MESOSPHERIC MODELS AND RELATED EXPERIMENTS

ASTROPHYSICS AND SPACE SCIENCE LIBRARY

A SERIES OF BOOKS ON THE RECENT DEVELOPMENTS

OF SPACE SCIENCE AND OF GENERAL GEOPHYSICS AND ASTROPHYSICS

PUBLISHED IN CONNECTION WITH THE JOURNAL

SPACE SCIENCE REVIEWS

VOLUME 25

MESOSPHERIC MODELS AND RELATED EXPERIMENTS

PROCEEDINGS OF THE FOURTH ESRIN-ESLAB SYMPOSIUM
HELD IN FRASCATI, ITALY, 6–10 JULY, 1970

Edited by

G. FIOCCO

European Space Research Institute, Frascati (Rome), Italy

D. REIDEL PUBLISHING COMPANY

DORDRECHT-HOLLAND

The Symposium was jointly sponsored by the Space Science Department (ESLAB), Noordwijk, The Netherlands, and the European Space Research Institute (ESRIN), Frascati, Italy, of the European Space Research Organisation (ESRO)

Library of Congress Catalog Card Number 70–154737

ISBN-13: 978-94-010-3116-5   e-ISBN-13: 978-94-010-3114-1
DOI: 10.1007/978-94-010-3114-1

FOREWORD

In July 1967 ESRIN and ESLAB, the two ESRO laboratories primarily concerned with basic research, held a joint symposium on Satellite and Rocket Measurements of Corpuscular Radiation from Outer Space. This was followed in September 1968 by a second symposium: Low-Frequency Waves and Irregularities in the Ionosphere; and in September 1969 by a third: Intercorrelated Satellite Observations Related to Solar Events.

A fourth symposium, on upper atmospheric models and related experiments, took place in Frascati, 6–10 July 1970. The main aim of the symposium was to assess current experimental work – both field and laboratory – related to mesospheric structure and composition, in the light of theoretical work on atmospheric models and atomic and molecular processes. To foster an interdisciplinary approach, the meeting brought together approximately 50 scientists working with different techniques but having a common interest in the interpretation of mesospheric phenomena.

Recent work such as that presented at the symposium has deepened our knowledge of upper atmospheric composition, and increased our understanding of upper atmospheric temperature and dynamics. In particular, it has shown that the concentration of minor species in the atmosphere is somewhat different than expected. Theoretical models have not yet considered many of the minor constituents and have generally not included dynamical effects. It is now time to reconsider these models and elaborate them in the light of recent knowledge.

Professor E. Hesstvedt and Dr A. Pedersen helped in defining the aims and scientific programme of the symposium, and Miss M. Sachs in the preparation of the meeting and the publication of its proceedings. Her further contribution was to prod the editor into brief but intense bursts of activity.

G. FIOCCO

Frascati, 18 December 1970

TABLE OF CONTENTS

AERONOMIC REACTIONS OF HYDROGEN AND OZONE

M. NICOLET

Institut d'Aéronomie Spatiale de Belgique, Brussels, Belgium

Introduction

Forty years ago Chapman (1930) showed that the dissociation of molecular oxygen was important above 100 km and, therefore, that the aeronomic behavior of ozone was related to the dissociation of oxygen below 100 km (Chapman, 1943). Furthermore, twenty years ago, vibrational-rotational bands of the hydroxyl radical OH, which appear in the airglow (Meinel, 1950), aroused interest in the photochemistry of hydrogen-oxygen compounds (Bates and Nicolet, 1950a; Herzberg, 1951) and, in particular, of methane (Bates and Nicolet, 1950b) and of water vapor (Bates and Nicolet, 1950c).

The photochemistry of atmospheric water vapor, which was studied in considerable detail by Bates and Nicolet (1950c), was difficult owing to the grievous lack of reliable basic data. Our knowledge concerning the experimental rate coefficients has increased rapidly in recent years and systematic accounts can be found in several review papers published since 1963: three-body reactions by Barth (1964); reactions involving nitrogen and oxygen by Schiff (1964, 1969) and reactions involving hydrogen and other minor constituents by Kaufman (1964, 1969). A complete application to aeronomy must still await other laboratory investigations under controlled conditions which are of fundamental importance for a useful interpretation of space observations.

As far as the solar radiation and its absorption are concerned, much progress has been made since the analysis of the H_2O photodissociation in a nitrogen-oxygen atmosphere by Bates and Nicolet (1950c), who adopted a solar flux corresponding to 5000 K with the possibility of local variations of 300 K in different parts of the spectral range concerned. The influence of the N_2 absorption is negligible in the mesosphere and, in the present connection, only molecular oxygen by the two systems, the Herzberg and the Schumann-Runge, plays the essential role. Recently, new results (Ackerman, Biaumé and Kockarts, this symposium) have been obtained in the region of 2100 Å to 1750 Å and these have shown that our knowledge of solar radiation and relevant absorption cross sections must be improved.

1. Ozone Formation and Dissociation

When the dissociation of molecular oxygen occurs it is followed by the three-body recombination

$$(k_1); \quad O + O + M \rightarrow O_2 + M + 118 \text{ kcal} \tag{1}$$

Fiocco (ed.), Mesospheric Models and Related Experiments, 1–51. All Rights Reserved.
Copyright © 1971 by D. Reidel Publishing Company, Dordrecht-Holland.

or by ozone formation

$$(k_2);\quad O + O_2 + M \rightarrow O_3 + M + 24\ \text{kcal}.\qquad(2)$$

Ozone is destroyed by the bimolecular process

$$(k_3);\quad O + O_3 \rightarrow O_2 + O_2^* + 94\ \text{kcal}.\qquad(3)$$

A precise knowledge of the rate coefficients k_1, k_2 and k_3 is required in order to determine the chemical rate coefficients for the formation and destruction of atmospheric ozone. The value of k_1 seems to be known with sufficient precision for aeronomic purposes. A recent experimental analysis (Campbell and Thrush, 1967) indicates that the molecular oxygen formation in the presence of molecular nitrogen $(M \equiv N_2)$ has a rate coefficient varying with temperature. It can be expressed as

$$k_1 = 3 \times 10^{-33}\,(300/T)^3\,n(N_2)\ \text{cm}^3\ \text{s}^{-1}\qquad(4)$$

in order to represent values such as about 4×10^{-33} cm^3 s^{-1} at 273 K and 1×10^{-32} cm^3 s^{-1} at 195 K.

As far as k_2 and k_3 are concerned, the experimental values are not yet sufficiently precise for aeronomic purposes. The thermal decomposition of ozone has been studied in terms of the following mechanisms:

$$O_2 + O \underset{k_{-2}}{\overset{k_2}{\rightleftharpoons}} O_3$$

and

$$O + O_3 \overset{k_3}{\rightarrow} 2O_2.$$

From the usual Arrhenius plot for an *ozone* atmosphere between 343 K and 383 K, the values of k_{-2} and $k_{-2}\,k_3/k_2$ have been deduced by Benson and Axworthy (1957) and lead to

$$k_2 n(M)/k_3 = (1.4 \pm 0.14) \times 10^{-24} e^{(3300 \pm 300)/T} n(M)\qquad(5)$$

in which $n(M)$ denotes the total concentration in the homosphere.

Adopting the result

$$n(M)\,k_2 = 3.3 \times 10^{-35} e^{950/T} n(M)\ \text{cm}^3\ \text{s}^{-1}\qquad(6)$$

from measurements by Clyne *et al.* (1962), the adopted expression for k_3 is, from (5) and (6),

$$k_3 = 2.4 \times 10^{-11}\,e^{-2350/T}\ \text{cm}^3\ \text{s}^{-1}.\qquad(7)$$

The numerical values obtained from (5), (6) and (7) are in agreement with experimental data at 300 K (Kaufman and Kelso, 1964 and 1967; Mathias and Schiff, 1964; Hochanadel *et al.*, 1968). However, there is still some uncertainty in the values of the above coefficients particularly at low temperatures which correspond to those at stratospheric and mesospheric levels. Keeping the same activation energies, k_2 should be considered as a maximum and k_3 as a minimum value. There is perhaps a possibility that $\frac{1}{2}k_2$ may correspond to a minimum value and $2k_3$ to a maximum value. An aeronomic calculation made with these two alternatives shows that the differences are not negligible. However, more important difficulties occur in the mesosphere when only a pure oxygen-nitrogen atmosphere is considered since the vertical distribution of ozone and its absolute concentration cannot be explained without the introduction of other minor constituents.

Since it is convenient to use as basic parameters the particle concentrations in the chemical problems the adoption of an atmospheric model is required. This is represented in Table I and gives the essential properties for consistent calculations, but

TABLE I
Atmospheric parameters in the stratosphere and mesosphere

Altitude (km)	Scale height (km)	Temperature (K)	Pressure (T)	Total concentration	Oxygen concentration
15	6.20	211	8.5×10^1	3.9×10^{18}	8.1×10^{17}
20	6.45	219	3.9×10^1	1.7×10^{18}	3.6×10^{17}
25	6.70	227	1.8×10^1	7.7×10^{17}	1.6×10^{17}
30	6.95	235	8.6	3.6×10^{17}	7.4×10^{16}
35	7.45	252	4.3	1.7×10^{17}	3.5×10^{16}
40	7.95	268	2.2	8.1×10^{16}	1.7×10^{16}
45	8.15	274	1.2	4.3×10^{16}	8.9×10^{15}
50	8.15	274	6.6×10^{-1}	2.3×10^{16}	4.8×10^{15}
55	8.15	273	3.6×10^{-1}	1.3×10^{16}	2.6×10^{15}
60	7.54	253	1.9×10^{-1}	7.2×10^{15}	1.5×10^{15}
65	6.93	232	9.4×10^{-2}	3.9×10^{15}	8.2×10^{14}
70	6.32	211	4.4×10^{-2}	2.0×10^{15}	4.2×10^{14}
75	5.82	194	1.9×10^{-2}	9.6×10^{14}	2.0×10^{14}
80	5.32	177	7.9×10^{-3}	4.2×10^{14}	9.0×10^{13}
85	4.82	160	2.9×10^{-3}	1.9×10^{14}	3.7×10^{13}

not for a study of seasonal or latitudinal variations. The total number of molecules in a vertical column is given by the product of the concentration and the scale height, i.e.

$$\int_z^\infty n \, dz \simeq n\mathrm{H}. \tag{8}$$

With the adopted values of the atmospheric parameters given in Table I, the rate coefficients in which ozone and oxygen are involved are given in Table II. However, before beginning a discussion of these numerical values, it is necessary to consider the photodissociation rate coefficients of O_2 and O_3.

It can be shown that the calculated values of the O_3 concentrations are too large, particularly in the mesosphere, and that the introduction of ozone-hydrogen reactions changes the situation. However, since at least superficial agreement is obtained by taking rate coefficients which fit the observational data, it seems to us that an adopted

TABLE II

Stratospheric and mesospheric rate coefficients for ozone and oxygen reactions

Altitude (km)	$k_1 n$ (M) (cm^3 s^{-1})	$k_2 n$ (M) (cm^3 s^{-1})	k_3 (cm^3 s^{-1})	$k_2 n$ (M) n (O$_2$) (s^{-1})	$k_2 n$ (M) n (O$_2$)/k_3 (cm^{-3})
15	1.1×10^{-14}	1.2×10^{-14}	3.5×10^{-16}	9.5×10^{3}	2.7×10^{19}
20	5.5×10^{-15}	5.2×10^{-15}	5.2×10^{-16}	1.6×10^{3}	2.9×10^{18}
25	2.1×10^{-15}	1.7×10^{-15}	7.7×10^{-16}	2.7×10^{2}	3.5×10^{17}
30	9.8×10^{-16}	6.7×10^{-15}	1.1×10^{-15}	5.0×10^{1}	4.5×10^{16}
35	4.6×10^{-16}	2.4×10^{-16}	2.1×10^{-15}	8.3×10^{0}	3.9×10^{15}
40	2.2×10^{-16}	9.3×10^{-17}	3.8×10^{-15}	1.6×10^{0}	4.2×10^{14}
45	1.2×10^{-16}	4.5×10^{-17}	4.6×10^{-15}	4.0×10^{-1}	8.7×10^{13}
50	6.4×10^{-17}	2.4×10^{-17}	4.5×10^{-15}	1.2×10^{-1}	2.6×10^{13}
55	3.5×10^{-17}	1.3×10^{-17}	4.5×10^{-15}	3.5×10^{-2}	7.8×10^{12}
60	2.0×10^{-17}	1.0×10^{-17}	2.2×10^{-15}	1.5×10^{-2}	6.9×10^{12}
65	1.1×10^{-17}	7.8×10^{-18}	9.6×10^{-16}	7.6×10^{-3}	6.7×10^{12}
70	5.6×10^{-18}	6.0×10^{-18}	3.5×10^{-16}	2.5×10^{-3}	7.2×10^{12}
75	2.7×10^{-18}	4.2×10^{-18}	1.3×10^{-16}	8.5×10^{-4}	6.4×10^{12}
80	1.2×10^{-18}	3.0×10^{-18}	4.2×10^{-17}	2.7×10^{-4}	6.5×10^{12}
85	4.9×10^{-19}	2.2×10^{-18}	1.0×10^{-17}	8.1×10^{-5}	7.9×10^{12}

model for the vertical distribution of ozone could more adequately describe the aeronomic processes which are involved.

In Figures 1a and 1b the profile adopted for the distribution of the ozone concentration in the mesosphere and stratosphere corresponds to conditions which are consistent with observational data. For the height range 40–80 km, such a vertical distribution leads to a simple density distribution $e^{-0.23z(\text{km})}$ giving $n(\text{O}_3) = 10^{12}$ at

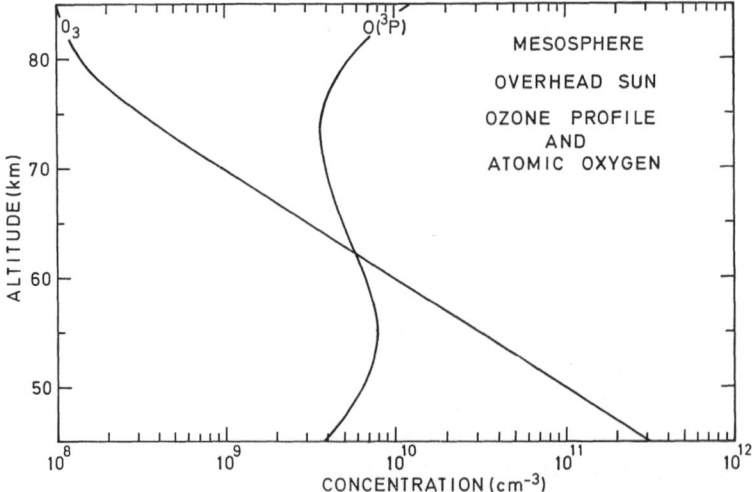

Fig. 1a. Atomic oxygen concentration for daytime conditions in the mesosphere when the vertical distribution of the ozone concentration corresponds to 10^{11} cm^{-3}, 10^{10} cm^{-3} and 10^{9} cm^{-3} at 50 km, 60 km and 70 km, respectively.

40 km and 10^8 cm^{-3} at 80 km. From such distribution it is possible to deduce average conditions in the stratosphere and mesosphere.

When an analysis of the ozone photodissociation is made, it is necessary to distinguish between the various products in order to know the production rate of excited atoms or molecules. For $\lambda \leqslant 11\,800$ Å

$$O_3 + hv\,(\lambda \leqslant 11\,800\,\text{Å}) \rightarrow O_2\,(^3\Sigma_g) + O\,(^3P) \tag{9}$$

with O_2 and O in their normal states. In the spectral range of the Chappuis band (with predissociation) it is logical to consider that (9) represents the normal process even if the following processes, with spin changes, are energetically possible

$$O_3 + hv\,(\lambda \leqslant 6110\,\text{Å}) \rightarrow O\,(^3P) + O\,(^1\varDelta_g), \tag{10a}$$

$$O_3 + hv\,(\lambda \leqslant 4630\,\text{Å}) \rightarrow O_2\,(^1\Sigma_g^+) + O\,(^3P). \tag{10b}$$

However, (10a) seems to be an important process in the Huggins band (Jones and Wayne, 1969) while

$$O_3 + hv\,(\lambda < 3100\,\text{Å}) \rightarrow O\,(^1\varDelta_g) + O\,(^1D) \tag{11}$$

leads to a complete production of $O\,(^1D)$ atoms for $\lambda < 3000$ Å (De More and Raper, 1966) and to 0.4 ± 0.15 of the total production at $\lambda = 3130$ Å.

Other processes leading to $O_2\,(^1\Sigma_g^+)$ and to $O\,(^1S)$ at shorter wavelengths should also be considered and require an experimental analysis in order to determine the exact fraction of their production by absorption in the Hartley band.

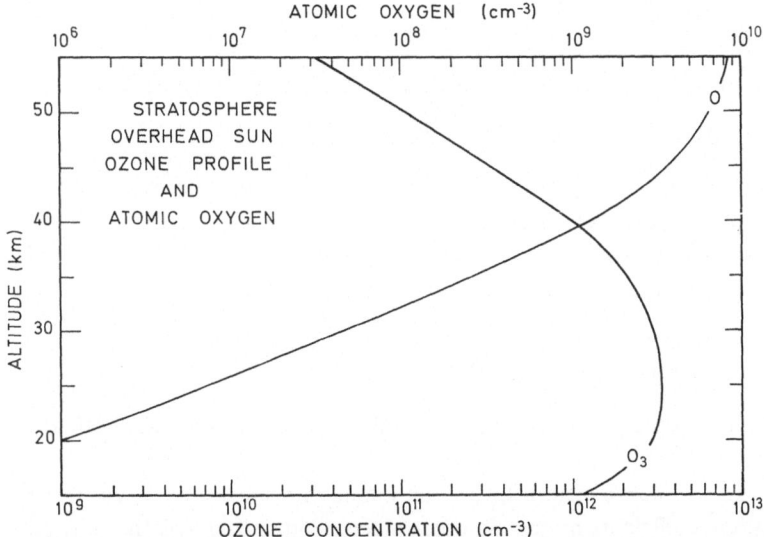

Fig. 1b. Atomic oxygen concentration in the stratosphere for overhead sun conditions when the ozone concentration is given.

For example, the spin forbidden photodissociation at $\lambda < 2360$ Å

$$O_3 + h\nu \, (\lambda \leqslant 2360 \text{ Å}) \rightarrow O_2 \, (X^3\Sigma_g^-) + O \, (^1S) \tag{12a}$$

cannot have an important role since it corresponds only to a small fraction of the total photodissociation. But the spin allowed transition

$$O_3 + h\nu \, (\lambda \leqslant 1995 \text{ Å}) \rightarrow O_2 \, (a^1\Delta_g) + O \, (^1S) \tag{12b}$$

may be considered at wavelengths less than 2000 Å. However, it can be shown that the production of $O(^1S)$ atoms by (12) is small compared with that of $O(^1D)$ atoms since the solar flux is greater at 3000 Å than at 2000 Å.

The small absorption cross-section of the Chappuis band (Vigroux, 1953; Tanaka *et al.*, 1953; Inn and Tanaka, 1953), which reaches only 5×10^{-21} cm^2 at the absorption peak near 6000 Å, leads to a photodissociation coefficient

$$J_{O_3} (\text{Chappuis}) = 3 \times 10^{-4} \text{ s}^{-1} \tag{13}$$

for zero optical depth.

The absorption cross-section in the Huggins bands increases from about 10^{-22} cm^2 between 3500 Å and 3600 Å up to 10^{-19} cm^2 at 3100 Å. This spectral region leads to a photodissociation coefficient

$$J_{O_3} (\text{Huggins}) = 1 \times 10^{-4} \text{ s}^{-1} \tag{14}$$

for zero optical depth. However, since the absorption cross-section increases rapidly with decreasing wavelength, the photodissociation coefficient for the spectral range $\lambda\lambda \, 3100$–3000 Å is

$$J_{O_3} (3100\text{–}3000 \text{ Å}) = 2 \times 10^{-4} \text{ s}^{-1} \tag{15}$$

for zero optical depth.

The total photodissociation coefficient for zero optical depth is

$$J_{O_3} = 10^{-2} \text{ s}^{-1} \tag{16}$$

corresponding practically to the absorption of solar radiation of $\lambda > 2300$ Å for which the absorption cross-section in the Hartley band is greater than 5×10^{-18} cm^2. Since the continuous absorption of molecular oxygen begins at $\lambda < 2420$ Å, it can be said that at least 90% of the photodissociation coefficient of ozone depends on its own absorption, i.e.

$$J(O_3) \simeq J_\infty (O_3) \, e^{-\tau(O_3)} \tag{17}$$

where $J_\infty (O_3)$ is the photodissociation coefficient for zero optical depth and $\tau(O_3)$ the optical depth.

With the adopted distribution of the ozone concentration (Figure 1) the various dissociation coefficients are easily determined. Figure 2 shows the photodissociation coefficients in the stratosphere where the optical depth becomes important. The total dissociation coefficient $J(O_3)$ decreases from about 10^{-2} s^{-1} in the mesosphere to about 3×10^{-4} s^{-1} in the lower stratosphere, where the photodissociation is due to the Chappuis band. The direct production of excited oxygen molecules in the $^1\Delta_g$ state is almost equivalent to the total photodissociation except in the lower stratosphere. As far as the production of excited oxygen in the ^1D and ^1S states is concerned, only the O(^1D) production is sufficiently important to be considered. However, the photo-dissociation coefficient decreases rapidly in the lower stratosphere due to the important absorption in the Hartley band.

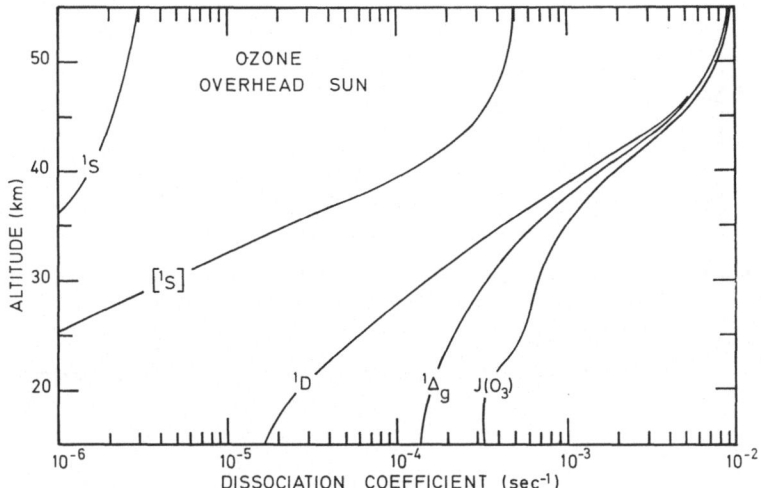

Fig. 2. Photodissociation coefficients of stratospheric ozone for overhead sun conditions. Curve $J(O_3)$ is the total photodissociation involving all processes. Curve $^1\Delta_g$ is the production coefficient of oxygen molecules in the first excited electronic level $^1\Delta_g$. Curve ^1D is the production coefficient of oxygen atoms in the metastable state ^1D. Curve [^1S] would correspond to a maximum production of O(^1S) atoms if there were no restriction with such spin forbidden processes. Curve ^1S corresponds to the simultaneous maximum production of O(^1S) and O$_2$($^1\Delta_g$).

The production of such excited atoms and molecules varies with the solar zenith angle which leads to a rapid decrease from the stratopause to the tropopause. The production of O(^1D) atoms and O$_2$($^1\Delta_g$) molecules by the ozone photodissociation for various solar zenith angles is illustrated in Figure 3 and Figure 4. Molecular oxygen in the first excited state is produced in the whole of the stratosphere during the day with a rate between 10^8 and 10^9 molecules cm^{-3} s^{-1}. With a quenching coefficient

$$k_Q(^1\Delta_g) = 4.4 \times 10^{-19} \text{ cm}^3 \text{ s}^{-1}$$

corresponding to the quenching by molecular oxygen (Clark and Wayne, 1969), the photoequilibrium concentration is very high. With the ozone profile given in Figure 1 and the production rates illustrated in Figure 4, the photoequilibrium concentrations are shown in Figure 5. For daytime conditions the concentration peak in the neighborhood of the stratopause has a value between 3×10^{10} and 7×10^{10} O$_2$($^1\Delta_g$) molecules cm^{-3}. At the horizon a concentration of the order of 10^{10} cm^{-3} near 60 km

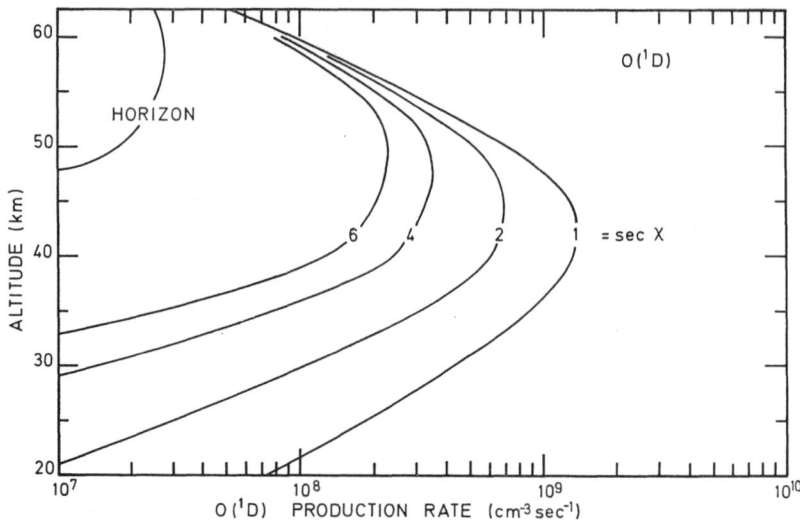

Fig. 3.   Production rate of oxygen atoms in their first excited state ^1D for various solar zenith angles. From $\sec\chi = 1$, overhead sun conditions, to horizon, $\chi = 90°$.

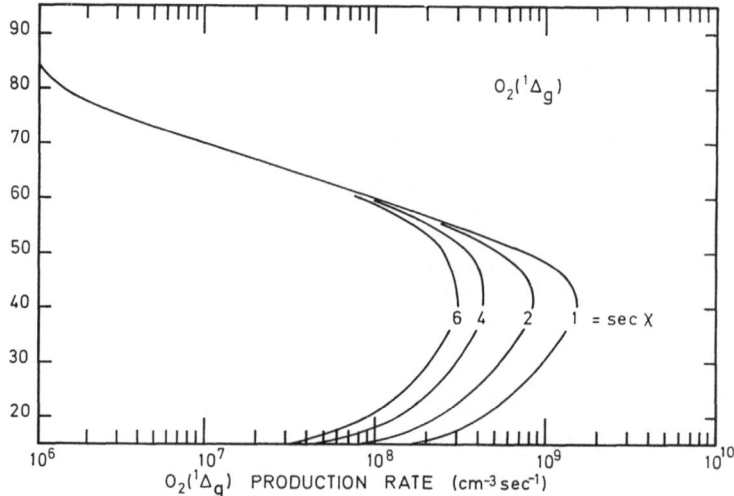

Fig. 4.   Production rate of molecular oxygen in the first excited electronic level $^1\Delta_g$ for various solar zenith angles.

corresponds to a maximum for photoequilibrium conditions. Such theoretical values which correspond to observational data (Evans and Llewellyn, 1970) show that the photolysis of ozone in the Hartley band (reaction 11) is the primary process as original-ly proposed by Vallance Jones and Gattinger (1963). Above the mesopause there is an increase of the $O_2(^1\Delta_g)$ concentration which is due to an increase, in the lower ther-mosphere, of the ozone concentration in photoequilibrium with atomic oxygen in a sunlit atmosphere, and in chemical equilibrium with atomic hydrogen in nighttime conditions.

The production of the $O(^1D)$ atoms by the photolysis process of the stratospheric and mesospheric ozone is also important (Figure 3). However, the variation with the solar zenith angle is more important for the production of $O(^1D)$ atoms than for that of $O_2(^1\Delta_g)$ molecules. Furthermore, the quenching rate is also more important.

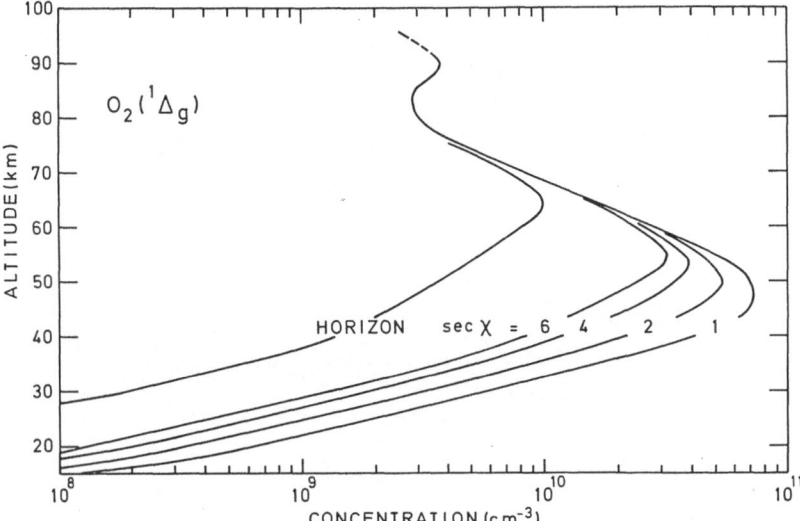

Fig. 5. Concentrations of $O_2(^1\Delta_g)$ molecules corresponding to *photoequilibrium* conditions for various solar zenith angles.

Several experimental determinations have been published (De More and Raper, 1964; Snelling and Bair, 1967; Young and Black, 1967; Young *et al.*, 1968; Noxon, 1970; Biedenkapp and Bair, 1970; De More, 1970). The following working value for the whole homosphere is adopted here

$$k_Q(^1D) = 5 \times 10^{-11} \text{ cm}^3 \text{ s}^{-1}.$$

This is an average value between the maximum value 10^{-10} cm^3 s^{-1} and the minimum value 2×10^{-11} cm^3 s^{-1}. It must be pointed out that molecular oxygen is excited to its $b^1\Sigma_g^+$ state in a reaction with $O(^1D)$ atoms

$$O(^1D) + O_2(A^3\Sigma_g^-) \rightarrow O(^3P) + O_2(^1\Sigma_g^+)$$

with a near unitary efficiency (Young and Black, 1967; Noxon, 1970; Clark, 1970).

With the average value $k_Q(^1D) = 5 \times 10^{-11}$ cm^3 s^{-1}, the production rate of $O(^1D)$ atoms illustrated in Figure 3 leads to the photoequilibrium concentrations for various solar zenith angles shown in Figure 6. The peak concentration of more than 10^2 cm^{-3} occurs near the stratopause for solar zenith angles where sec $\chi \leqslant 6$. At the mesopause, the concentration of $O(^1D)$ begins to increase again in consequence of the photo-dissociation of O_2 in the Schumann-Runge continuum.

As far as the $O(^1S)$ atoms are concerned, their low collisional deactivation with N_2 (Filseth et al., 1970)

$$k_Q(^1S, N_2) < 2 \times 10^{-16} \text{ cm}^3 \text{ s}^{-1},$$

and their moderate deactivation by O_2 (Filseth et al., 1970; Black et al., 1969) are not sufficient to lead to important $O(^1S)$ concentrations since the production rate is too low. In any case, an oxidizing reaction with $O(^1S)$ atoms would require a rate coefficient much too great compared with the normal rate coefficient with $O(^1D)$.

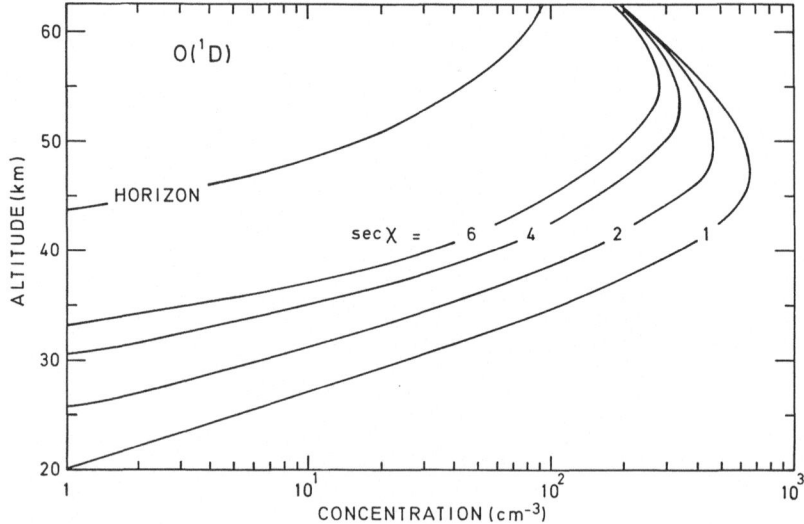

Fig. 6.   Concentrations of $O(^1D)$ atoms for various solar zenith angles.

2. The Oxygen Dissociation

As far as molecular oxygen is concerned, its photodissociation must be considered for aeronomic purposes as follows:

$$O_2(\lambda < 2424 \text{ Å}) \rightarrow 2 \, O(^3P) \tag{18}$$

with two oxygen atoms in the normal state. The absorption cross-section in the Herzberg continuum is very small near the threshold, being only of the order of 10^{-24} cm^2 and becoming 10^{-23} cm^2 near 2050 Å (Vassy, 1941; Ditchburn and Young, 1962; Shardanand, 1970). The contribution from the Herzberg continuum to the photo-dissociation rate is essentially constant in the mesosphere since there is practically no absorption; it is

$$J_{O_2}(\text{Herzberg}) = 1.3 \times 10^{-9} \text{ s}^{-1}. \tag{19a}$$

In the spectral range of the Schumann-Runge bands ($v' > 3$), where predissociation occurs, the absorption is extremely variable. The relevant cross-sections have been determined by Ackerman *et al.* (1970). In using the results, allowance must, of course, be made for the effect of temperature (cf. Kockarts, this symposium, for the optical depth in the stratosphere and mesosphere and Ackerman for solar radiation data). At the mesopause, the photodissociation coefficient for the spectral range of the Schumann-Runge bands is

$$J_{O_2}(\text{Schumann-Runge bands})_{85\ km} = 1.7 \times 10^{-8}\ s^{-1}, \tag{19b}$$

and at the stratopause,

$$J_{O_2}(\text{Schumann-Runge bands})_{50\ km} = 1.7 \times 10^{-10}\ s^{-1} \tag{19c}$$

for an overhead sun.

The continuum to which the Schumann-Runge system converges

$$O_2 + h\nu\,(\lambda < 1759\ \text{Å}) \rightarrow O\,(^3P) + O\,(^1D) \tag{20}$$

leads to an excited atom ^1D. For zero optical depth the photodissociation coefficients are

$$J_{O_2}(\text{Schumann-Runge continuum}) = 3.7 \times 10^{-6}\ s^{-1}, \tag{21a}$$

and at 100 km

$$J_{O_2}(\text{Schumann-Runge continuum})_{100\ km} = 3.7 \times 10^{-7}\ s^{-1}. \tag{21b}$$

With such small values for the photodissociation coefficient it is certain that molecular oxygen cannot be in photochemical equilibrium in the thermosphere (Nicolet and Mange, 1954; Nicolet, 1954).

Finally, the absorption of Lyman-α at 1215.7 Å by molecular oxygen must be considered since its energy is of the order of several ergs $cm^{-2}\ s^{-1}$. With a flux of 3×10^{11} photon $cm^{-2}\ s^{-1}$ and an absorption cross-section of $10^{-20}\ cm^2$ (Watanabe, 1958), the rate is

$$J_{O_2}(\text{Lyman-}\alpha) = 3 \times 10^{-9}\ s^{-1}. \tag{22}$$

In the thermosphere (see Figure 7a), the principal process in the O_2 photodissociation is due to the absorption in the Schumann-Runge continuum; the effects of Lyman-α and of the Herzberg continuum are negligible. The predissociation process in the Schumann-Runge bands begins to play a role near the mesopause and becomes very important in the mesosphere. Figure 7b shows the transition from a dissociation at the mesopause through the predissociation of the bands of the Schumann-Runge system to a photodissociation by absorption in the Herzberg continuum at the stratopause. For this reason, the calculation of the photodissociation of molecular oxygen in the mesosphere requires the precise absorption cross-sections determined by Ackerman *et al.* (1970). It is not possible to use a constant absorption cross-section of the Schumann-Runge bands in order to determine the exact effect of the dissociation of molecular

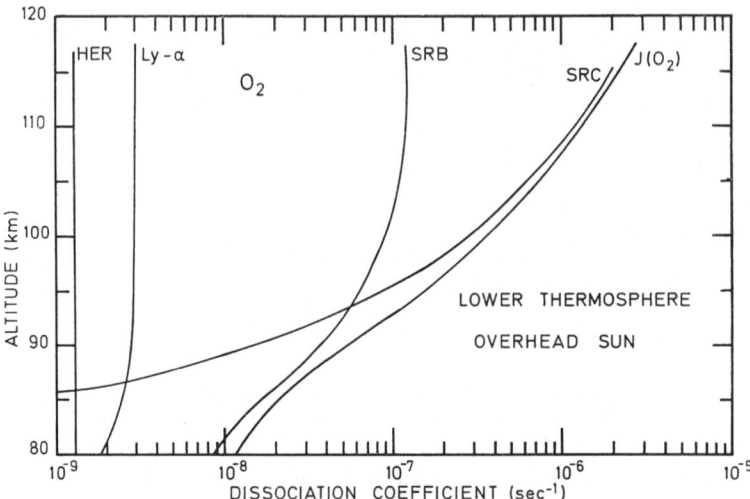

Fig. 7a. Photodissociation coefficient of oxygen molecules in the lower thermosphere. The total photodissociation [curve $J(O_2)$ in the whole of the thermosphere depends on the absorption in the Schumann-Runge continuum (curve SRC). Near the mesopause, the spectral region of the Schumann-Runge bands (curve SRB) begins to play the principal role. The Herzberg continuum (curve HER), and Lyman-α (curve Ly-α) do not play any practical role.

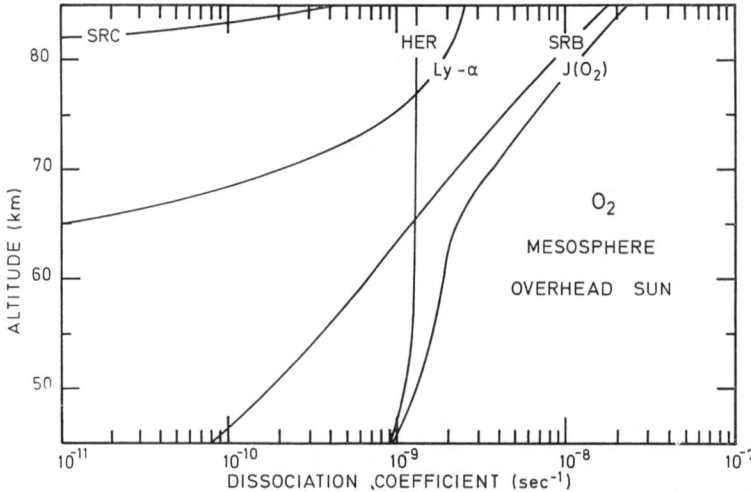

Fig. 7b. Photodissociation coefficient of molecular oxygen in the mesosphere for overhead sun conditions. In the upper part of the mesosphere the major role is played by the region of the Schumann-Runge bands (curve SRB). Above the stratopause the Herzberg continuum is involved (curve HER).

oxygen and other constituents such as H_2O. However, in the specific case of O_2, a constant cross-section corresponding to an optical depth of about 1 can fit the dissociation rate in the mesosphere while average cross-sections corresponding to $\tau > 1$ lead to too small photodissociation coefficients.

The photodissociation of molecular oxygen reaches a peak near the stratopause (Figure 8) with a production rate of about 10^7 cm^{-3} s^{-1} at 40 km for an overhead sun, and of about 10^6 cm^{-3} s^{-1} at 60 km when the sun is at a solar zenith angle of about 90°. Figure 8 shows the importance of variation of the O_2 dissociation in the stratosphere with the variation of the solar zenith angle.

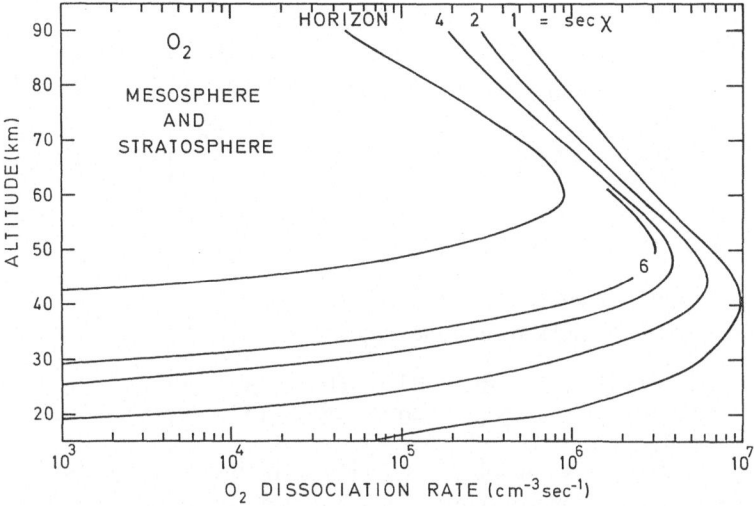

Fig. 8. Photodissociation rate of molecular oxygen for various solar zenith angles. From overhead sun conditions, $\sec \chi = 1$, to $\chi = 90°$, horizon.

3. Conditions in a Pure Oxygen Atmosphere

Considering now the theoretical problem of oxygen dissociative and ozone formation, the equations governing the rates of change of the concentrations $n(O)$ and $n(O_3)$ are

$$dn(O)/dt + 2k_1 n(M) n^2(O) + k_2 n(M) n(O_2) n(O) + k_3 n(O_3) n(O) =$$
$$= 2n(O_2) J_2 + n(O_3) J_3 \quad (23)$$

$$dn(O_3)/dt + n(O_3) J_3 - k_2 n(M) n(O_2) n(O) + k_3 n(O) n(O_3) = 0 \quad (24)$$

which lead to the general form

$$dn(O)/dt + dn(O_3)/dt + 2k_1 n(M) n^2(O) + 2k_3 n(O_3) n(O) =$$
$$= 2n(O_2) J_2, \quad (25)$$

where J_2 and J_3 are the photodissociation coefficients of O_2 and O_3, respectively.

In the thermosphere and mesosphere where the photodissociation coefficient of O_3 is high: $10^{-3} \leqslant J \leqslant 10^{-2}$ s^{-1}, there is a daytime photoequilibrium for ozone. Thus, from (24),

$$n(O_3) [J_3 + k_3 n(O)] = k_2 n(M) n(O_2) n(O). \quad (26a)$$

With $k_3 \leqslant 10^{-15}$ cm^3 s^{-1} (cf. Table II), the following approximation can be made

$$n(O_3)/n(O) = k_2 n(M) n(O_2)/J_3 \qquad (26b)$$

since $n(O) \leqslant 10^{12}$ cm^{-3}. At the mesopause level, Equation (25) becomes, using (26b),

$$\frac{dn(O)}{dt} + 2\left[k_1 n(M) + \frac{k_3 k_2 n(M) n(O_2)}{J_3} \right] n^2(O) = 2n(O_2) J_2. \qquad (27)$$

The time τ_0 which is required in a sunlit atmosphere to reach about 50% of the photochemical equilibrium value (or to reach 80% from 50%) is given by

$$\tau_0(50\%) = 0.275 \left\{ \left[k_1 n(M) + \frac{k_3 k_2 n(M) n(O_2)}{J_3} \right] n(O_2) J_2 \right\}^{1/2} \qquad (28)$$

or

$$\tau_0(50\%) = 0.275 n_*(O)/n(O_2) J_2 \qquad (29)$$

if $n_*(O)$ is the photochemical equilibrium concentration.

Application of (29) leads to the following times

Altitude (km)	90	85	80	75	70
Time (h)	200	90	32	12	4

Thus, at the mesopause and above, in the lower thermosphere, the vertical distribution of the atomic oxygen concentration will depend on the relative importance of atmospheric transport conditions as compared with chemical equilibrium conditions. The introduction of ozone-hydrogen reactions can change the situation (Bates and Nicolet, 1950c, 1965; Nicolet 1964, 1966) and various studies (Hampson, 1966; Hunt, 1966; Hesstvedt, 1968; Crutzen, 1969) show how the ozone behavior in the mesosphere is affected by hydrogen compounds.

TABLE III

Atmospheric parameters in the stratosphere and mesosphere (overhead sun conditions)

Altitude (km)	$n(O_2)$ (cm^{-3})	$n(O_2) J_2$ (cm^{-3} s^{-1})	$n(O_3)$ (cm^{-3})	$n(O_3) J_3$ (cm^{-3} s^{-1})	$n[O(^3P)]$ (cm^{-3})	$n[O(^1D)]$ (cm^{-3})
85	3.7×10^{13}	7.3×10^5	1.0×10^8	1.0×10^6	3.0×10^{10}	1.2×10^2
80	9.0×10^{13}	1.2×10^6	1.4×10^8	1.4×10^6	1.4×10^{10}	7.0×10^1
75	2.0×10^{14}	1.5×10^6	3.2×10^8	3.2×10^6	3.8×10^9	6.7×10^1
70	4.2×10^{14}	1.7×10^6	1.0×10^9	1.0×10^7	4.0×10^9	9.6×10^1
65	8.2×10^{14}	2.2×10^6	3.2×10^9	3.2×10^7	5.0×10^9	1.5×10^2
60	1.5×10^{15}	3.0×10^6	1.0×10^{10}	9.9×10^7	6.5×10^9	2.6×10^2
55	2.6×10^{15}	4.3×10^6	3.2×10^{10}	2.9×10^8	8.4×10^9	4.4×10^2
50	4.8×10^{15}	6.6×10^6	1.0×10^{11}	7.7×10^8	6.5×10^9	6.1×10^2
45	8.9×10^{15}	9.0×10^6	3.2×10^{11}	1.5×10^9	3.7×10^9	6.0×10^2
40	1.7×10^{16}	9.7×10^6	1.0×10^{12}	1.9×10^9	1.2×10^9	3.3×10^2
35	3.5×10^{16}	8.1×10^6	2.0×10^{12}	1.9×10^9	2.4×10^8	1.0×10^2
30	7.4×10^{16}	6.7×10^6	2.9×10^{12}	2.0×10^9	3.9×10^7	2.5×10^1
25	1.6×10^{17}	2.9×10^6	3.2×10^{12}	1.8×10^9	6.7×10^6	5.0×10^0
20	3.6×10^{17}	8.0×10^5	2.9×10^{12}	1.4×10^9	9.4×10^5	9.0×10^{-1}

With the numerical values used in Tables I and II and the results given in various figures, it is possible to adopt the various parameters for an idealized atmosphere. Since such an atmosphere exhibits the general aeronomic features found in the stratosphere, mesosphere and lower thermosphere, Table III gives data which are required in the subsequent discussion of the reaction of the various hydrogen-oxygen compounds.

4. Hydrogen-Oxygen Atmosphere

A hydrogen-oxygen atmosphere is very complicated. However, when the less probable reactions are eliminated, it is possible to have a general idea of the role of the principal reactions.

The reactions involving only a single hydrogen atom (free or combined) and one of the allotropic forms of oxygen are as follows:

The three-body reaction of atomic hydrogen with atomic oxygen

$$(a_0); \quad H + O + M \rightarrow OH + M + 103 \text{ kcal} \tag{30}$$

is unimportant as compared with

$$(a_1); \quad H + O_2 + M \rightarrow HO_2 + M + 46 \text{ kcal} \tag{31a}$$

which has a rate coefficient (Clyne and Thrush, 1963b) taken for aeronomic purposes as

$$a_1 n(M) = 3.3 \times 10^{-33} e^{800/T} n(N_2, O_2) \text{ cm}^3 \text{ s}^{-1}. \tag{31b}$$

Another important loss process for atomic hydrogen is

$$(a_2); \quad H + O_3 \rightarrow O_2 + OH^*_{v \leqslant 9} + 77 \text{ kcal}. \tag{32a}$$

It is the Bates-Nicolet process, introduced for the interpretation of the OH airglow, which is observed in the lower thermosphere and perhaps also in the upper mesosphere. Its rate coefficient (Philips and Schiff, 1962; Kaufman, 1964, 1969) with practically no activation energy is taken as

$$a_2 = 1.5 \times 10^{-12} T^{1/2} \text{ cm}^3 \text{ s}^{-1}. \tag{32b}$$

The production of hydroperoxyl radicals by

$$(a_3); \quad H + O_3 \rightarrow O + HO_2 + 22 \text{ kcal} \tag{33}$$

with a rate coefficient a_3 which has not been measured, and by a three-body association

$$(a_4); \quad OH + O + M \rightarrow M + HO_2 + 63 \text{ kcal} \tag{34}$$

with a conventional value of the three-body rate coefficient a_4, is negligible compared with reaction (31).

An important reaction which forms a chain leading to the formation of oxygen molecules in conjunction with (32) is the bimolecular process

$$(a_5); \quad OH + O \rightarrow H + O_2 + 16.6 \text{ kcal}. \tag{35a}$$

The rate coefficient (Clyne and Thrush, 1963a; Kaufman, 1964, 1969)

$$a_5 = 3 \times 10^{-12} T^{1/2} \text{ cm}^3 \text{ s}^{-1} \tag{35b}$$

leads to the value $5 \times 10^{-11} \text{ cm}^3 \text{ s}^{-1}$ at 300K. Thus, reactions (32) and (35) show that hydrogen acts as a catalyst for the destruction of odd oxygen atoms and will affect the atomic oxygen (and ozone) distributions.

Other reactions involve hydroxyl radicals and ozone. The reaction

$$(a_6); \quad \text{OH} + \text{O}_3 \rightarrow \text{HO}_2 + \text{O}_2 + 39 \text{ kcal} \tag{36a}$$

competes with (35) where atomic oxygen has too small a concentration (stratosphere) as compared with the ozone concentration. No direct measurement of the rate coefficient has been reported. Since an upper limit at room temperature should be of the order of $5 \times 10^{-13} \text{ cm}^3 \text{ s}^{-1}$ (Kaufman, 1964, 1969) an activation energy of only 3 kcal leads to

$$a_6 = 1.5 \times 10^{-12} T^{1/2} e^{-1500/T} \text{ cm}^3 \text{ s}^{-1} \tag{36b}$$

which gives $10^{-13} \text{ cm}^3 \text{ s}^{-1}$ at the stratopause and $7.5 \times 10^{-15} \text{ cm}^3 \text{ s}^{-1}$ at the tropopause (190 K). An experimental analysis of this reaction is required since it has been introduced in various calculations to explain the photolysis of ozone in the presence of water vapor. Furthermore, another reaction has been introduced by various authors

$$(a_{6c}); \quad \text{HO}_2 + \text{O}_3 \rightarrow \text{OH} + 2\text{O}_2 + 31 \text{ kcal} \tag{36c}$$

in which ozone and hydroperoxyl radicals react in order to explain the OH-catalyzed chain decomposition of ozone. Such a reaction requires the simultaneous breaking of the bonds $\text{OH}-\text{O}$ and O_2-O and cannot be introduced in aeronomic calculations (Nicolet, 1970).

The principal reaction leading to OH involves atomic oxygen (Kaufman, 1964)

$$(a_7); \quad \text{O} + \text{HO}_2 \rightarrow \text{O}_2 + \text{OH}^*_{v \leqslant 6} + 55 \text{ kcal} \tag{37a}$$

for which a rate coefficient

$$a_7 = 3 \times 10^{-12} T^{1/2} \text{ cm}^3 \text{ s}^{-1} \tag{37b}$$

identical to a_5 is adopted in order to simplify the aeronomic analysis, even if it is perhaps too high by a factor of two.

In addition to these various collision processes, there is a possibility of dissociation of the hydroxyl and hydroperoxyl radicals.

$$(a_8); \quad \text{OH} + hv \rightarrow \text{O} + \text{H}. \tag{38}$$

Such a reaction will be ignored here ($a_8 = 0$), even if there is a possibility of a dissociation due to the predissociation of the $A^2 \Sigma^+$ state for levels $v' > 0$ (Smith, 1970).

Nothing is known about the photodissociation of perhydroxyl. If a comparison is made with H_2O_2 for which the photodissociation rate for zero optical depth is

$J_{H_2O_2} = 1.4 \times 10^{-4}$ s^{-1}, and with NO$_2$ for which $J_{NO_2} = 3.5 \times 10^{-3}$ s^{-1},

$$(a_9); \quad HO_2 + h\nu \rightarrow OH + O \tag{39a}$$

is possible, and a working value such as

$$10^{-3} \leqslant a_9 \equiv J_{HO_2} \leqslant 10^{-4} \text{ s}^{-1} \tag{39b}$$

may be considered. It is almost certain that the photodissociation process is less important than reaction (37) above 30 km in the stratosphere and in the whole mesosphere.

As for the photodissociation process

$$(a_{10}); \quad HO_2 + h\nu \rightarrow H + O_2 \tag{40}$$

in which the bonds between H and O atoms are difficult to break, it can be ignored and $a_{10} = 0$.

Expressions for the equilibrium ratios of $n(OH)/n(H)$ and $n(HO_2)/n(H)$ can be easily obtained if it is assumed, as a first approximation, that only (30) to (40) are involved. Thus, if reactions with rate coefficients $a = 0$ are ignored,

$$\frac{n(OH)}{n(H)} = \frac{\left[a_1 n(M)\, n(O_2) + a_3 n(O_3)\right]\left[a_7 n(O) + J_{HO_2}\right] + a_7 n(O)\, a_2 n(O_3)}{a_5 n(O)\left[a_7 n(O) + J_{HO_2}\right]} \tag{41a}$$

In this equation $a_3 n(O_3) < a_1 n(M)\, n(O_2)$; in the atmospheric region where $a_7 n(O) > J_{HO_2}$, (41a) becomes

$$\frac{n(OH)}{n(H)} = \frac{a_1 n(M)\, n(O_2) + a_2 n(O_3)}{a_5 n(O)} \tag{41b}$$

which is the simple expression for the ratio of $n(OH)/n(H)$ in the whole mesosphere.

In the same way the expression for the equilibrium ratio of $n(HO_2)/n(H)$ is obtained

$$\frac{n(HO_2)}{n(H)} = \frac{a_1 n(M)\, n(O_2) + a_3 n(O_3)}{a_7 n(O) + J_{HO_2}} + \frac{a_1 n(M)\, n(O_2) + a_2 n(O_3)}{a_7 n(O) + J_{HO_2}}$$
$$\times \frac{a_6 n(O_3)}{a_5 n(O)} \tag{41c}$$

and, where $a_7\, n(O) > J_{HO_2}$ and $a_3\, n(O_3) < a_1 n(M)\, n(O_2)$,

$$\frac{n(HO_2)}{n(H)} = \frac{a_1 n(M)\, n(O_2)}{a_7 n(O)} + \frac{a_6 n(O_3)}{a_5 n(O)} \times \frac{a_1 n(M)\, n(O_2) + a_2 n(O_3)}{a_7 n(O)} \tag{41d}$$

Expressions (41b) and (41d), assuming $a_5 = a_7$, lead to

$$\frac{n(HO_2)}{n(OH)} = \frac{a_1 n(M)\, n(O_2)}{a_1 n(M)\, n(O_2) + a_2 n(O_3)} + \frac{a_6 n(O_3)}{a_7 n(O)} \tag{41e}$$

which can be accepted where $a_7\, n(O) > J_{HO_2}$.

Numerical values show that

$$\frac{n(HO_2)}{n(OH)} = \frac{a_1 n(M) n(O_2)}{a_1 n(M) n(O_2) + a_2 n(O_3)} \tag{41f}$$

is a sufficient approximation in the lower thermosphere and mesosphere. In the stratosphere, the term $a_6 n(O_3)/a_7 n(O)$ could play a role. In the same way

$$\frac{n(HO_2)}{n(H)} = \frac{a_1 n(M) n(O_2)}{a_7 n(O)} \tag{41g}$$

is the practical ratio in the whole mesosphere. Numerical values are given in Table IV. The essential result is that the atomic hydrogen concentration is greater than the hydroxyl and hydroperoxyl concentrations in the upper mesosphere. The ratio of $n(HO_2)/n(OH)$, which is about 0.8, indicates that both radicals have almost the same concentrations in the whole of the mesosphere.

TABLE IV

Hydroxyl and hydroperoxyl radicals in the mesosphere. Overhead sun conditions.

Altitude	$n(OH)/n(H)$	$n(HO_2)/n(H)$	$n(HO_2)/n(OH)$
85	4.5×10^{-3}	2.8×10^{-3}	0.63
80	2.6×10^{-2}	2.1×10^{-2}	0.81
75	2.9×10^{-1}	2.5×10^{-1}	0.86
70	8.3×10^{-1}	7.0×10^{-1}	0.85
65	1.8×10^{0}	1.4×10^{0}	0.82
60	3.5×10^{0}	2.7×10^{0}	0.78
55	6.7×10^{0}	4.9×10^{0}	0.73
50	2.9×10^{1}	2.3×10^{1}	0.79

When hydrogen compounds are considered, the differential Equation (23) relating to atomic oxygen becomes

$$\frac{dn(O)}{dt} + 2k_1 n(M) n^2(O) + [k_2 n(M) n(O_2) + k_3 n(O_3) + a_5 n(OH)$$
$$+ a_7 n(HO_2)] n(O) = 2n(O_2) J_2 + n(O_3) J_3 \tag{42}$$

where the reactions with the a coefficients are the additional processes in which atomic oxygen is involved in a hydrogen-oxygen atmosphere. From this equation, one sees that oxygen atoms may disappear by reaction with OH and HO_2. Since $a_5 \simeq a_7 \simeq$ $\simeq 4 \times 10^{-11}$ cm^3 s^{-1}, a direct effect is not important at low altitudes where the term $k_2 n(M) n(O_2)$ (see Table II) reaches its highest values. However, in the mesosphere and particularly in its upper part, it is clear that $k_2 n(M) n(O_2)$ can be less than $a_5 n(OH) + a_7 n(HO_2)$.

Instead of (26a), for the upper mesosphere and lower thermosphere we must write

$$\frac{n(O_3)}{n(O)} = \frac{k_2 n(M) n(O_2)}{J_3 + k_3 n(O) + a_2 n(H)} \equiv \frac{k_2 n(M) n(O_2)}{J_{3a}} \tag{43}$$

with

$$J_{3a} = J_3 + k_3 n(O) + a_2 n(H) \simeq J_3 + a_2 n(H). \tag{44}$$

By introducing (43) into (42), the variation of atomic oxygen is given by

$$\frac{dn(O)}{dt} + 2n^2(O)\left[k_1 n(M) + \frac{k_3 k_2 n(M) n(O_2)}{J_{3a}}\right]$$
$$+ n(O)\left[a_2 n(H) \frac{k_2 n(M) n(O_2)}{J_{3a}} + a_5 n(OH) + a_7 n(HO_2)\right] =$$
$$= 2n(O_2) J_2. \tag{45}$$

Such an equation corresponds to photochemical conditions in the lower thermosphere and in the mesosphere. Numerical calculations near the mesopause level show (see Table II) that

$$k_1 n(M) = 1 \times 10^{-18} \text{ cm}^3 \text{ s}^{-1} \simeq k_3 k_2 n(M) n(O_2/J_3) \tag{46}$$

at 80 km; i.e. that the direct recombination rate of oxygen atoms is of the same order of magnitude as their association rate to form ozone. However, such rate coefficients are too small (see Equation 29) to lead to photochemical equilibrium conditions. As far as the term involving H, OH and HO_2 is concerned, the effect is also too small in the thermosphere and chemical equilibrium cannot be reached for atomic oxygen. Its vertical distribution in the lower thermosphere is subject to the transport conditions controlled by eddy diffusion.

In such conditions, if we adopt the conventional value $n(H) = 3 \times 10^7$ cm^{-3} at 100 km corresponding to

$$n(H) = n(N_2)/2 \times 10^5 \text{ cm}^{-3} \tag{47}$$

with the hypothesis that atomic hydrogen in the lower thermosphere is practically uniformly mixed, its concentration reaches 7×10^8 cm^{-3} at the mesopause (see Table V). A possible approximation for Equation (45) is

$$\frac{dn(O)}{dt} + \left[a_2 n(H) \frac{k_2 n(M) n(O_2)}{J_{3a}} + a_5 n(OH) + a_7 n(HO_2)\right] n(O) =$$
$$= 2n(O_2) J_2 \tag{48}$$

or even

$$\frac{dn(O)}{dt} + \left[a_2 n(H) \frac{k_2 n(M) n(O_2)}{J_3 + a_2 n(H)}\right] n(O) = 2n(O_2) J_2 \tag{49}$$

at the mesopause. With the following time τ_0

$$\tau_0 = \frac{J_3 + a_2 n(H)}{2 a_2 n(H) k_2 n(M) n(O_2)} \tag{50}$$

TABLE V
Adopted parameters in the lower thermosphere. Overhead sun conditions.

Altitude (km)	$n(O)$ (cm^{-3})	$n(H)$ (cm^{-3})	$n(O_3)$ (cm^{-3})	$n[O(^1D)]$ (cm^{-3})
120	1.0×10^{11}	4×10^6	6.0×10^2	5.1×10^3
115	1.5×10^{11}	6×10^6	4.0×10^3	5.1×10^3
110	2.3×10^{11}	9×10^6	2.7×10^4	4.1×10^3
105	2.9×10^{11}	1.5×10^7	2.0×10^5	2.2×10^3
100	3.2×10^{11}	3.5×10^7	1.6×10^6	1.2×10^3
95	3.3×10^{11}	8.7×10^7	1.3×10^7	5.0×10^2
90	3.0×10^{11}	2.3×10^8	1.1×10^8	4.2×10^2
85	3.0×10^{10}	6.9×10^8	1.0×10^8	1.2×10^2

necessary to reach a sufficient fraction of the equilibrium value, it is found that τ_0 at 85 km is only 8.4×10^3 sec. Thus, when atomic hydrogen plays a role at the mesopause level, an equilibrium value of $n(O)$ can be reached in about two hours. Adopting all the numerical parameters used here and for an overhead sun,

$$n(O)_{85\ km} \simeq 3 \times 10^{10}\ cm^{-3} \tag{51}$$

and, with (49) and (44),

$$n(O_3)_{85\ km} \simeq 10^8\ cm^{-3}. \tag{52}$$

In the lower thermosphere there is a transition region where the atomic oxygen concentration must increase rapidly since at 90 km the time to reach photoequilibrium is not less than 10^5 s.

After sunset, in the lower thermosphere,

$$\frac{dn(O_3)}{dt} + n(O_3)\,a_2 n(H) = k_2 n(M)\,n(O_2)\,n(O). \tag{53}$$

Provided $n(O)$ does not vary appreciably, the relevant solution to the differential Equation (53) is simply

$$n(O_3) = n(O_3)_{t=0} e^{-a_2 n(H)t} + \frac{k_2 n(M)\,n(O_2)\,n(O)}{a_2 n(H)}\left[1 - e^{-a_2 n(H)t}\right]. \tag{54}$$

Since $a_2 n(H)\,t = 1$ after less than 100 s at 100 km, the ozone concentration can be conveniently written in the nighttime thermosphere as

$$n(O_3) = \frac{k_2 n(M)\,n(O_2)\,n(O)}{a_2 n(H)} \tag{55}$$

while the daytime value is

$$n(O_3) = \frac{k_2 n(M)\,n(O_2)\,n(O)}{a_2 n(H) + J_3}. \tag{56}$$

The Bates-Nicolet process (32) which leads to the OH airglow spectrum occurs in the lower thermosphere according to (56),

$$\frac{dn(OH^*)_{v\leqslant 9}}{dt} = a_2n(H)\,n(O_3) = a_2n(H)\frac{k_2n(M)\,n(O_2)\,n(O)}{J_3 + a_2n(H)} \tag{57}$$

or for nighttime conditions, (55),

$$\frac{dn(OH^*)}{dt} = k_2n(M)\,n(O_2)\,n(O) = +\frac{dn(O_3)}{dt} \tag{58}$$

which corresponds to the ozone formation. Thus, in the lower thermosphere, the number of excited molecules OH* which are produced is almost 10^7 cm^{-3} s^{-1}, i.e. a total production which is not less than 5×10^{12} OH* molecules cm^{-2} s^{-1}.

In the region where atomic oxygen is in photochemical equilibrium, the conventional stratospheric equation

$$\frac{dn(O_3)}{dt} + 2n^2(O_3)\frac{J_3k_3}{k_2n(M)\,n(O_2)} = 2n(O_2)\,J_2 \tag{59}$$

indicates that photochemical equilibrium values can be reached even in the upper stratosphere. Below 35 km, the times required become too long and departures from equilibrium conditions are observed. However, (59) must be replaced in a hydrogen-oxygen atmosphere by

$$\frac{dn(O_3)}{dt} + 2n^2(O_3)\frac{J_3k_3}{k_2n(M)\,n(O_2)} + n(O_3)\,a_2n(H) +$$
$$+ n(O_3)\left\{\left[\frac{a_5J_3}{k_2n(M)\,n(O_2)} + a_6\right]n(OH) +\right.$$
$$\left. + \left[\frac{a_7J_3}{k_2n(M)\,n(O_2)} + a_{6c}\right]n(HO_2)\right\} = 2n(O_2)\,J_2 \tag{60}$$

in which the effect of H, OH and HO$_2$ reactions is represented by the symbol a from reactions (32), (35), (36) and (37). If (60) is compared with (59), it is clear that the effect of hydrogen compounds is equivalent to an increase in the photodissociation coefficient J_3. In other words, the ozone dissociation is increased by additional processes involving oxygen atoms so that the equivalent O$_3$ photodissociation coefficient is written

$$J_{3A} = J_3[1 + A] \tag{61}$$

where A is given by

$$A = \frac{a_2n(H)\,k_2n(M)\,n(O_2)}{2J_3k_3n(O)} + \frac{a_5n(OH)}{2k_3n(O_3)}\left[1 + \frac{a_6}{a_5}\cdot\frac{k_2n(M)\,n(O_2)}{J_3}\right] +$$
$$+ \frac{a_7n(HO_2)}{2k_3n(O_3)}\left[1 + \frac{a_{6c}}{a_5}\frac{k_2n(M)\,n(O_2)}{J}\right]. \tag{62}$$

Using (60) and (61), the conventional equation for equilibrium conditions in the stratosphere is replaced by

$$n(O_3) = \left[\frac{k_2}{k_3} n(M) n^2(O_2) \frac{J_2}{J_3(1+A)}\right]^{1/2}. \tag{63}$$

The correction term $(1+A)^{1/2}$ in the stratosphere where $n(H)$ is negligible, and when a_6 is neglected and $a_7 = a_5$, can be written

$$[1+A]^{1/2} = \left[1 + \frac{a_7\{n(OH) + n(HO_2)\}}{2k_3 n(O_3)}\right]^{1/2}. \tag{64}$$

It can be shown that a very high mixing ratio $n(H_2O)/n(M)$ is required in order to have an important effect on the ozone concentration in the stratosphere. For example, a mixing ratio $n(H_2O)/n(M) = 6.5 \times 10^{-6}$ leads to a value $(1+A)^{1/2} = 1.5$ in the stratosphere. Such a modification is only important from a theoretical point of view. It seems that almost impossibly high mixing ratios $(5 \times 10^{-5}, 10^{-4})$ would be required in order to affect in any significant way the ozone concentration in the lower stratosphere.

In the mesosphere there is an important difference between a pure oxygen atmosphere and a hydrogen-oxygen atmosphere. The correction term which applies in the mesosphere

$$[1+A]^{1/2} = \left[1 + \frac{a_2 n(H) k_2 n(M) n(O_2)/J_3 + a_5 n(OH) + a_7 n(HO_2)}{2k_3 n(O_3)}\right]^{1/2} \tag{65}$$

increases to about 100 at the mesopause if it is 1.5 at the stratopause. There is a considerable decrease in the atomic oxygen concentration and consequently a parallel decrease in the ozone concentration, Figure 9 illustrates the difference between an oxygen mesosphere and a hydrogen-oxygen mesosphere.

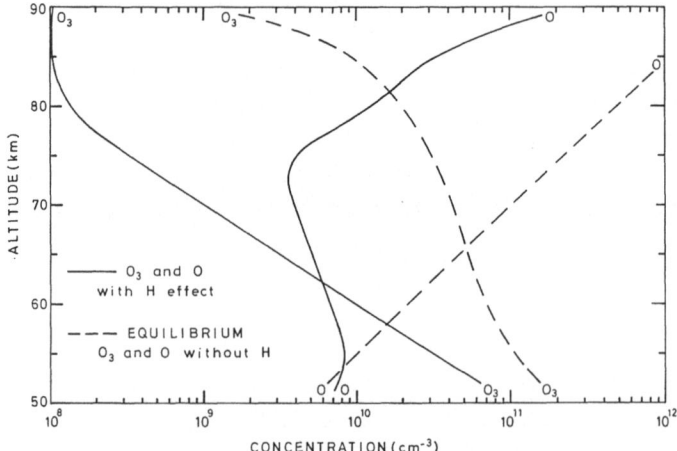

Fig. 9.   Equilibrium conditions in the mesosphere of ozone and atomic oxygen in a pure oxygen-nitrogen atmosphere and in an atmosphere where hydrogen is involved.

In the thermosphere, atomic hydrogen is the most important hydrogen constituent, $n(\text{H})/n(\text{OH}) > 1$ and $n(\text{H})/n(\text{HO}_2) > 1$, and its vertical distribution (Table V) has been assumed to be a mixing distribution. Atomic oxygen concentrations are also fixed according to arbitrary eddy diffusion coefficients. Dynamic effects acting on the vertical distribution of atomic oxygen will modify the ozone distribution in the lower thermosphere. Below the mesopause, 85 to 80 km, the aeronomic conditions are such that they correspond to a transition region between mixing and photochemical equilibrium. In the mesosphere, therefore, it is necessary to determine the production of hydroxyl and hydroperoxyl radicals.

5. Methane in the Stratosphere and Mesosphere

Methane, which has been found as a permanent constituent of the troposphere, has continuous sources at ground level. Its total amount by volume is 1.5×10^{-6} of the major gases N_2 and O_2. Methane can be dissociated by ultraviolet radiation. However, since the absorption cross-section is very small at $\lambda > 1500$ Å, the aeronomic photodissociation of CH_4 arises principally from Lyman-α with a photodissociation coefficient at zero optical depth

$$J_{CH_4}(\lambda = 1216 \text{ Å}) = 5 \times 10^{-6} \text{ s}^{-1}. \tag{66a}$$

The total photodissociation coefficient is

$$J_{CH_4} = 7 \times 10^{-6} \text{ s}^{-1}, \tag{66b}$$

and it must be considered only above 100 km. Thus, the photodissociation coefficient of CH_4 is a decreasing function of the optical depth in the mesosphere and is negligible in the lower mesosphere (Figure 10). The photolysis of CH_4 due to Lyman-α may be

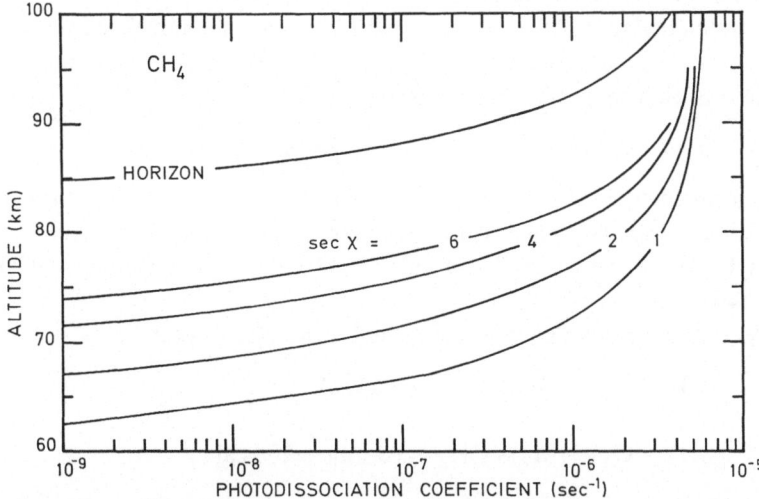

Fig. 10. Photodissociation coefficient of methane for various solar zenith angles. From overhead sun conditions, $\chi = 0$, to $\chi = 90°$, horizon.

represented by (see a discussion by Strobel, 1969),

$$(75\%), \quad CH_4 + Ly\text{-}\alpha \rightarrow CH_2 + H_2 \tag{67a}$$

$$(25\%), \quad CH_4 + Ly\text{-}\alpha \rightarrow CH_3 + H. \tag{67b}$$

But photodissociation is not the only disruptive process of CH_4. The oxidation of methane according to

$$O(^3P) + CH_4 \rightarrow OH + CH_3 - 2 \text{ kcal}, \tag{68}$$

being endothermic by about 2 kcal, is a slow reaction. Its activation energy is not less than 7 kcal, according to various laboratory measurements (see for example, Brown and Thrush, 1967; Westenberg and de Haas, 1969; Wong and Potter, 1967). Adopting the minimum activation energy, the rate coefficient is written

$$a_{CH_4-O} = 3.5 \times 10^{-13} \, T^{1/2} \, e^{-3650/T} \text{ cm}^3 \text{ s}^{-1}. \tag{69}$$

Other oxidation processes may be considered; with OH the following reaction is possible

$$OH + CH_4 \rightarrow CH_3 + H_2O + 15 \text{ kcal}. \tag{70}$$

This reaction, according to recent laboratory measurements (Greiner, 1970), has a rate coefficient which can be written

$$a_{OH-CH_4} = 1.8 \times 10^{-13} \, T^{1/2} \, e^{-1750/T} \text{ cm}^3 \text{ s}^{-1}. \tag{71}$$

A reaction such as

$$HO_2 + CH_4 \rightarrow CH_3 + H_2O_2 - 15 \text{ kcal} \tag{72}$$

is endothermic by about 15 kcal and it is difficult to see how hydroperoxyl radicals could be responsible for an attack on methane.

When the methyl radical CH_3 is produced, i.e. when the removal of a single hydrogen atom from a methane molecule takes place, it is certain that there is a permanent destruction of that molecule. The following reaction

$$O(^3P) + CH_3 \rightarrow H + HCHO + 67 \text{ kcal} \tag{73}$$

yielding formaldehyde (Niki et $al.$, 1968) can be considered as the most important process since its rate coefficient is not less than $3 \times 10^{-13} \text{ cm}^3 \text{ s}^{-1}$ at 300K. Reaction (73) may be followed by $CH_2O + O \rightarrow CHO + OH$ and can end by $O + HCO \rightarrow OH + CO + 84$ kcal.

Methyl radicals react also with ozone and molecular oxygen

$$CH_3 + O_3 \rightarrow CH_3O + O_2 + 66 \text{ kcal} \tag{74}$$

and

$$CH_3 + O_2 (+ M) \rightarrow CH_3O_2 (+ M) + 26 \text{ kcal}. \tag{75}$$

If we consider that atomic oxygen can be in the electronically excited ^1D state, the reaction $O(^1D) + CH_4$ is possible (Cadle, 1964) with no activation energy (De More and

Raper, 1967). It proceeds by the three paths

$$O(^1D) + CH_4 + M \rightarrow CH_3OH + M \tag{76}$$

$$O(^1D) + CH_4 \rightarrow CH_2O + H_2O \tag{77}$$

$$O(^1D) + CH_4 \rightarrow CH_3 + OH \tag{78}$$

which, from an aeronomic point of view, correspond to a dissociation process of CH_4. As (78) seems to be the most important reaction (De More and Raper, 1967), we conclude that

$$O(^1D) + CH_4 \rightarrow CH_3 + OH\,(v \leqslant 4) + 43.5 \text{ kcal} \tag{79}$$

with the following working value for the rate coefficient

$$a^*_{CH_4-O^*} = 10^{-10} \text{ cm}^3 \text{ s}^{-1}. \tag{80}$$

Such a reaction will play a role in a sunlit atmosphere where the ozone photodissociation is important, i.e. in the stratosphere and lower mesosphere. A reaction such as

$$O(^1S) + CH_4 \rightarrow CH_3 + OH \tag{81}$$

cannot be important since the quenching rate coefficient, which is about 5×10^{-14} cm^3 s^{-1} (Filseth et al., 1970) at 298 K, is too small.

From the foregoing discussion it is apparent that the mechanisms responsible for the dissociation of CH_4 can be identified with an oxidation process by $O(^1D)$ atoms throughout the stratosphere, by $O(^3P)$ atoms near the stratopause level and by photodissociation by Lyman-α in the mesosphere.

Using the numerical values of $O(^1D)$ and $O(^3P)$ atoms which are given in Table III for overhead sun conditions, the dissociation of CH_4 can be determined and is particularly important near the stratopause level. With a mixing distribution such as is indicated in Table VI, the average diffusion flow which would be required at the stratosphere should be

$$F_{50 \text{ km}}(CH_4) > 10^9 \text{ cm}^2 \text{ s}^{-1}. \tag{82}$$

Under mixing conditions, the average molecular diffusion flow F_D in the stratosphere is

$$F_D(CH_4) = 7 \times 10^6 \text{ cm}^2 \text{ s}^{-1} \tag{83}$$

which is a negligible fraction of the possible loss by an oxidation process. Since the tropospheric mixing time is short enough to lead to a uniform distribution of CH_4 in the whole troposphere, the injection rate of CH_4 into the stratosphere is the essential process which must be considered. In other words, the conditions of the tropopause must determine the behavior of the stratospheric CH_4. Assuming that the CH_4 is injected with no restriction into the stratosphere, it is possible to determine the conditions which are required to maintain a certain vertical distribution of a constituent which is continuously lost, such as CH_4.

The vertical diffusive speed w_D of a minor constituent of mass m_1 and concentration n_1 relative to the general mass m is of the form (see e.g. Nicolet, 1968)

$$w_D = - D_{12} \left[\frac{1}{n_1} \frac{\partial n_1}{\partial r} + \frac{m_1}{mH} + \frac{1}{T} \frac{\partial T}{\partial r} \right] \tag{84}$$

where D_{12} is the coefficient of molecular diffusion. H is the atmospheric scale height

$$H = kT/mg \tag{85}$$

where k is the Boltzmann constant, T is the temperature and g denotes the gravitational acceleration. The properties of the gas are taken to vary in only one direction, namely along the Earth's radius; r is the geocentric height.

Since the differentiation of the gas law in the form $p = nkT$, where p and n denote the total pressure and total concentration respectively, leads to

$$\frac{dp}{p} = \frac{dn}{n} + \frac{dT}{T} = - \frac{dr}{H}, \tag{86}$$

if the static law $dp = -gnm\, dr$ is used, it is possible to use the scale height $\mathcal{H}_1$ for the constituent of mass m_1 defined by

$$\frac{dp_1}{p_1} = \frac{dn_1}{n_1} + \frac{dT}{T} = - \frac{dr}{\mathcal{H}_1}. \tag{87}$$

With (87) the formula (84) for molecular diffusion becomes

$$w_D = \frac{D_{12}}{H} \left[\frac{H}{\mathcal{H}_1} - \frac{m_1}{m} \right]. \tag{88}$$

When eddy diffusion is involved, the vertical speed is defined by

$$w_{\mathcal{D}} = \frac{\mathcal{D}_{12}}{H} \left[\frac{H}{\mathcal{H}_1} - 1 \right] \tag{89}$$

where $\mathcal{D}_{12}$ is the eddy diffusion coefficient. Since the coefficient of molecular diffusion in the stratosphere is

$$2 \text{ cm}^2 \text{ s}^{-1} \leqslant D_{CH_4} \leqslant 4 \times 10^2 \text{ cm}^2 \text{ s}^{-1} \tag{90}$$

it is clear that eddy diffusion must be involved in order to maintain CH_4 in the stratosphere. Considering that the methane which is destroyed chemically in a sunlit atmosphere by reaction with $O(^1D)$ oxygen atoms is given approximately by

$$F_{CH_4} = \tfrac{1}{2} a^*_{O^*-CH_4} n(O^1D)\, n(CH_4)\, H(CH_4) \text{ cm}^2 \text{ s}^{-1} \tag{91}$$

and since the loss must correspond to the diffuse upward current of CH_4 molecules by eddy diffusion

$$F_{CH_4} = n(CH_4)\, w_{CH_4} = \frac{n(CH_4)\, \mathcal{D}_{12}}{H} \left[\frac{H}{\mathcal{H}_1} - 1 \right] \text{cm}^2 \text{ s}^{-1}, \tag{92}$$

it is possible to determine for a certain ratio H_{CH_4}/H, the eddy coefficient $\mathscr{D}_{12}$ which is required. With $H_{CH_4}=0.9H$, in order to indicate a small departure from mixing conditions, and $H_{CH_4}=0.5H$, to show that there is a definite decrease in the mixing ratio in the stratosphere, the following expressions are obtained:

$$\mathscr{D}_{12}(0.9H) = 4a^*n^*H^2 \tag{93}$$

if $H_{CH_4}=0.9H$, and

$$\mathscr{D}_{12}(H/2) = a^*n^*H^2/4 \tag{94}$$

if $H_{CH_4}=H/2$. Results are given in Table VI and are also illustrated in Figure 11. An eddy diffusion coefficient greater than 10^5 cm^2 s^{-1} would be required in the upper stratosphere in order to maintain an approximately constant volume ratio of

TABLE VI

Dissociation of CH$_4$ under mixing conditions in the stratosphere for an overhead sun

Altitude (km)	Concentration (cm^{-3})	O(^1D) oxidation (cm^{-3} s^{-1})	O(^1D) oxidation (cm^{-2} s^{-1})	Eddy diffusion coefficient (cm^{-2} s^{-1})
15	5.8×10^{12}	5.3×10^1	–	1.4×10^1
20	2.6×10^{12}	2.3×10^2	8.0×10^7	1.5×10^2
25	1.2×10^{12}	5.7×10^2	2.9×10^8	8.9×10^2
30	5.3×10^{11}	1.3×10^3	7.7×10^8	4.8×10^3
35	2.5×10^{11}	2.5×10^3	1.8×10^9	2.3×10^4
40	1.2×10^{11}	4.0×10^3	3.9×10^9	8.3×10^4
45	6.4×10^{10}	3.9×10^3	5.5×10^9	1.6×10^5
50	3.5×10^{10}	2.1×10^3	6.9×10^9	1.6×10^5

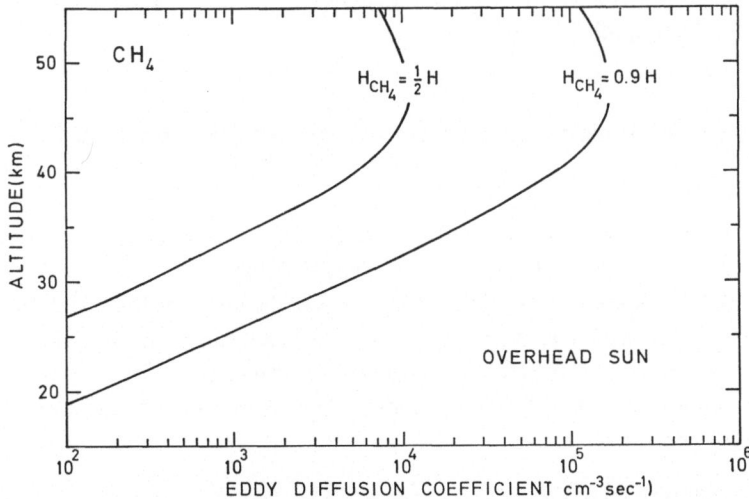

Fig. 11. Eddy diffusion coefficient which would be required in the stratosphere to maintain a vertical distribution of CH$_4$ with a scale height H_{CH_4} corresponding to $\frac{1}{2}$ and to 9/10 of the atmospheric scale height H.

$n(\mathrm{CH_4})/n(\mathrm{M})$. Even for a scale height $H_{\mathrm{CH_4}} = \frac{1}{2}H(\mathrm{M})$, an eddy diffusion coefficient not less than 10^4 cm^2 s^{-1} is required below the stratopause. It seems, therefore, that it is not possible to maintain a constant mixing ratio of methane in the stratosphere. Recent observations (Bainbridge and Heidt, 1966; Scholz et al., 1970) show that there is a decrease of the CH$_4$ mixing ratio in the stratosphere and, therefore, that the eddy diffusion coefficient in the upper stratosphere is certainly less than 10^4 cm^2 s^{-1}. Thus, the oxidation of the stratospheric methane leads to a continuous production of stratospheric water vapor with the formation of CO and CO$_2$. Further measurements of the stratospheric CH$_4$ distribution are needed to determine the scale height of CH$_4$ in the stratosphere and, therefore, to provide information on its injection from the troposphere into the stratosphere and on the upward diffuse current resulting from eddy diffusion. Finally, since the oxidation of methane in the stratosphere is a very important process, its presence in the mesosphere will not be sufficient to permit it to play a leading role in the various reactions in which hydrogen compounds are involved.

6. Water Vapor in the Stratosphere and Mesosphere

The water vapor content is very small in the stratosphere; 3×10^{-6} is not an unreasonable value to adopt for the order of magnitude of the fractional volume concentration of water vapor (Williamson and Houghton, 1965; Mastenbrook, 1968; Sissenwine et al., 1968; McKinnon and Morewood, 1970). Since the amount of methane by volume is 1.5×10^{-6} and that of molecular hydrogen is not far from 0.5×10^{-6}, the total amount of hydrogen atoms cannot be far from 10^{-5} above the tropopause; the equivalent H$_2$O amount should correspond to 6.5×10^{-6}.

An effect caused by oxygen atoms in their normal state ^3P on H$_2$O has not been detected since the reaction is endothermic, but atoms in their first excited state ^1D lead to (Engleman, 1965)

$$\mathrm{O}(^1D) + \mathrm{H_2O} \rightarrow \mathrm{OH} + \mathrm{OH^*}\,(v \leqslant 2) + 28.8 \text{ kcal} \tag{95}$$

for which the following round figure for the rate coefficient is adopted

$$a_{\mathrm{H_2O}}^* = 10^{-10} \text{ cm}^3 \text{ s}^{-1} \tag{96}$$

while the possible variation is $(1 \pm 0.5) \times 10^{-10}$ cm^3 s^{-1}. With the O(^1D) concentrations which are used here (Table III), it is clear that there is permanent destruction of H$_2$O in the stratosphere. The dissociation rate coefficient is of the order 10^{-10} s^{-1} in the lower stratosphere (see Figure 12) for overhead sun conditions and reaches a maximum of more than 5×10^{-8} s^{-1} near the stratopause.

Water vapor can be dissociated by sunlight and its photodissociation coefficient for zero optical depth is

$$J_{\mathrm{H_2O}} = 10^{-5} \text{ s}^{-1} \tag{97}$$

corresponding to

$$\mathrm{H_2O} + h\nu \rightarrow \mathrm{OH} + \mathrm{H} \quad \text{or} \quad \mathrm{H_2} + \mathrm{O}. \tag{98}$$

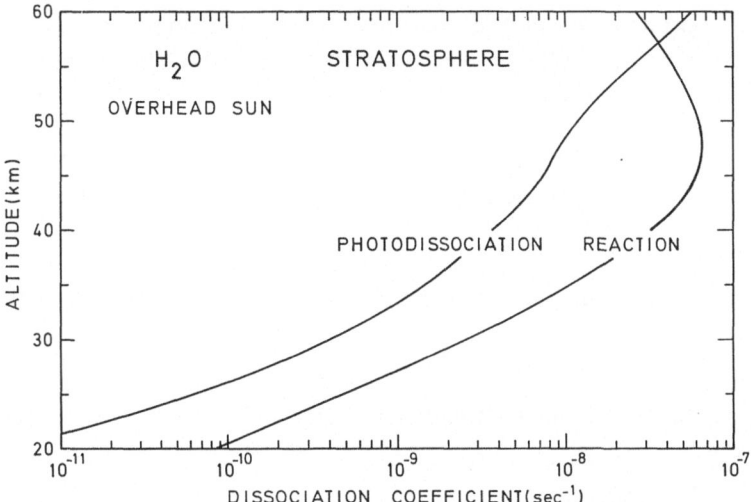

Fig. 12.   Dissociation coefficient of the water vapor in the stratosphere. The effect of the reaction of H_2O with $O(^1D)$ is more important than the direct photodissociation.

The photodissociation in the upper mesosphere is due mainly to the effect of Lyman-α and the effect of the spectral range of the Schumann-Runge bands plays the leading role in the lower mesosphere. The result of the calculation of the photodissociation is given in Figure 13 for overhead sun conditions. The photodissociation rate coefficient decreases in the mesosphere by a factor of the order of 1000, since it reaches at 50 km about

$$J_{H_2O}(50 \text{ km}) \simeq 10^{-8} \text{ s}^{-1}. \tag{99}$$

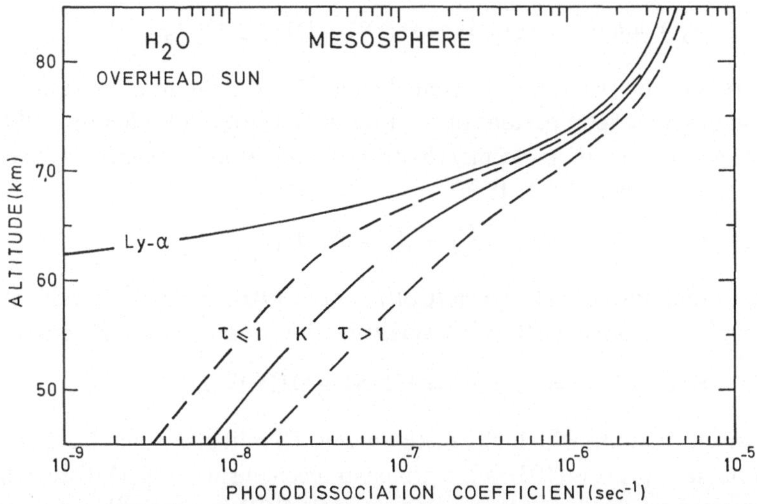

Fig. 13.   Photodissociation coefficient of H_2O in the mesosphere. In the upper part of the mesosphere Lyman-α (curve Ly-α) plays the principal role. In the lower part the dissociation comes from the solar radiation of the spectral region involving the Schumann-Runge bands.

In the stratosphere, the photodissociation process is less important than the oxidation process (see Figure 12) due to excited oxygen atoms $O(^1D)$. As before, the effect of $O(^1S)$ atoms is considered negligible since their production rate is too small as compared with other production processes.

It must be pointed out that the photodissociation process which corresponds to Lyman-α leads to a dissociation coefficient J_{H_2O} for zero optical depth

$$J_{H_2O}(\text{Ly-}\alpha) = 4.3 \times 10^{-6}\,\text{s}^{-1} \tag{100a}$$

which is identical to the photodissociation coefficient in the Schumann-Runge continuum

$$J_{H_2O}(\text{SRC}) = 4.3 \times 10^{-6}\,\text{s}^{-1}. \tag{100b}$$

The values in the O_2 Schumann-Runge band region and in the O_2 Herzberg continuum are respectively,

$$J_{H_2O}(\text{SRB}) = 1.5 \times 10^{-6}\,\text{s}^{-1} \tag{100c}$$

and

$$J_{H_2O}(\text{HER}) = 1.2 \times 10^{-9}\,\text{s}^{-1}. \tag{100d}$$

Since the H_2O photodissociation in the lower mesosphere depends on its absorption in the Schumann-Runge band region, the optical depth is related to the band structure. In Figure 13, the results of the exact computation with the detailed rotational structure and the data obtained with mean coefficient adapted to $\tau \leqslant 1$ and $\tau \geqslant 1$ show that an error of a factor of 2 is involved even when appropriate constant cross-sections for the absorption of molecular oxygen are used.

At the mesopause level, the photodissociation of H_2O by Lyman-α is the most important process. At 1216 Å the following process occurs

$$H_2O + h\nu(\lambda = 1216\,\text{Å}) = H(^2S) + OH(^2\Sigma^+)_{v' \leqslant 2} \tag{101}$$

where OH is in the electronically excited state $^2\Sigma^+$ leading to a fluorescence process corresponding to several percent of the total dissociation (Carrington, 1964).

Direct production of molecular hydrogen and atomic oxygen is also possible (McNesby et al., 1962; Stief, 1966)

$$(a_{12}); \quad H_2O + h\nu(\lambda = 1216\,\text{Å}) = H_2 + O(^1D). \tag{102}$$

The exact process is not yet known (Cottin et al., 1966; Welge and Stuhl, 1967) since the probability of process (102) is a small fraction of that of the general process

$$(a_{13}); \quad H_2O + h\nu(\lambda = 1216\,\text{Å}) = H(^2S) + OH(X^2\Pi) \tag{103}$$

which leads to H atoms with high kinetic energy. Thus, it seems that from an aeronomic point of view, process (103) is the principal mechanism for H_2O dissociation in the upper mesosphere and lower thermosphere while (101) represents between 1 and 10% of the total. As far as the photodissociation (102) leading to H_2 is concerned, it is not certain whether it is a sufficiently large fraction to be considered in the upper mesosphere (Cottin *et al.*, 1966). In any case, such a process can be neglected in the aeronomic analysis of the H_2O photolysis, but perhaps not in the production of molecular hydrogen at mesospheric levels.

When photodissociation of water vapor proceeds through (103) its re-formation may occur through

$$(a_{14}); \quad H + OH + M \rightarrow H_2O + M + 118 \text{ kcal} \tag{104}$$

which is not important compared to other possible aeronomic processes.

The reaction of hydrogen atoms and perhydroxyl radicals leads to two hydroxyl radicals

$$(a_{15}); \quad H + HO_2 \rightarrow OH + OH^*_{v \leqslant 4} + 39 \text{ kcal} \tag{105a}$$

a reaction which has been observed (Cashion and Polanyi, 1959). With

$$a_{15} = 1.5 \times 10^{-12} \, T^{1/2} \text{ cm}^3 \text{ s}^{-1} \tag{105b}$$

the experimental conditions can be followed (Kaufman 1964).

The reaction between two hydroxyl radicals leads to the re-formation of water vapor

$$(a_{16}); \quad OH + OH \rightarrow H_2O + O + 17 \text{ kcal} \tag{106a}$$

with an activation energy not greater than 2 kcal. All the measurements have been made at 300 K and the correct value is not yet known with sufficient precision (Dixon-Lewis *et al.*, 1966; Kaufman, 1969; Breen and Glass, 1970). Thus, the following rate coefficient, which is adopted for stratospheric and mesospheric conditions,

$$a_{16} = 5 \times 10^{-12} \, T^{1/2} \, e^{-1000/T} \text{ cm}^3 \text{ s}^{-1}. \tag{106b}$$

leads to about 10^{-13} cm^3 s^{-1} near the mesopause (170 K) and 2×10^{12} cm^3 s^{-1} near the stratopause (270 K).

In the same way, the reaction between hydroxyl and perhydroxyl radicals leads to H_2O

$$(a_{17}); \quad OH + HO_2 \rightarrow H_2O + O_2 + 72 \text{ kcal} \tag{107a}$$

with a small activation energy (~ 1 kcal). There is no direct measurement (Kaufman, 1969), but

$$a_{17} = 3 \times 10^{-12} \, T^{1/2} \, e^{-500/T} \text{ cm}^3 \text{ s}^{-1} \tag{107b}$$

which is adopted for aeronomic applications, leads to 1.7×10^{-12} cm^3 s^{-1} at the mesopause and 8×10^{-12} cm^3 s^{-1} at the stratopause.

Other reactions can lead to the formation of H_2O, but they are not important in the mesosphere and stratosphere compared with those reactions leading to its re-formation. The three-body association

$$(a_{18}); \quad H_2 + O + M \rightarrow H_2O + M + 116 \text{ kcal} \tag{108}$$

should have a small rate coefficient since it is spin forbidden when normal oxygen atoms are involved. Such a reaction is, therefore, neglected.

The exothermic reaction between H_2 and OH

$$(a_{19}); \quad H_2 + OH \rightarrow H_2O + H + 15 \text{ kcal} \tag{109a}$$

requires a relatively high activation energy. With a steric hindrance factor of about 0.01, the measurements at 300–500 K (Greiner 1969) lead to

$$a_{19} = 2.0 \times 10^{-18} \ T^{1/2} \ e^{-1800/T} \text{ cm}^3 \text{ s}^{-1}. \tag{109b}$$

Finally, in addition to (105), the reaction between hydrogen atoms and perhydroxyl radicals may lead to the re-formation of H_2O

$$(a_{20}); \quad H + HO_2 \rightarrow H_2O + O + 55 \text{ kcal} \tag{110a}$$

with a rate coefficient less important than a_{15}. We adopt the following working value

$$a_{20} = 1.5 \times 10^{-12} \ T^{1/2} \ e^{-2000/T} \text{ cm}^3 \text{ s}^{-1}. \tag{110b}$$

If only (105) to (110) are involved in the production of H_2O, equilibrium conditions are

$$n(H_2O) \{J_{H_2O} + a^*_{H_2O} n[O(^1D)]\} = a_{20} n(HO_2) \, n(H) +$$
$$+ [a_{16} n(OH) + a_{17} n(HO_2)] \, n(OH) + a_{19} n(OH) \, n(H_2). \tag{111A}$$

An approximation for mesospheric conditions can be

$$n(H_2O) \{J_{H_2O} + a^*_{H_2O} n[O(^1D)]\} \simeq a_{17} n(HO_2) \, n(OH). \tag{111B}$$

Since the ratio of $n(HO_2)/n(OH)$ (see Table IV) is about 0.8 in the mesosphere, it is possible to obtain an idea of the possible concentration of the hydroxyl and perhydroxyl radicals if the water vapor concentration is known. With the mixing ratio 6.5×10^{-6}, numerical values of $n(HO_2) \, n(OH)$ can be obtained; they are not less than 10^{15} cm^{-6} and lead to OH and HO_2 concentrations greater than 10^7 cm^{-3}. However, in the upper mesosphere (111B) does not represent the real conditions since hydrogen is involved in the chemical equilibrium of hydroxyl and hydroxyl radicals (Equations (41)). In any case at the stratopause the concentrations, which are $n(OH) \simeq n(HO_2) = 5 \times 10^7$ cm^{-3}, correspond to a total dissociation rate of 10^4 H_2O molecules cm^{-3} s^{-1}, and are practically photochemical equilibrium values.

Thus there is a clear indication that in the whole stratosphere the dissociation of water vapor does not modify its vertical distribution if the reaction between OH and HO_2 giving H_2O plays a leading role. Furthermore, the diffuse upward current due to eddy diffusion, which is needed in order to maintain CH_4 at the stratopause, is not needed by H_2O. Considering the destruction of H_2O in the stratosphere principally by its reaction with $O(^1D)$ atoms (cf. Figure 12) it is possible to consider the eddy diffusion coefficients which would be needed (cf. CH_4) with no re-formation of water vapor. Figure 14 shows the increasing ratios of the eddy diffusions coefficients which would be necessary in the stratosphere. Since there is a continuous re-formation of H_2O, no transport is needed to maintain its normal distribution with altitude. However, in the mesosphere, where the OH and HO_2 concentration decreases, the re-formation of water vapor is too slow and a diffuse upward current of H_2O is required.

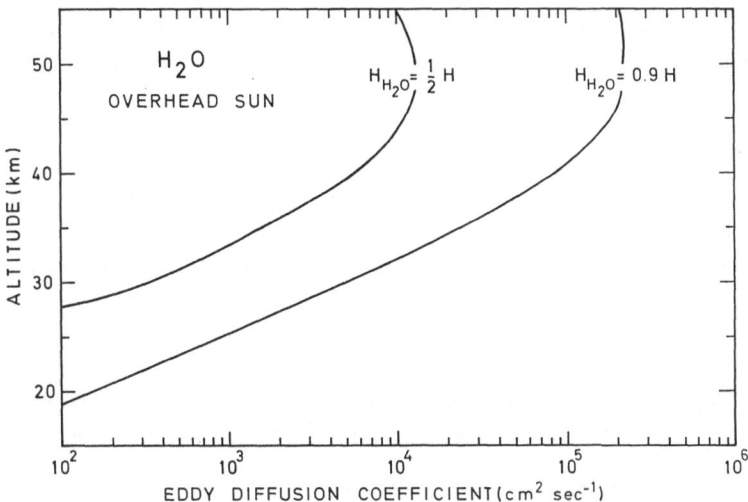

Fig. 14. Eddy diffusion coefficients which would be necessary in the stratosphere to maintain H_2O assuming that there is no re-formation. See Figure 11.

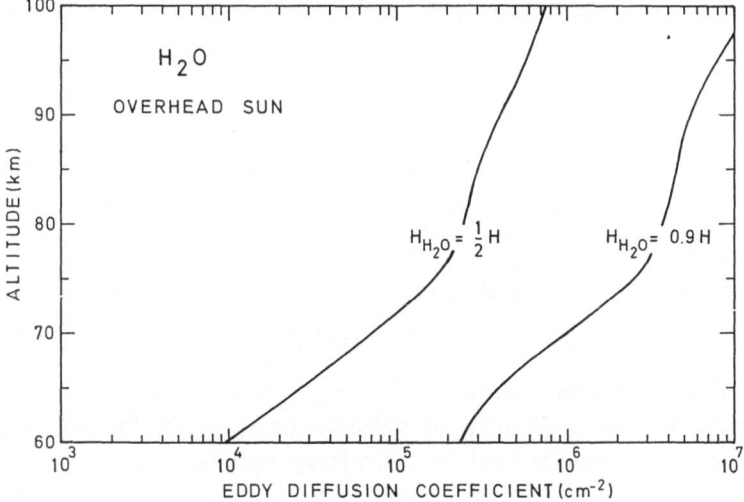

Fig. 15. Eddy diffusion coefficients which are required in the mesosphere and lower thermosphere to maintain H_2O which is subject to photodissociation.

Figure 15 shows that the mesospheric photodissociation requires an eddy diffusion coefficient greater than 10^6 cm^2 s^{-1} to maintain a mixing distribution at the mesopause. The eddy diffusion coefficient increases up to 10^7 cm^2 s^{-1} at 100 km if mixing of water vapor is adopted at 100 km.

Thus, the aeronomic problem of water vapor in the mesosphere depends on the photochemical conditions that are adopted at the stratopause and on the vertical distribution of the eddy diffusion coefficient above the stratopause.

A coefficient less than 2×10^5 cm^2 s^{-1} (see Figure 15) would lead to a significant reduction of the mixing ratio since $H_{H_2O} \leqslant \frac{1}{2} H_M$, the H$_2$O scale height being less than half of the atmospheric scale height. Values reaching 4×10^6 cm^2 s^{-1} are necessary to maintain an almost constant mixing ratio up to the mesopause. It is clear that the vertical distribution of H$_2$O in the neighbourhood of the mesopause and in the lower mesosphere, and also the aeronomic conditions for other minor constituents, will depend on the choice made for the eddy diffusion coefficient in the calculations of the diffuse upward current. Special care should be taken to avoid unrealistic values for the diffusion coefficients.

7. Molecular Hydrogen

Molecular hydrogen, which is a permanent constituent of the troposphere, may be produced at high altitude by the three-body association of two hydrogen atoms

$$(a_{21}); \quad H + H + M \rightarrow H_2 + M + 103.2 \text{ kcal} \tag{111a}$$

with the conventional value (Larkin and Thrush, 1964; Bennett and Blackmore, 1968) for mesospheric conditions

$$a_{21} = 1 \times 10^{-32} \text{ cm}^6 \text{ s}^{-1}. \tag{111b}$$

The bimolecular reaction

$$(a_{22}); \quad H + OH \rightarrow H_2 + O + 1.9 \text{ kcal} \tag{112a}$$

requires a high activation energy (see its reverse reaction 114). Its rate coefficient must be between

$$a_{22} = 5 \times 10^{-13} T^{1/2} e^{-4400/T} \text{ cm}^3 \text{ s}^{-1} \tag{112b}$$

with the highest activation energy, and

$$a_{22} = 2 \times 10^{-13} T^{1/2} e^{-3400/T} \text{ cm}^3 \text{ s}^{-1} \tag{112c}$$

with the lowest activation energy.

In fact a more important mode of molecular hydrogen formation occurs through the reaction of hydrogen atoms with perhydroxyl radicals

$$(a_{23}); \quad H + HO_2 \rightarrow H_2 + O_2 + 57 \text{ kcal} \tag{113a}$$

which has a coefficient of about 3×10^{-12} cm^3 s^{-1} at 300 K (Kaufman, 1964, 1969). Thus, it is possible to assume an activation energy not greater than 2 kcal in order to write

$$a_{23} = 5 \times 10^{-12} T^{1/2} e^{-1000/T} \text{ cm}^3 \text{ s}^{-1}. \tag{113b}$$

The reaction

$$(a_{24}); \quad H_2 + O \rightarrow OH + H - 1.9 \text{ kcal}, \tag{114a}$$

which is endothermic by about 1.9 kcal, cannot be neglected as a loss process in the thermosphere where the temperature is relatively high. From the experimental observations (Clyne and Thrush, 1963; Wong and Potter, 1965; Ripley and Gardiner, 1966; Westenberg and De Haas, 1967; Campbell and Thrush, 1968)

$$a_{24} = 2.5 \times 10^{-12} \, T^{1/2} \, e^{-5000/T} \, \text{cm}^3 \, \text{s}^{-1} \tag{114b}$$

with the highest activation energy or

$$a_{24} = 5 \times 10^{-13} \, T^{1/2} \, e^{-4400/T} \, \text{cm}^3 \, \text{s}^{-1} \tag{114c}$$

with the lowest activation energy. It is difficult to make a choice between the various values.

The photodissociation of H_2 (Dalgarno and Allison, 1969; Mentall and Gentieu, 1970) can be ignored in the mesosphere. The threshold for a possible dissociation by ultraviolet absorption which lies at 1109 Å (Stecher and Williams, 1967) and that for direct photodissociation at 845 Å are in a spectral region where radiation is absorbed in the thermosphere. The photodissociation coefficient will be taken for the calculation in the mesosphere as $a_{25} = 0$. In fact the most important process is the reaction of molecular hydrogen with excited oxygen atoms

$$(a_{24}^*); \quad H_2 + O\,(^1D) \rightarrow H + OH^*\,(v \leqslant 4) + 43.5 \, \text{kcal} \tag{114d}$$

for which we adopt the following round figure

$$a_{24}^* = 10^{-10} \, \text{cm}^3 \, \text{s}^{-1}$$

since it is not yet possible to deduce an exact value for the various quenching rate coefficients of $O\,(^1D)$ deduced from laboratory measurements. Values $a_{24}^* = (1.5 \pm 1) \times \times 10^{-10} \, \text{cm}^3 \, \text{s}^{-1}$ are possible.

Thus the chemical conditions for molecular hydrogen are written as follows

$$\frac{dn\,(H_2)}{dt} + n\,(H_2)\,[\{a_{18}n\,(M) + a_{24}\}\,n\,(O) + a_{19}n\,(OH) + a_{24}^*n^*\,(O)] =$$
$$= n\,(H_2O)\,J_{H_2-O} +$$
$$+ n\,(H)\,[a_{21}n\,(M)\,n\,(H) + a_{22}n\,(OH) + a_{23}n\,(HO_2)]. \tag{115a}$$

Considering the numerical values of the various rate coefficients (115a) can be written, with a sufficiently good approximation,

$$\frac{dn\,(H_2)}{dt} + n\,(H_2)\,[a_{24}n\,(O) + a_{24}^*n^*\,(O)] = n\,(H)\,n\,(HO_2)\,a_{23}. \tag{115b}$$

In the stratosphere, the reaction between $O\,(^1D)$ and H_2 plays a role whereas that between $O\,(^3P)$ and H_2 is not negligible in the thermosphere where the temperature is relatively high. The H_2 production is particularly important in the upper mesosphere where the product $n\,(H)\,n\,(HO_2)$ is maximum.

If the hypothesis of a constant mixing of H_2 in the stratosphere is adopted, it requires high eddy diffusion coefficients (identical to those in Figure 11) which cannot be accepted. Alternatively it must be assumed that the total dissociation, which is

2.6×10^9 H_2 molecules $cm^{-2}\, s^{-1}$ for overhead sun conditions, is replaced by an identical downward transport of H_2 molecules formed in the upper mesosphere.

Thus more molecular hydrogen is produced below the mesopause than is destroyed there; molecular hydrogen from this layer flows downwards into the upper stratosphere where process (114d) converts it into atomic hydrogen and hydroxyl radicals. Such a process which occurs only in a sunlit atmosphere should lead to not less than 10^9 dissociations $cm^{-2}\, s^{-1}$ if the mixing ratio $n(H_2)/n(M) = 5 \times 10^{-7}$ is maintained. The exact distribution depends on the eddy diffusion coefficient which is introduced into the continuity equation

$$\frac{\partial n(H_2)}{\partial t} + \frac{\partial [n(H_2)\, w]}{\partial z} + a_{24}^* n^*(O) + a_{24} n(O) = a_{23} n(H)\, n(OH) \quad (116)$$

where w is the diffusion velocity. A reliable estimate of the concentration of molecular hydrogen cannot be made without an exact knowledge of the values of the eddy diffusion coefficient. In any case, it seems that the stratospheric concentration of molecular hydrogen depends on its production in the upper mesosphere and on its downward transport.

Various reactions with nitrogen oxides in which H, OH, and HO_2 are involved will be introduced in another section. However, a reaction such as

(a_{26}); $HO_2 + NO \rightarrow OH + NO_2$,

for which the rate coefficient is not known, cannot be forgotten in the stratosphere where the reaction with atomic oxygen (a_7) is not important enough.

8. Hydrogen Peroxide

In the foregoing discussion hydrogen peroxide was considered to be of minor importance and its reactions were not listed. However, it is produced by a two-body process

(a_{27}); $HO_2 + HO_2 \rightarrow H_2O_2 + O_2 + 42\, kcal$ $\qquad\qquad\qquad$ (117a)

with a rate coefficient which does not require a high activation energy (Foner and Hudson, 1962). With $a_{27} = 3 \times 10^{-12}$ $cm^3\, s^{-1}$ at 300 K, the following value which is similar to a_{16} can be adopted

$$a_{27} = 5 \times 10^{-12}\, T^{1/2}\, e^{-1000/T}\, cm^3\, s^{-1}. \qquad\qquad (117b)$$

The photodissociation of H_2O_2 is known from laboratory measurements (Urey et al., 1929; Holt et al., 1948). The photodissociation coefficient for zero optical depth is

$$J_{H_2O_2} = 1.2 \times 10^{-4}\, s^{-1} \qquad\qquad\qquad (118a)$$

The essential process in the stratosphere is

(a_{28}); $H_2O_2 + hv \rightarrow 2OH$ $\qquad\qquad\qquad\qquad\qquad\qquad$ (118b)

since the photolysis cannot occur in the far ultraviolet and below 2100 Å the photo-

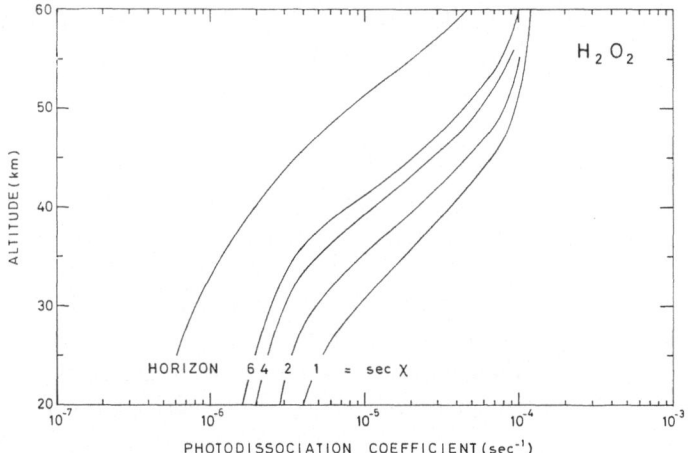

Fig. 16. Photodissociation coefficient of hydrogen peroxide in the stratosphere for various solar zenith angles. From overhead sun conditions, $\chi = 0$, to $\chi = 90°$, horizon.

dissociation coefficient is less than 10^{-5} s^{-1}. The variation of $J_{H_2O_2}$ with the solar zenith angle is shown in Figure 16. In the lower stratosphere the photodissociation coefficient is less than 10^{-5} s^{-1}.

The reaction with atomic hydrogen leads to

(a_{29a}); $\quad H + H_2O_2 \rightarrow H_2O + OH + 69\,kcal$ \hfill (119a)

(a_{29b}); $\quad H + H_2O_2 \rightarrow H_2 + HO_2 + 14\,kcal.$ \hfill (119b)

Experimental evidence (Foner and Hudson, 1962; Baldwin *et al.*, 1964; Hata and Giguere, 1966) seems to indicate that the activation energies of (119a) and (119b) are relatively high, and that $a_{29a} > a_{29b}$. We write

$$a_{29} = 5 \times 10^{-12}\, T^{1/2}\, e^{-3000/T}\ \text{cm}^3\ \text{s}^{-1} \hfill (119c)$$

$$a_{29b} = 5 \times 10^{-12}\, T^{1/2}\, e^{-4500/T}\ \text{cm}^3\ \text{s}^{-1} \hfill (119d)$$

which show that the reaction between H and H_2O_2 is not rapid in aeronomic conditions. But the reaction between a hydroxyl radical and hydrogen peroxide

(a_{30}); $\quad OH + H_2O_2 \rightarrow H_2O + HO_2 + 30\,kcal$ \hfill (120a)

is more important since its activation energy is of the order of 1 kcal (Greiner, 1968). Its rate coefficient is about

$$a_{30} = 3 \times 10^{-13}\, T^{1/2}\, e^{-500/T}\ \text{cm}^3\ \text{s}^{-1}. \hfill (120b)$$

Finally, the reaction between oxygen atoms and hydrogen peroxide molecules must be introduced since atomic oxygen is an important minor constituent

(a_{31a}); $\quad O + H_2O_2 \rightarrow H_2O + O_2 + 86\,kcal$ \hfill (121a)

(a_{31b}); $\quad O + H_2O_2 \rightarrow OH + HO_2 + 15\,kcal.$ \hfill (121b)

It appears (Foner and Hudson, 1962b) that $a_{31a} > a_{31b}$. With an activation energy of the order of 4 kcal,

$$a_{31a} = 1.5 \times 10^{-13} \, T^{1/2} \, e^{-2000/T} \, \text{cm}^3 \, \text{s}^{-1} \tag{121c}$$

leads to a rate coefficient of about $3 \times 10^{-15} \, \text{cm}^3 \, \text{s}^{-1}$ at 293 K.

Experimental evidence does not rule out $O(^1D) - H_2O_2$ reactions such as

$$(a_{31}^*) \, O(^1D) + H_2O_2 \rightarrow OH + O_2 + 60 \, \text{kcal} \tag{121d}$$

which could have no activation energy and could occur at rate similar to the reactions of O^* with O_3, H_2, H_2O or CH_4. However, the photodissociation coefficient $J_{H_2O_2}$ is greater than $a^* \, n^*(O)$ and remains the most important loss process of H_2O_2.

If nitric oxide is involved, hydrogen peroxide decomposes according to overall reactions such as

$$NO + H_2O_2 \rightarrow H_2O + NO_2 \tag{121e}$$

$$2NO + H_2O_2 \rightarrow 2HNO_3 \tag{121f}$$

which may come, for example, from

$$(a_{32}); \quad NO + H_2O_2 \rightarrow HNO_2 + OH + 11 \, \text{kcal} \tag{122}$$

$$(a_{33}); \quad NO_2 + H_2O_2 \rightarrow HNO_3 + OH + 4 \, \text{kcal} \tag{123}$$

for which the rate coefficients are not known. NO_2 and NO concentrations of the order of $10^9 \, \text{cm}^{-3}$ in the stratosphere would lead to a rate coefficient less than $10^{-5} \, \text{s}^{-1}$ if $a_{32} \geqslant 10^{-14} \, \text{cm}^3 \, \text{s}^{-1}$.

If all reactions from (117) to (122) are considered in the production and destruction of H_2O_2, the differential equation for H_2O_2 is

$$\frac{dn(H_2O_2)}{dt} + n(H_2O_2) \left[J_{H_2O_2} + \{a_{29a} + a_{29b}\} \, n(H) + a_{30}n(OH) + \right.$$
$$\left. + \{a_{31a} + a_{31b}\} \, n(O) + a_{32}n(NO) + a_{33}n(NO_2) \right] =$$
$$= a_{27}n^2(HO_2). \tag{124a}$$

Since $J_{H_2O_2}$ is, for overhead sun conditions, about $10^{-4} \, \text{s}^{-1}$ and $4 \times 10^{-6} \, \text{s}^{-1}$ at the stratopause and near 20 km, respectively, it is the principal loss term of hydrogen peroxide compared with its reactions with H and OH; the reaction with atomic oxygen, which reaches a maximum of about $10^{-5} \, \text{s}^{-1}$ near the stratopause, is practically negligible compared with the photodissociation. As far as the reactions with nitrogen oxides are concerned, it is not possible to neglect them in the lower stratosphere if their rate coefficients are not less than $10^{-14} \, \text{cm}^3 \, \text{s}^{-1}$. Therefore, an idea of the behavior of hydrogen peroxide can be obtained only with the oversimplified equation

$$\frac{dn(H_2O_2)}{dt} + n(H_2O_2) \, J_{H_2O_2} = a_{27}n^2(HO_2). \tag{124b}$$

Photoequilibrium conditions at the stratopause lead to

$$\frac{n^2(HO_2)}{n(H_2O_2)} = \frac{10^{-4}}{2 \times 10^{-12}} = 5 \times 10^7 \; cm^{-3} \tag{125a}$$

and at 20 km to

$$\frac{n^2(HO_2)}{n(H_2O_2)} = \frac{4 \times 10^{-6}}{8 \times 10^{-13}} = 5 \times 10^6 \; cm^{-3}. \tag{125b}$$

They give a clear indication that the hydrogen peroxide concentration is greater than that of the hydroperoxyl radical at 20 km and is of the same order of magnitude at the stratopause.

Since all the reactions, in which H_2O_2 is involved, except (123), are reactions occurring in a sunlit atmosphere, it is important to know the possible action of nitrogen dioxide. In any case, special experimental data are necessary before the role of hydrogen peroxide in the stratosphere can be determined.

9. Nitrogen Oxides

In the foregoing discussion on the reactions in which hydrogen peroxide is involved, a possible action of nitrogen oxide has been indicated. However, other aeronomic reactions with nitrogen oxides can play a role in the stratosphere and mesosphere (Nicolet, 1965, 1970).

A rapid reaction (Phillips and Schiff, 1962)

$$(a_{34}); \quad H + NO_2 \rightarrow OH + NO + 30 \; kcal \tag{126a}$$

with a rate coefficient

$$a_{34} = 2 \times 10^{-12} \; T^{1/2} \; cm^3 \; s^{-1} \tag{126b}$$

is not important since the reaction $H+O_3 \rightarrow OH+O_2$ is also rapid and since $O+NO_2 \rightarrow NO+O_2$ also plays a leading role.

In the mesosphere, the effect of NO on HO_2 and H_2O_2 is also without great importance since $O+HO_2 \rightarrow OH+O_2$ and similar reactions are more important. However, in the stratosphere their action may become important. The reaction

$$(a_{26}); \quad NO + HO_2 \rightarrow OH + NO + 9 \; kcal \tag{127a}$$

is fast relative to other reactions of HO_2 in similar circumstances (Tyler, 1962). No data are available on the rate coefficient but, assuming two extreme values, with activation energy reaching 2.5 kcal,

$$a_{26a} = 3 \times 10^{-12} \; T^{1/2} \; cm^3 \; s^{-1} \tag{127b}$$

in order to reach not less than $5 \times 10^{-11} \; cm^3 \; s^{-1}$ at 273 K, and

$$a_{26b} = 3 \times 10^{-12} \; T^{1/2} \; e^{-1250/T} \; cm^3 \; s^{-1} \tag{127c}$$

leading to not less than 10^{-11} cm^3 s^{-1} at 500 K, it is possible to see that a leading role can be played by (127a) in the lower stratosphere. The ratio $n(HO_2)/n(OH)$ given by (41c) must be modified when (127) is introduced; it becomes

$$\frac{n(HO_2)}{n(OH)} = \left\{ \frac{a_1 n(M) n(O_2) a_5 n(O)}{a_1 n(M) n(O_2) + a_2 n(O_3)} + a_6 n(O_3) \right\} \times$$
$$\times \{a_7 n(O) + a_{26} n(NO)\}. \tag{128}$$

Since $n(O)$ decreases rapidly from about 10^9 cm^{-3} at 40 km to 4×10^7 cm^{-3} at 30 km and to 10^6 cm^{-3} at 20 km, when its value is obtained by (26), it is clear that NO concentrations of the order of 10^9 cm^{-3} can play a role in the stratosphere if the rate coefficient a_{26a} is used. With the lowest value a_{26c}, NO can still play a role at 20 km if its concentration reaches 10^9 cm^{-3}. Consequently, the ratio $n(HO_2)/n(OH)$ is not yet defined with sufficient precision in the lower stratosphere. It depends on the exact value of the rate coefficient a_6, which has not been measured, and on the nitric oxide concentration.

If nitric oxide is present in the lower stratosphere, it is clear that nitrogen dioxide must be considered. The ratio (Nicolet, 1965) which corresponds to photochemical equilibrium conditions is

$$n(NO_2)/n(NO) = 1 \text{ to } 4 \tag{129}$$

between 20 and 40 km. In addition to the loss process involving O_3,

$$(b_4); \quad NO + O_3 \rightarrow NO_2 + O_2 + 48 \text{ kcal}, \tag{130a}$$

which has been observed in the laboratory (Johnson and Crossby, 1954; Phillips and Schiff, 1962a; Clough and Thrush, 1967) with a rate coefficient which may correspond to
$$b_4 = 5 \times 10^{-14} \, T^{1/2} \, e^{-1200/T} \text{ cm}^3 \text{ s}^{-1}, \tag{130b}$$

it is necessary to consider the following reaction

$$(b_3); \quad O + NO_2 \rightarrow NO + O_2 + 46 \text{ kcal} \tag{131a}$$

which is relatively rapid (Schiff, 1964; Klein and Herron, 1964) with a rate coefficient
$$b_3 = 1.5 \times 10^{-12} \, T^{1/2} \, e^{-500/T}. \tag{131b}$$

This reaction must be compared with the reaction of atomic oxygen (cf. formula (7)) with O_3

$$(k_3); \quad O + O_3 \rightarrow O_2 + O_2 + 94 \text{ kcal}. \tag{132}$$

and with OH and HO$_2$ (reactions 35 and 37)

$$(a_5); \quad O + OH \rightarrow H + O_2 + 17 \text{ kcal} \tag{133}$$

$$(a_7); \quad O + HO_2 \rightarrow OH = O_2 + 55 \text{ kcal}. \tag{134}$$

Instead of (62) the correction term A becomes

$$A = \left\{ a_2 n\,(\mathrm{H}) \frac{k_2 n\,(\mathrm{M})\,n\,(\mathrm{O}_2)}{J_3} + a_5 n\,(\mathrm{OH}) \left[1 + \frac{a_6 k_2 n\,(\mathrm{M})\,n\,(\mathrm{O}_2)}{a_5 \quad J_3} \right] + \right.$$

$$+ a_7 n\,(\mathrm{HO}_2) \left[1 + \frac{a_{6c}\,k_2 n\,(\mathrm{M})\,n\,(\mathrm{O}_2)}{a_5 \quad J_3} \right] +$$

$$\left. + 2 b_3 n\,(\mathrm{NO}_2) \right\} \bigg/ 2 k_3 n\,(\mathrm{O}_3) \tag{135}$$

Since $2 k_3\,n\,(\mathrm{O}_3) > 10^{-3}\ \mathrm{s}^{-1}$ in the major part of the stratosphere, the correction term A plays a role when the various terms in which hydroxyl and hydroperoxyl radicals and nitrogen oxides are involved are greater or of the same order of magnitude. With the mixing ratio which has been used here for $\mathrm{H}_2\mathrm{O}$, an effect due to OH and HO_2 occurs in the upper stratosphere since $a_7\,n\,(\mathrm{HO}_2) > 10^{-3}\ \mathrm{s}^{-1}$. As far as nitrogen dioxide is concerned an effect requires NO_2 concentrations greater than $10^9\ \mathrm{cm}^{-3}$ since its rate coefficient b_3 is of the order of $(3 \pm 1) \times 10^{-12}\ \mathrm{cm}^3\ \mathrm{s}^{-1}$. The effect of nitric oxide is involved in the term $2 b_3 n\,(\mathrm{NO}_2)$ through the factor of 2 since (Nicolet, 1965)

$$\frac{n\,(\mathrm{NO}_2)}{n\,(\mathrm{NO})} = \frac{b_4 n\,(\mathrm{O}_3)}{b_3 n\,(\mathrm{NO}) + J_{\mathrm{NO}_2}} \tag{136}$$

where $J_{\mathrm{NO}_2} = 3.5 \times 10^{-3}\ \mathrm{s}^{-1}$ is the photodissociation coefficient which leads to the photoequilibrium between NO_2 and NO. It is extremely important to determine the exact concentration of NO_2 in the stratosphere. A first measurement by Ackerman and Frimout (1969) would indicate a mixing ratio reaching 10^{-8} which seems to be very large, since it affects the photochemistry of the whole stratosphere.

Finally, the effect of $\mathrm{N}_2\mathrm{O}$ which has been introduced in the mesosphere (Hesstvedt and Jansson, 1969; Shimazaki and Laird, 1970) cannot be considered since its formation from N_2 and O in their normal states is not possible. The only possibility of an aeronomic effect due to $\mathrm{N}_2\mathrm{O}$ is the dissociation of nitrous oxide (Bates and Hays, 1967) in the lower stratosphere.

$\mathrm{N}_2\mathrm{O}$ is subject to various photodissociation processes in the stratosphere. The principal aeronomic process is

$$\mathrm{N}_2\mathrm{O} + h\nu\,(\lambda < 3400\ \text{Å}) \rightarrow \mathrm{N}_2\,(X^1\Sigma) + \mathrm{O}\,(^1D). \tag{137}$$

The photodissociation coefficient increases from about $1.5 \times 10^{-9}\ \mathrm{s}^{-1}$, at the beginning of the stratosphere, up to about $10^{-6}\ \mathrm{s}^{-1}$ at the stratopause. With such a coefficient and no rapid re-formation of $\mathrm{N}_2\mathrm{O}$, its concentration decreases rapidly in the stratosphere. $\mathrm{N}_2\mathrm{O}$ is practically absent in the mesosphere and does not belong to the system of nitrogen oxides in which NO and NO_2 are involved. However, in the lower stratosphere, the reactions

$$(b_{38});\quad \mathrm{O}\,(^1D) + \mathrm{N}_2\mathrm{O} \rightarrow \mathrm{N}_2 + \mathrm{O}_2 \tag{138}$$

$$(b_{39});\quad \mathrm{O}\,(^1D) + \mathrm{N}_2\mathrm{O} \rightarrow \mathrm{NO} + \mathrm{NO} \tag{139}$$

which have been studied by Greenberg and Heicklen (1970) lead to $k_{38}/k_{39} = 1$. Thus, reaction (139) leads to a formation of nitric oxide in the lower stratosphere where N_2O is present. The total production depends on the eddy diffusion in the stratosphere which is necessary to sustain N_2O subject to photodissociation.

In any case, the problem of nitrogen oxides in the lower stratosphere must be considered in association with nitrogen dioxide and nitrous and nitric acids (Nicolet, 1965). Nitrous acid is formed in the stratosphere by association of NO with OH and H_2O_2 (Figure 17) while nitric acid (Figure 18) seems to be subject to an important production not only from association of NO and NO_2 with HO_2 and OH respectively, but also by a reaction between H_2O_2 and NO_2.

NITROUS ACID

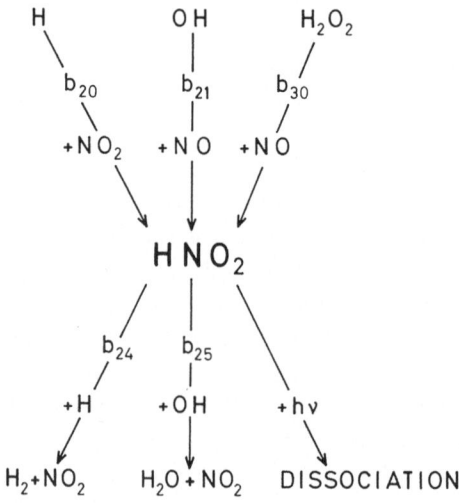

Fig. 17.   Principal reactions in which nitrous acid may be involved in the mesosphere and stratosphere.

Three-body associations leading to nitrous and nitric acid must be considered. They are

(b_{20}); $NO_2 + H + M \rightarrow HNO_2 + M + 80$ kcal   (140)

(b_{21}); $NO + OH + M \rightarrow HNO_2 + M + 60$ kcal   (141)

(b_{22}); $NO_2 + OH + M \rightarrow HNO_3 + M + 50$ kcal   (142)

(b_{23}); $NO + HO_2 + M \rightarrow HNO_3 + M + 63$ kcal.   (143)

The rate coefficients for (140–143) are not known, but typical values for such three-body rate coefficients are between 10^{-32} and 10^{-31} cm^6 s^{-1}. Reaction (140) leading to nitrous acid is not important since it requires H and NO_2, but it is clear that reactions (141, (142) and (143) may be important in the stratosphere.

The photodissociation must be included with the loss processes

(b_{24}); $H + HNO_3 \rightarrow H_2 + NO_2 + 23$ kcal (144)

(b_{25}); $OH + HNO_2 \rightarrow H_2O + NO_2 + 38$ kcal (145)

(b_{26}); $H + HNO_3 \rightarrow H_2O + NO_2 + 66$ kcal (146)

(b_{27}); $OH + HNO_3 \rightarrow H_2O + NO_3 + 15$ kcal. (147)

NITRIC ACID

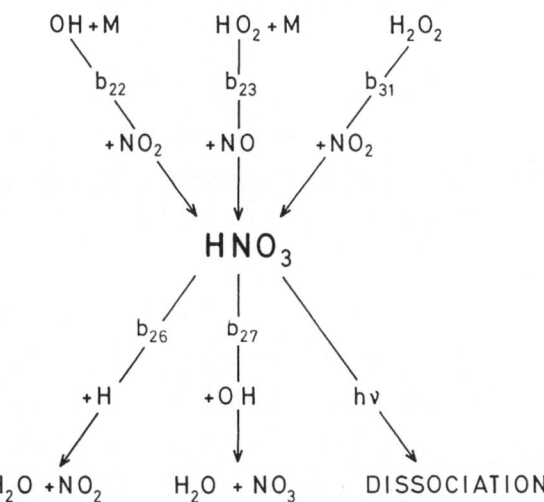

Fig. 18. Principal reactions in which nitric acid may be involved in the mesosphere and stratosphere.

The reactions (122) and (123) which have been considered before cannot be neglected in the aeronomic discussion. Thus

$$NO + H_2O_2 \rightarrow OH + HNO_2 + 11 \text{ kcal} \qquad (148)$$

and

$$NO_2 + H_2O_2 \rightarrow OH + HNO_3 + 4 \text{ kcal} \qquad (149)$$

must be added as production processes of nitrous and nitric acids. All these reactions are exothermic, but their activation energies are not known. Since the reaction of HNO_3 with OH is perhaps the principal loss process, it seems possible that the fraction of nitric acid which is not destroyed after its formation in the stratosphere leads to the dissappearance of NO. Thus, it is extremely important to measure the concentration of nitrogen dioxide (Ackerman and Frimout, 1969, 1970) or nitric oxide (Pontano and

Hale, 1970) simultaneously with that of nitric acid (Murcray *et al.*, 1969; Rhine *et al.*, 1969) in order to know the proportions of different nitrogen oxides which are definitively converted into more complex compounds going from the stratosphere into the troposphere. The nitrate plus nitrite content of South Polar snow (Wilson and House, 1965) corresponds to an infall of nitrogen oxide of the order of 1.5×10^7 molecules cm^{-2} s^{-1}. In any case, a special study of the photodecomposition of nitrogen trioxide and of nitric acids should be made since their decomposition is of special interest in the aeronomic processes of the stratosphere.

10. General Conclusions

When all reactions dealing with H, OH and HO_2 are considered the following general equations can be written

$$\frac{dn(H_2O)}{dt} + n(H_2O)\left[J_{OH-H} + J_{H_2-O} + a^*_{H_2O}n(O^*)\right] =$$
$$= n(H)\left[a_{17}n(M)\,n(OH) + a_{20}n(HO_2) + a_{29a}n(H_2O_2)\right] +$$
$$+ n(OH)\left[a_{16}n(OH) + a_{17}n(HO_2) + a_{19}n(H_2) + a_{30}n(H_2O_2)\right] +$$
$$+ n(O)\left[a_{18}n(M)\,n(H_2) + a_{31}n(H_2O_2)\right] \tag{150}$$

$$\frac{dn(H_2)}{dt} + n(H_2)\left[\{a_{18}n(M) + a_{24}\}\,n(O) + a_{19}n(OH) + a^*_{24}n(O)\right] =$$
$$= n(H)\left[a_{21}n(M)\,n(H) + a_{22}n(OH) + a_{23}n(HO_2) +\right.$$
$$\left.+ a_{29b}n(H_2O_2)\right] + n(H_2O)\,J_{H_2-O} \tag{151}$$

$$\frac{dn(H_2O_2)}{dt} + n(H_2O_2)\left[J_{H_2O_2} + \{a_{29a} + a_{29b}\}\,n(H) + a_{30}n(OH) +\right.$$
$$+ \{a_{31a} + a_{31b}\}\,n(O) + a_{32}n(NO) + a_{33}n(NO_2) +$$
$$\left.+ a_{37}n(CO)\right] = a_{27}n^2(HO_2) \tag{152}$$

$$\frac{dn(H)}{dt} + n(H)\left[a_0n(M)\,n(O) + a_1n(M)\,n(O_2) + \{a_2 + a_3\}\,n(O_3) +\right.$$
$$+ \{a_{14}n(M) + a_{22}\}\,n(OH) + \{a_{15} + a_{20} + a_{23}\}\,n(HO_2) +$$
$$+ 2a_{21}n(M)\,n(H) + \{a_{29a} + a_{29b}\}\,n(H_2O_2)\right] = n(H_2O)\times$$
$$\times J_{OH-H} + \left[a_5n(O) + a_{19}n(H_2)\right]n(OH) +$$
$$+ n(H_2)\left[a_{24}n(O) + a^*_{24}n(O^*)\right] \tag{153}$$

$$\frac{dn(OH)}{dt} + n(OH)\left[\{a_4n(M) + a_5\}\,n(O) + a_6n(O_3) +\right.$$
$$+ \{a_{14}n(M) + a_{12}\}\,n(H) + 2a_{16}n(OH) + a_{17}n(HO_2) +$$
$$\left.+ a_{19}n(H_2) + a_{30}n(H_2O_2) + a_{36}n(CO)\right] =$$

$$= n(H) [a_0 n(M) n(O) + a_2 n(O_3) + a_{15} n(HO_2)] +$$
$$+ n(HO_2) [J_{HO_2} + a_{6b} n(O_3) + a_7 n(O) + a_{15} n(H) +$$
$$+ a_{26} n(NO)] + n(H_2O_2) [2J_{H_2O_2} + a_{29a} n(H) +$$
$$+ a_{31b} n(O) + a_{32} n(NO) + a_{33} n(NO_2)] +$$
$$+ n(H_2) [a_{24} n(O) + a_{24}^* n(O^*)] +$$
$$+ n(H_2O) [J_{H-OH} + 2a_{H_2O}^* n(O^*)] +$$
$$+ 4n(CH_4) [a_{CH_4} n(O) + a_{CH_4}^* n(O^*)] \qquad (154)$$

$$\frac{dn(HO_2)}{dt} + n(HO_2) [J_{HO_2} + a_{6b} n(O_3) + a_7 n(O) +$$
$$+ \{a_{15} + a_{20} + a_{23}\} n(H) + a_{17} n(OH) + 2a_{27} n(HO_2) +$$
$$+ a_{26} n(NO)] = [a_1 n(M) n(O_2) + a_3 n(O_3)] n(H) +$$
$$+ [a_4 n(M) n(O) + a_6 n(O_3)] n(OH) +$$
$$+ [a_{29b} n(H) + a_{30} n(OH) + a_{31b} n(O)] n(H_2O_2). \qquad (155)$$

All the reactions which were discussed in the preceding sections have been introduced in Equations (150) to (155) with the addition of two reactions of OH and HO_2 with CO which may play a role in the lower stratosphere.

The reaction

$$(a_{36}); \quad OH + CO \to CO_2 + H + 24 \text{ kcal} \qquad (156a)$$

has been observed between 300 K and 500 K (Greiner, 1969) with practically no activation energy. The rate coefficient would be about

$$a_{36} = 10^{-13} \text{ cm}^3 \text{ s}^{-1}. \qquad (156b)$$

An analysis should be made of the reactions between carbon monoxide with hydroperoxyl radicals and hydrogen peroxide in order to determine the exact chemical

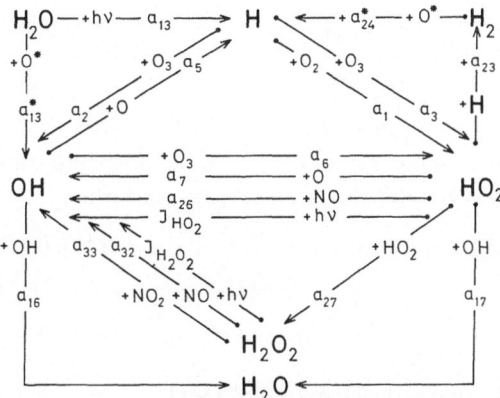

Fig. 19.   Fundamental H_2O reaction scheme. Initiating and terminating reactions depend on the production and destruction of hydroxyl radicals. Nitrogen oxides must be involved in the lower stratosphere.

behavior of the minor constituents in the lower stratosphere. For example, a study should be made of

$$(a_{37}); \quad CO + HO_2 \rightarrow CO_2 + OH + 62 \text{ kcal} \tag{157}$$

for which the rate coefficient a_{37} is not known.

In order to give an idea of the reactions which play a real role in the aeronomic behavior of H_2O and H_2 and the various radicals H, OH, HO_2 and H_2O_2, Figure 19 shows a general scheme. It is clear that H, OH and HO_2 may be in equilibrium when reactions with atomic oxygen are sufficiently rapid. From (153), (154) and (155), we can write

$$\frac{dn(H)}{dt} + \frac{dn(OH)}{dt} + \frac{dn(HO_2)}{dt} + 2n(H)\left[\{a_{15}n(M) + a_{22}\}\,n(OH) + \right.$$
$$+ \{a_{20} + a_{23}\}\,n(HO_2) + a_{21}n(M)\,n(H)] +$$
$$+ 2n^2(OH)\,a_{16} + 2a_{17}n(OH)\,n(HO_2) + 2n^2(HO_2)\,a_{27} =$$
$$= 2n(H_2O)\left[J_{OH-H} + a_{H_2O}^*n(O^*)\right] +$$
$$+ 2n(H_2O_2)\left[J_{H_2O_2} + a_{31b}n(O)\right] +$$
$$+ 2n(H_2)\left[a_{24}n(O) + a_{24}^*n(O^*)\right] +$$
$$+ 4n(CH_4)\left[a_{CH_4}n(O) + a_{CH_4}^*n(O^*)\right] \tag{158}$$

If $a_7\,n(O) + J_{HO_2}$ is greater than 10^{-3} s^{-1} equilibrium conditions can be considered in (155), and $dn(HO_2)/dt = 0$. Considering the essential reactions (Figure 19), we can write

$$n(HO_2)\left[J_{HO_2} + a_7n(O) + a_{26}n(NO)\right] = a_1n(M)\,n(O_2)\,n(H) +$$
$$+ a_6n(O_3)\,n(OH) + a_{31b}n(O)\,n(HO_2). \tag{159a}$$

Since $a_7\,n(O) > 10^{-3}$ s^{-1} above 30 km for overhead sun conditions (see Figure 20), (159a) is certainly correct in the upper stratosphere and in the whole mesosphere and lower thermosphere. At the stratopause and in the mesosphere (159a) can be simplified and is written (cf. 41g)

$$\frac{n(HO_2)}{n(H)} = \frac{a_1n(M)\,n(O_2)}{a_7n(O)}. \tag{159b}$$

In the same way, when $a_1\,n(M)\,n(O_2) + a_2\,n(O_3) > 10^{-3}$ s^{-1}, i.e. (Figure 20) below the mesopause in the whole mesosphere and stratosphere, $dn(H)/dt = 0$, and (153) becomes (see Figure 19), with the essential reactions only,

$$n(H)\left[a_1n(M)\,n(O_2) + a_2n(O_3)\right] = a_5n(O)\,n(OH) +$$
$$+ n(H_2O)\,J_{OH-H} + n(H_2)\left[a_{24}^*n(O^*)\right]. \tag{160a}$$

In the major part of the stratosphere and the mesosphere (160a) becomes practically

$$\frac{n(OH)}{n(H)} = \frac{a_1n(M)\,n(O_2) + a_2n(O_3)}{a_5n(O)} \tag{160b}$$

which is (41b).

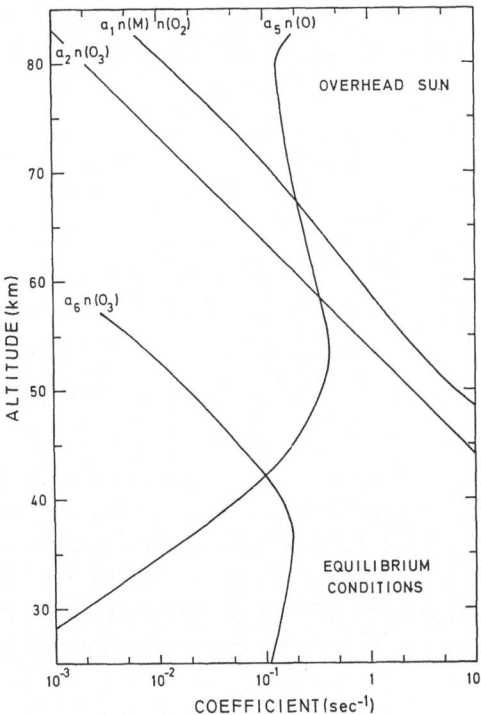

Fig. 20.   Reaction coefficients indicating that the times which are involved are less than 10^{-3} s^{-1} above 30 km. $a_6 n$ (O_3) is probably less than 10^{-5} s^{-1}.

Thus, at the stratopause equilibrium conditions can be used, and

$$\frac{n(\text{OH})}{n(\text{HO}_2)} = \frac{a_1 n(\text{M}) \, n(\text{O}_2) + a_2 n(\text{O}_3)}{a_1 n(\text{M}) \, n(\text{O}_2)} \tag{161}$$

is the equation which must be used in the mesosphere and lower thermosphere.

Equation (158) can be applied to the upper mesosphere or to the lower mesosphere if different conditions are considered. It is clear (see Figure 20 and Table IV) that the ratios $n(\text{HO}_2)/n(\text{H})$ and $n(\text{OH})/n(\text{H})$ increase from values < 1 at the mesopause to values > 1 at the stratopause.

At the mesopause and above (158) becomes

$$\frac{\mathrm{d}n(\text{H})}{\mathrm{d}t} + 2n(\text{H}) \left[a_{23} n(\text{HO}_2) \right] = P(\text{H}) \tag{162a}$$

where $P(\text{H})$ corresponds to all possible production processes, but principally the photodissociation of water vapor.

With (159b), (162a) is written

$$\frac{\mathrm{d}n(\text{H})}{\mathrm{d}t} + 2n^2(\text{H}) \, a_{23} \frac{a_1 n(\text{M}) \, n(\text{O}_2)}{a_7 n(\text{O})} = P(\text{H}) \tag{162b}$$

which shows that several days are required at the mesopause to reach equilibrium conditions. For this reason, mixing conditions lead to the best approximation for the vertical distribution of atomic hydrogen in the lower thermosphere. The introduction of eddy and molecular diffusion in the aeronomic problem of atomic hydrogen requires special attention.

Below 60 km $a_5 n(O) < a_1 n(M) \, n(O_2) + a_2 n(O_3)$, and it can be said that the stratopause conditions are, from (158),

$$\frac{dn(OH)}{dt} + \frac{dn(HO_2)}{dt} + 2n^2(OH) \, a_{16} + 2a_{17} n(OH) \, n(HO_2) +$$
$$+ 2n^2(HO_2) \, a_{27} = 2n(H_2O) \left[J_{OH-H} + a^*_{H_2O} n(O^*) \right] +$$
$$+ 2n(H_2O_2) \left[J_{H_2O_2} + a_{31b} n(O) \right]. \tag{163}$$

The essential fact is that equilibrium conditions can be applied in the major part of the stratosphere and mesosphere, since a close approach to photochemical equilibrium can be reached in less than three hours. Furthermore, there is a variation between day and night (Nicolet, 1964, 1970) which leads to a re-formation of H_2O. It can be concluded that there is no difficulty in maintaining a normal vertical distribution for H_2O. Nevertheless, the role of hydrogen peroxide in the stratosphere should be defined when exact rate coefficients are known. A complete simplification of (152) which would lead to

$$\frac{dn(H_2O_2)}{dt} + n(H_2O_2) \, J_{H_2O_2} = a_{27} n^2(HO_2) \tag{164}$$

indicates that equilibrium conditions for (163) can be written in the following form

$$a_{16} n^2(OH) + a_{17} n(OH) \, n(HO_2) = n(H_2O) \left[J_{OH-H} + a^*_{H_2O} n(O^*) \right], \tag{165a}$$

and with (161)

$$n^2(OH) \left[\frac{a_1 n(M) \, n(O_2)}{a_1 n(M) \, n(O_2) + a_2 n(O_3)} a_{17} + a_{16} \right] =$$
$$= n(H_2O) \left[J_{OH-H} + a^*_{H_2O} n(O^*) \right]. \tag{165b}$$

Such an equation leads to the following average concentrations

$n(OH)$	40 km	50 km	60 km
	8×10^7 cm^{-3}	6×10^7 cm^{-3}	3×10^7 cm^{-3}

which can be accepted as approximate values representing average daytime concentrations.

As far as molecular hydrogen is concerned a special analysis is required. Keeping the essential reactions in (151), the simplified equation is (cf. Figure 19)

$$\frac{dn(H_2)}{dt} + n(H_2) \left[a_{24} n(O) + a^*_{24} n(O^*) \right] =$$
$$= n(H) \, a_{23} n(HO_2) + n(H_2O) \, J_{H_2-O} \tag{166}$$

and it shows that there is an important production process near the mesopause and two important loss processes near the stratopause and in the thermosphere, respectively. Eddy and molecular diffusion processes must play an important role in the vertical distribution of molecular hydrogen.

Finally, Equation (150) shows that in the lower thermosphere the re-formation of H_2O is not important, and that its photodissociation must be associated with eddy and molecular diffusion processes which determine its vertical distribution.

References

Ackerman, M., Biaume, F., and Kockarts, G.: 1970, *Planetary Space Sci.* **18**, 1639.
Ackerman, M. and Frimout, D.: 1969, *Bull. Acad. Roy. Belg., Cl. Sc.* **55**, 948.
Bainbridge, A. E. and Heidt, L. E.: 1966, *Tellus* **18**, 221.
Baldwin, R. R., Jackson, D., Walker, R. W., and Webster, S. J.: 1965, 'The Use of the Hydrogen-Oxygen Reaction in Evaluating Velocity Constants', *10th Symp. Comb.*
Barth, C. A.: 1964, *Ann. Geophys.* **20**, 182.
Bates, D. R. and Hays, P. B.: 1967, *Planetary Space Sci.* **15**, 189.
Bates, D. R. and Nicolet, M.: 1950a, *Compt. Rend. Acad. Sci. Paris* **230**, 1943.
Bates, D. R. and Nicolet, M.: 1950b, *Publ. Astron. Soc. Pacific* **62**, 106.
Bates, D. R. and Nicolet, M.: 1950c, *J. Geophys. Res.* **55**, 301.
Bennett, J. E. and Blackmore, D. R.: 1968, *Proc. Roy. Soc.* **A305**, 553.
Benson, S. W. and Axworthy, A. E., Jr.: 1957, *J. Chem. Phys.* **26**, 1718.
Biedenkapp, D. and Bair, E. J.: 1970, *J. Chem. Phys.* **52**, 6119.
Black, G., Slanger, T. G., St John, G. A., and Young, R. A.: 1969, *J. Chem. Phys.* **51**, 116.
Breen, J. E. and Glass, G. P.: 1970, *J. Chem. Phys.* **52**, 1082.
Brown, J. M. and Thrush, B. A.: 1967, *Trans. Faraday Soc.* **63**, 630.
Cadle, R. D.: 1964, *Disc. Faraday Soc.* **37**, 66.
Campbell, I. M. and Thrush, B. A.: 1967, *Proc. Roy. Soc.* **A296**, 222.
Carrington, T.: 1964, *J. Chem. Phys.* **41**, 2012.
Cashion, J. K. and Polanyi, J. C.: 1959, *J. Chem. Phys.* **30**, 316.
Chapman, S.: 1930, *Phil. Mag.* **10**, 369.
Chapman, S.: 1943, *Rep. Prog. Phys.* **9**, 92.
Clark, I. D.: 1970, *Chem. Phys. Letters* **5**, 317.
Clark, I. D. and Wayne, R. P.: 1969, *Proc. Roy. Soc.* **A314**, 111.
Clough, P. N. and Thrush, B. A.: 1967, *Trans. Faraday Soc.* **63**, 915.
Clyne, M. A. A., McKenney, D. J., and Thrush, B. A.: 1965, *Trans. Faraday Soc.* **61**, 2701.
Clyne, M. A. A. and Thrush, B. A., 1963a, *Proc. Roy. Soc.* **A275**, 544.
Clyne, M. A. A. and Thrush, B. A.: 1963b, *Proc. Roy. Soc.* **A275**, 559.
Cottin, M., Masanet, J., and Vermeil, C.: 1966, *J. Chim. Phys., Paris* **63**, 959.
Crutzen, P. J.: 1969, *Tellus* **21**, 368.
Crutzen, P. J.: 1970, *Quart. J. Roy. Meteorol. Soc.* **96**, 320.
Dalgarno, A. and Allison, A. C.: 1969, *J. Geophys. Res.* **74**, 4178.
De More, W. B.: 1970, *J. Chem. Phys.* **52**, 4309.
De More, W. B. and Raper, O. F.: 1966, *J. Chem. Phys.* **44**, 1780.
De More, W. B. and Raper, O. F.: 1967, *J. Chem. Phys.* **46**, 2500.
De More, W. B. and Raper, O. F.: 1968, *J. Chem. Phys.* **37**, 2048.
Ditchburn, R. W. and Young, P. A.: 1962, *J. Atmospheric Terrest. Phys.* **24**, 127.
Dixon-Lewis, G., Wilson, W. E., and Westenberg, A. A.: 1966, *J. Chem. Phys.* **44**, 2877.
Engleman, R.: 1965, *J. Amer. Chem. Soc.* **87**, 4193.
Evans, W. F. J. and Llewellyn, E. J.: 1970, *Ann. Geophys.* **26**, 1.
Filseth, S. V., Stuhl, F., and Welge, K. H.: 1970, *J. Chem. Phys.* **52**, 239.
Foner, S. N. and Hudson, R. L.: 1962, *J. Chem. Phys.* **36**, 2681.
Greenberg, R. I. and Heicklein, J. P.: 1970, *Int. J. Chem. Kinetics* **2**, 185.
Greiner, N. R.: 1968, *J. Phys. Chem.* **72**, 406.

Greiner, N. R.: 1969, *J. Chem. Phys.* **51**, 5049.
Greiner, N. R.: 1970, *J. Chem. Phys.* **53**, 1070.
Hampson, J.: 1966, in *Les problèmes météorologiques de la stratosphère et de la mésosphère*, Presses Universitaires de France, Paris, p. 393.
Hata, N. and Giguere, P. A.: 1966, *Can. J. Chem.* **44**, 869.
Herzberg, G.: 1951, *J. Roy. Astron. Soc. Canada* **45**, 100.
Hesstvedt, E.: 1968, *Geophys. Publik.* **27**, No. 4.
Hesstvedt, E. and Jansson, U. B.: 1969, *Aeronomy Report, Univ. Illinois*, No. 32, 190.
Hochanadel, C. J., Ghormley, J. A., and Boyle, J. W.: 1968, *J. Chem. Phys.* **48**, 2416.
Holt, R. K., McLane, C. R., and Oldenberg, O.: 1948, *J. Chem. Phys.* **16**, 225 and 638.
Hunt, B. G.: 1966, *J. Geophys. Res.* **71**, 1385.
Inn, E. C. Y. and Tanaka, Y.: 1953, *J. Opt. Soc. Amer.* **43**, 870.
Johnston, H. S. and Crosby, H. J.: 1954, *J. Chem. Phys.* **22**, 689.
Jones, I. T. N. and Wayne, R. P.: 1969, *J. Chem. Phys.* **51**, 3617.
Kaufman, F.: *Ann. Geophys.* **20**, 106.
Kaufman, F.: 1969, *Can. J. Chem.* **47**, 1917.
Kaufman, F. and Kelso, J. R.: 1964, *J. Chem. Phys.* **40**, 1162.
Kaufman, F. and Kelso, J. R.: 1967, *J. Chem. Phys.* **46**, 4541.
Klein, F. S. and Herron, J. T.: 1964, *J. Chem. Phys.* **41**, 1285.
Larkin, F. S. and Thrush, B. A.: 1964, *Disc. Faraday Soc.* **37**, 112.
McKinnon, D. and Morewood, H. W.: 1970, *J. Atmospheric Sci.* **27**, 483.
McNesby, J. R., Tanaka, I., and Okabe, H.: 1962, *J. Chem. Phys.* **36**, 605.
Mastenbrook, H. J.: 1968, *J. Atmospheric Sci.* **25**, 299.
Mathias, A. and Schiff, H. I.: 1964, *J. Chem. Phys.* **40**, 3118.
Meinel, A. B.: 1950, *Astrophys. J.* **111**, 207.
Mentall, J. E. and Gentien, E. P.: 1970, *J. Chem. Phys.* **52**, 5641.
Murcray, D. R., Kyle, T. G., Murcray, F. H., and Williams, W. J.: 1969, *J. Opt. Soc. Amer.* **59**, 1131.
Nicolet, M.: 1954, *J. Atmospheric Terrest. Phys.* **5**, 132.
Nicolet, M.: 1964, *Disc. Faraday Soc.* **37**, 7.
Nicolet, M.: 1965a, *J. Geophys. Res.* **70**, 679.
Nicolet, M.: 1965b, *Proc. Roy. Soc.* **A288**, 479.
Nicolet, M.: 1966, in *Les problèmes météorologiques dans la stratosphère et la mésosphère*, Les Presses Universitaires de France, Paris, p. 441.
Nicolet, M.: 1968, *Geophys. J. Roy. Astron. Soc.* **15**, 157.
Nicolet, M.: 1970, *Ann. Geophys.* **26**, 531.
Nicolet, M. and Mange, P.: 1954, *J. Geophys. Res.* **59**, 15.
Noxon, J. F.: 1970, *J. Chem. Phys.* **52**, 1852.
Phillips, L. F. and Shiff, H. I.: 1962a, *J. Chem. Phys.* **36**, 1509.
Phillips, L. F. and Schiff, H. I.: 1962b, *J. Chem. Phys.* **37**, 1233.
Pontano, B. A. and Hale, L. C.: 1970, *Space Res.* **10**, 208.
Rhine, P. E., Tubbs, L. D., and Williams, D.: 1969, *Appl. Opt.* **8**, 1500.
Ripley, D. L. and Gardiner, W. C., Jr.: 1966, *J. Chem. Phys.* **44**, 2285.
Schiff, H. I.: 1964, *Ann. Geophys.* **20**, 115.
Schiff, H. I.: 1969, *Can. J. Chem.* **47**, 1903.
Scholz, T. G., Ehhalt, D. H., Heidt, L. E., and Martell, E. A.: 1970, *J. Geophys. Res.* **75**, 3049.
Shardanand, X.: 1969, *Phys. Rev.* **186**, 5.
Shimazaki, T. and Laird, A. R.: 1970, *J. Geophys. Res.* **75**, 3221.
Sissenwine, N., Grantham, D. D., and Samela, H. A.: 1968, *J. Atmospheric Sci.* **25**, 1129.
Smith, W. H.: 1970, *J. Chem. Phys.* **53**, 792.
Snelling, R. D. and Bair, E. J.: 1968, *J. Chem. Phys.* **48**, 5737.
Stecher, T. P. and Williams, D. A.: 1967, *Astrophys. J.* **149**, L 29.
Stief, L. J.: 1966, *J. Chem. Phys.* **44**, 277.
Strobel, D. F.: 1969, *J. Atmospheric Sci.* **26**, 906.
Tanaka, Y., Inn, E. Y., and Watanabe, K.: 1953, *J. Chem. Phys.* **21**, 1651.
Tyler, B. J.: 1962, *Nature* **195**, 279.
Urey, H. C., Dawsey, L. C., and Rice, F. D.: 1929, *J. Amer. Chem. Soc.* **51**, 1371.
Vallance Jones, A. and Gattinger, R. L.: 1958, *J. Atmospheric Terrest. Phys.* **13**, 45.

Vassy, A.: 1941, *Ann. Phys. Paris* **16**, 145.
Vigroux, E.: 1951, *Ann. Phys.* **8**, 1.
Watanabe, K.: 1958, *Advances Geophys.* **5**, 154.
Welge, K. H. and Stuhl, F.: 1967, *J. Chem. Phys.* **46**, 2440.
Westenberg, A. A. and de Haas, N.: 1967, *J. Chem. Phys.* **47**, 4241.
Westenberg, A. A. and de Haas, N.: 1969, *J. Chem. Phys.* **50**, 2512.
Williamson, E. J. and Houghton, J.: 1965, *Quart. J. Roy. Meteor. Soc.* **91**, 330.
Wilson, A. T. and House, D. A.: 1965, *Nature* **205**, 793.
Wong, E. L. and Potter, A. E. Jr.: 1963, *J. Chem. Phys.* **39**, 2211.
Wong, E. L. and Potter, A. E. Jr.: 1967, *Can. J. Chem.* **45**, 367.
Young, R. A. and Black, G.: 1967, *J. Chem. Phys.* **47**, 2311.
Young, R. A., Black, G., and Slanger, T. G.: 1968, *J. Chem. Phys.* **49**, 4758.

A MERIDIONAL MODEL OF THE
OXYGEN-HYDROGEN ATMOSPHERE

EIGIL HESSTVEDT

University of Oslo, Oslo, Norway

Photochemical models containing reactions between oxygen and hydrogen compounds have been used as a most efficient tool in studies of the chemical composition of the mesosphere and lower thermosphere. Recent models are based upon the model presented by Bates and Nicolet (1950). Two important shortcomings of the latter – the distributions of atomic and molecular oxygen in the lower thermosphere, and water vapor above 70 km – were eliminated when vertical eddy mixing was introduced by Colegrove *et al.* (1965) for the pure oxygen model, and by Hesstvedt (1968) for the oxygen-hydrogen model.

Up to now the models have described the height distributions of the various components in a vertical column, and ignored the variation with latitude and season. It is the aim of this paper to analyse how the components are distributed in a meridional plane.

Such a study should include vertical mean transport as well as horizontal transport by mean motion and by eddies. However, our present knowledge about the circulation in the upper atmosphere does not permit us to take horizontal transport into consideration. Also the vertical transport is poorly known. A global mean profile is used for the vertical variation of the vertical eddy diffusion coefficient K_z. A variation with latitude and season is to be expected, and clearly our constant K_z profile leads to a considerable overestimate in high latitudes in the summer.

As computations are made for the height interval 65–105 km it has not been necessary to include molecular diffusion in the model.

The photochemistry in our model does not differ significantly from that used in earlier works. Ten components are considered: $O(^3P)$, $O(^1D)$, O_2, O_3, OH, HO_2, H_2O_2, H_2, H_2O, and H, and the following 38 reactions are assumed to take place:

(1) $O(^3P) + O(^3P) + M \rightarrow O_2 + M$ $\quad k_1 = 2.7 \times 10^{-33}$

(2) $O(^3P) + O_2 + M \rightarrow O_3 + M$ $\quad k_2 = 8.2 \times 10^{-35} \exp(0.89/RT)$

(3) $O(^3P) + O_3 \rightarrow 2O_2$ $\quad k_3 = 3.4 \times 10^{-11} \exp(-4.5/RT)$

(4a) $O_2 + hv \rightarrow O(^3P) + O(^3P)$ $\quad J_{2a}$ (1750 Å $< \lambda <$ 2424 Å) $\quad \left.\begin{array}{c} \\ \end{array}\right\} J_{2a} + J_{2b} = J_2$

(4b) $O_2 + hv \rightarrow O(^3P) + O(^1D)$ $\quad J_{2b}$ ($\lambda <$ 1750 Å)

(5a) $O_3 + hv \rightarrow O(^3P) + O_2$ $\quad J_{3a}$ ($\lambda >$ 3100 Å) $\quad \left.\begin{array}{c} \\ \end{array}\right\} J_{3a} + J_{3b} = J_3$

(5b) $O_3 + hv \rightarrow O(^1D) + O_2$ $\quad J_{3b}$ ($\lambda <$ 3100 Å)

(6) $OH + O(^3P) \rightarrow H + O_2$ $\quad k_6 = 5 \times 10^{-11}$

(7) $HO_2 + O(^3P) \rightarrow OH + O_2$ $\quad k_7 = 10^{-11}$

Fiocco (ed.), Mesospheric Models and Related Experiments, 52–64. All Rights Reserved.
Copyright © 1971 by D. Reidel Publishing Company, Dordrecht-Holland.

(8) $H + O_2 + M \rightarrow HO_2 + M$ $k_8 = 3 \times 10^{-32} (273/T)^{1.3}$

(9) $H + O_3 \rightarrow OH + O_2$ $k_9 = 2.6 \times 10^{-11}$

(10) $OH + HO_2 \rightarrow H_2O + O_2$ $k_{10} = 10^{-11}$

(11) $H_2O_2 + hv \rightarrow 2OH$ $J_{H_2O_2} (1875 \text{ Å} < \lambda < 3825 \text{ Å})$

(12) $O(^3P) + H_2O_2 \rightarrow OH + HO_2$ $k_{12} = 10^{-15}$

(13) $HO_2 + HO_2 \rightarrow H_2O_2 + O_2$ $k_{13} = 1.5 \times 10^{-12}$

(14) $OH + H_2O_2 \rightarrow H_2O + HO_2$ $k_{14} = 4 \times 10^{-13}$

(15) $OH + OH \rightarrow H_2O + O(^3P)$ $k_{15} = 2 \times 10^{-12}$

(16) $H_2O + hv \rightarrow OH + H$ $J_{H_2O} (1350 \text{ Å} < \lambda < 2375 \text{ Å} + Ly\alpha)$

(17) $H + HO_2 \rightarrow H_2O + O(^3P)$ $k_{17} = 2 \times 10^{-10} \exp(-4/RT)$

(18) $H + HO_2 \rightarrow H_2 + O_2$ $k_{18} = 3 \times 10^{-12}$

(19) $H + H + M \rightarrow H_2 + M$ $k_{19} = 1.2 \times 10^{-32} (273/T)^{0.7}$

(20) $O(^1D) + M \rightarrow O(^3P) + M$ $k_{20} = 3.8 \times 10^{-11}$

(21) $O(^1D) + H_2 \rightarrow OH + H$ $k_{21} = 10^{-11}$

(22) $O(^3P) + H_2 \rightarrow OH + H$ $k_{22} = 7 \times 10^{-11} \exp(-10.2/RT)$

(23) $HO_2 + O_3 \rightarrow OH + 2O_2$ $k_{23} = 10^{-14}$

(24) $OH + O_3 \rightarrow HO_2 + O_2$ $k_{24} = 5 \times 10^{-13}$

(25) $H + H_2O_2 \rightarrow H_2 + HO_2$ $k_{25} = 10^{-13}$

(26) $H + O(^3P) + M \rightarrow OH + M$ $k_{26} = 8 \times 10^{-33}$

(27) $H + O_3 \rightarrow HO_2 + O(^3P)$ $k_{27} = 2 \times 10^{-10} \exp(-4/RT)$

(28) $H + O_2 \rightarrow OH + O(^3P)$ $k_{28} = 1 \times 10^{-9} \exp(-16.8/RT)$

(29) $O(^1D) + H_2O \rightarrow 2OH$ $k_{29} = 10^{-11}$

(30) $H + HO_2 \rightarrow 2OH$ $k_{30} = 3 \times 10^{-12}$

(31) $H + OH + M \rightarrow H_2O + M$ $k_{31} = 2.5 \times 10^{-31}$

(32) $O(^3P) + OH + M \rightarrow HO_2 + M$ $k_{32} = 1.4 \times 10^{-31}$

(33) $H + OH \rightarrow H_2 + O(^3P)$ $k_{33} = 3 \times 10^{-11} \exp(-8.3/RT)$

(34) $H_2 + OH \rightarrow H_2O + H$ $k_{34} = 10^{-10} \exp(-5.9/RT)$

(35) $O(^1D) + O_3 \rightarrow 2O_2$ $k_{35} = 3 \times 10^{-10}$

(36) $HO_2 + hv \rightarrow OH + O(^3P)$ $J_{HO_2} = 1.4 \times 10^{-4}$

The dissociation rates are based upon values given by Ackerman (1971) for solar radiation and for the absorption cross-sections of O_3 and O_2, the latter with exception of the Schumann-Runge bands. In this region predissociation of O_2 is assumed and the cross-sections given by Huffman (1968) are used. Absorption cross-sections for H_2O_2 and H_2O are taken from Schumb et al. (1955) and Watanabe (1958). The dissociation rates of O_2 and H_2O are found to vary with height, latitude and season, while those for O_3 and H_2O_2 are constant in the height range considered. For J_3 a value of 10^{-2} is obtained, while $J_{H_2O_2}$ is found to take the value of 1.4×10^{-4}. No data are available for HO_2. In this study J_{HO_2} will be assumed to have the same value as $J_{H_2O_2}$.

Temperature and density data are taken mainly from Cole *et al.* (1965). A revision is made for high latitude summer, in accordance with more recent observations. Also, extrapolations are made for heights above 90 km.

The heat budget for the mesosphere and lower thermosphere is determined largely by infrared cooling due to CO_2, by absorption of solar radiation by O_2 and O_3, and by release of chemical energy by recombination of atomic oxygen $(O+O+M)$, of atomic and molecular oxygen $(O+O_2+M)$, and of atomic oxygen and ozone $(O+O_3)$. This represents a somewhat simplified picture, since part of the absorbed energy may be given off as emission rather than being converted to heat.

Summing up these sources and sinks of heat, we are left with local heatings or coolings which have to be compensated in order to maintain a steady state in the temperature distribution. Such a compensation may be caused by transport of air, vertically and horizontally, by mean motion and by eddy motion. (As mentioned above, horizontal transport is neglected in this paper. The temperature balance in our model is therefore maintained by vertical transport alone.) Johnson and Wilkins (1965) assumed that the heat transport was due to vertical mixing alone and computed a profile for the eddy diffusion coefficient, K_z. If, however, this procedure is applied to air columns at different latitudes, it is clear that we get the strongest diffusion in high latitude summer, where the heating due to absorption of solar radiation at 105 km is about 60 K/day, against 30 K/day over the equator and no heating in high latitudes, winter. Such a meridional variation in K_z cannot be accepted, since the eddy motion in

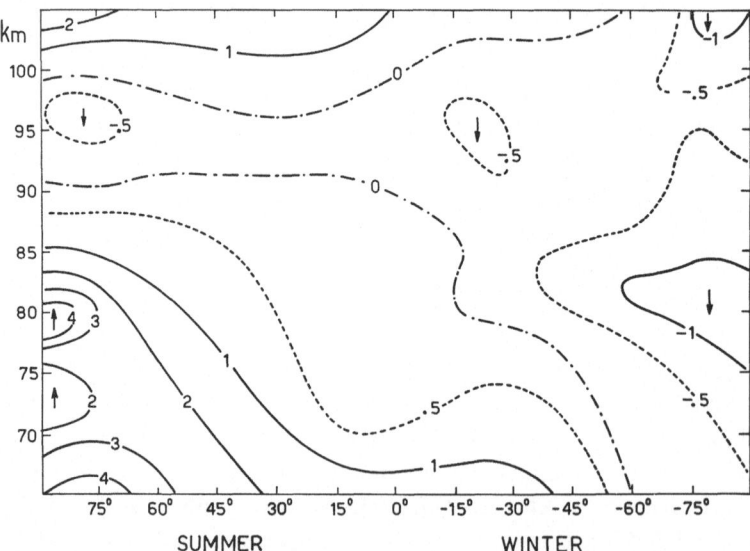

Fig. 1.   Vertical velocity (cm/s) computed on the assumption of steady state in the temperature field. Only vertical transport has been considered.

high latitudes, summer, is likely to be rather small. In this paper a K_z-profile which is meant to represent a mean for the globe is used for all latitudes, no attempt being made to account for a meridional variation. At 65 km a value of 2×10^5 cm²/s is taken for K_z, increasing to 5×10^5 cm²/s at the mesopause and to 8×10^6 cm²/s at 105 km. However, when the vertical eddy transport of heat is added to the heat sources mentioned above, we are still left with local heatings or coolings. We will assume that steady state in the temperature field is maintained, in addition to eddy mixing, by vertical mean motion. The result of the computations of the vertical velocity, w, is given in Figure 1.

The main feature of the w-field in our model is an upward motion over the summer pole and a downward motion over the winter pole. In low latitudes w is always very small, as one would expect. However, the model suffers from obvious shortcomings. The use of too high K_z-values in high latitudes, summer, locally leads to a net cooling where a heating is expected. This must in the model be compensated by a descent while an ascent is more likely. Our model satisfies the continuity of heat but disregards the continuity of mass. Therefore, Figure 1 should be used with the utmost care; it is meant only as a first approximation to the vertical velocity field.

The heat sources we consider have already been mentioned. The cooling rates due to infra-red radiative transfer by CO_2 are taken from Kuhn and London (1968). The absorption of solar radiation by O_2 and O_3 will lead partly to dissociation, partly to heating. The energy required for dissociation will be stored as chemical energy and transported to another region where it is converted to heat through a recombination process. We will assume that the energy in excess of that needed for dissociation goes

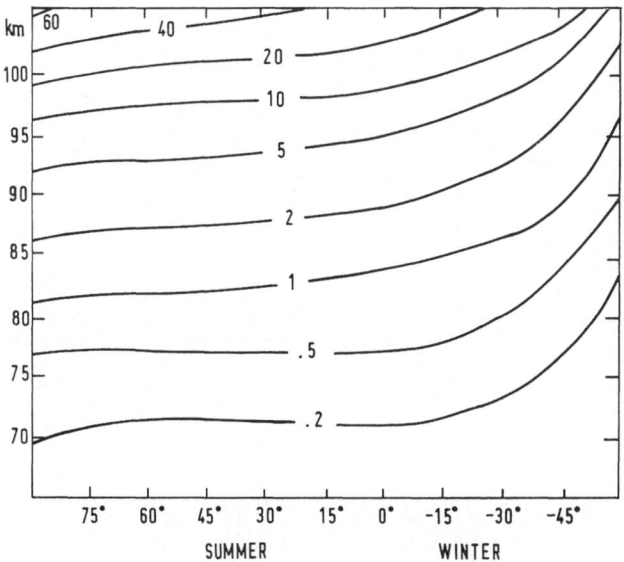

Fig. 2.   Heating by O_2 absorption (K/day).

to heating. (As mentioned, this may not be realistic, since part of the energy may be given off as emission.)

The heating rates due to absorption by O_2 and O_3 are shown in Figures 2 and 3. Absorption by O_2 is a very important heat source above 90 km. The heating rates at 105 km range from 60 K/day over the summer pole to 30 K/day over the equator and go to zero towards high latitudes in the winter hemisphere. Absorption by O_3 is important only below 95 km. It has a maximum at about 88 km with 6 K/day in high latitudes, summer. Below about 75 km, depending on latitude and season, the ozone heating becomes extremely important.

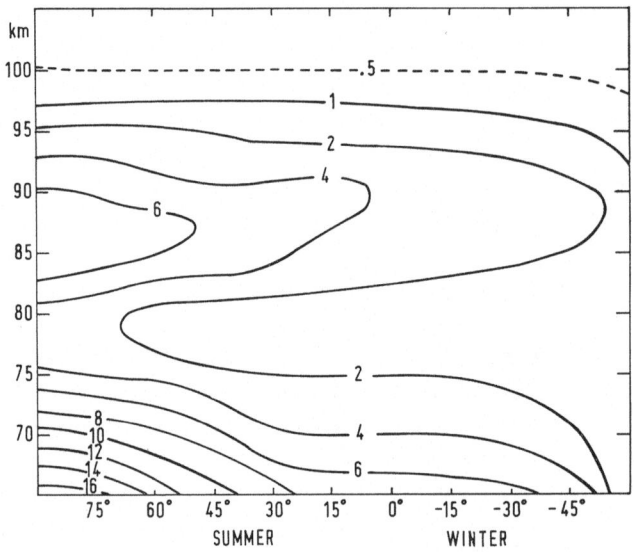

Fig. 3.   Heating by O_3 absorption (K/day).

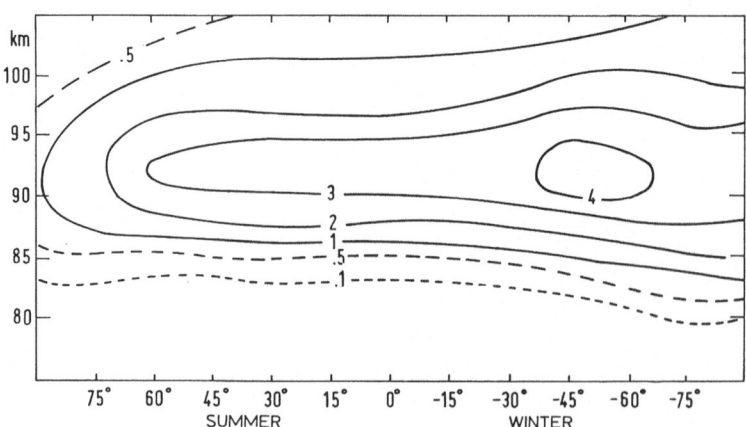

Fig. 4.   Heating by recombination $O + O + M \rightarrow O_2 + M$ (K/day).

Conversion of chemical energy to heat will also contribute significantly to the heat budget. It is well known that recombination of atomic oxygen $(O+O+M\rightarrow O_2+M)$ is an important heat source in the lower thermosphere. This heating has a maximum of about 4 K/day at 90–95 km, where we have the highest atomic oxygen number density. Figure 4 shows how the heating from this reaction is distributed. The reaction $O+O_3\rightarrow O_2+O_2$ is less important. It is seen from Figure 5 that the heating rates do not exceed 1 K/day. Much more important is the reaction $O+O_2+M\rightarrow O_3+M$. The heating from this reaction has a maximum at about 87 km, with values of about 5 K/day for most latitudes and seasons (see Figure 6).

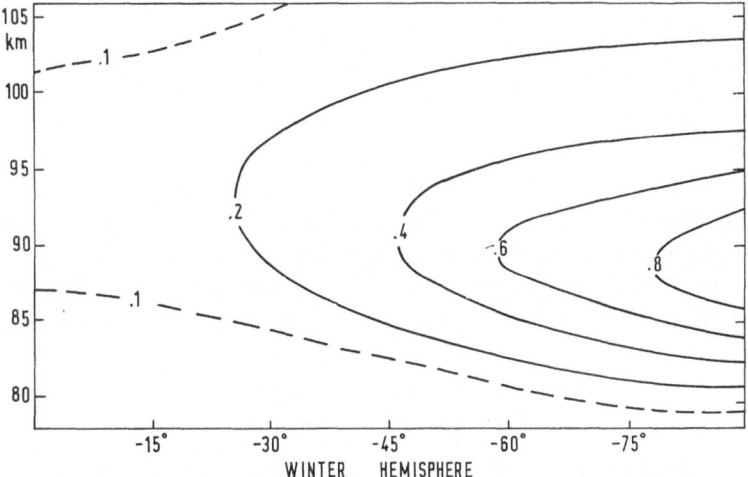

Fig. 5. Heating by recombination $O+O_3\rightarrow O_2+O_2$ (K/day). The heating rates in the summer hemisphere are 0.1 K/day or less.

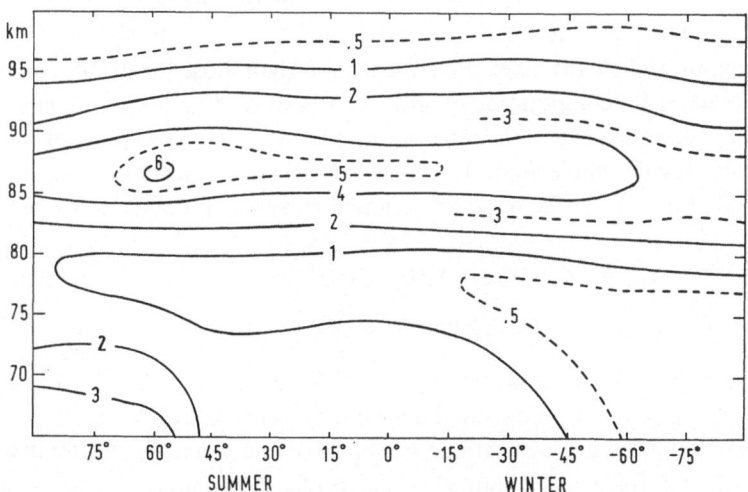

Fig. 6. Heating by recombination $O+O_2+M\rightarrow O_3+M$ (K/day).

Our model is a time-dependent one. Since it is assumed that the fractional amounts of oxygen (mixing ratio $=0.2095$) and hydrogen (mixing ratio $=7 \times 10^{-6}$) are the same for all points, we are left with eight partial differential equations. These are integrated by an explicit method over time steps of $\Delta t = 60$ s. The time variation of a component with number density x may be written in the following form

$$\partial x/\partial t = (P_{ph} - Q_{ph} \cdot x - R_{ph} \cdot x^2) + (P_{tr} - Q_{tr} \cdot x) = P - Q \cdot x - R \cdot x^2 \quad (1)$$

where subscript ph denotes photochemical terms and tr denotes transport terms (finite differences are introduced). In cases where P, Q, and R are nearly constant over the time-step, Equation (1) may be integrated

$$\frac{x - x_e}{x - x_e^*} = \frac{x_0 - x_e}{x_0 - x_e^*} \cdot \exp\left(- \sqrt{Q^2 + 4RP} \cdot \Delta t\right). \quad (2)$$

If $R=0$, or if the second degree term is an unimportant one, we obtain

$$x - x_e = (x_0 - x_e) \exp\left(- Q \cdot \Delta t\right) \quad (3)$$

Here x_e and x_e^* are the equilibrium values given by $\partial x/\partial t = 0$, x_0 and x are the number densities at the beginning and end of the time-step.

The advantage of using an explicit method is that it requires relatively short computer time. The drawback is that in many cases the integration is unstable. In the present model representative initial values for the height range 80–105 km are computed in a subroutine. This tends to decrease the possibilities of getting instability. But more important in this connection is the introduction of the functions

$$[\tilde{O}] = [O(^3P)] + [O(^1D)] + [O_3] - [OH] - 2 \cdot [HO_2] \quad (4)$$

$$[\tilde{H}] = [H] + [OH] + [HO_2N]. \quad (5)$$

These two functions have P, Q, and R terms which vary more slowly than those for atomic oxygen and atomic hydrogen. The expressions for $\partial[\tilde{O}]/\partial t$ and $\partial[\tilde{H}]/\partial t$ are therefore more suitable for an explicit integration than those for $\partial[O]/\partial t$ and $\partial[H]/\partial t$.

The subroutine for computation of initial values above 80 km is based on a simplified time-dependent model where the integration is made in two steps, one over the whole day and one over the whole night. If the number densities of a component at the end of the day and at the end of the night are denoted by subscripts d and n, respectively, we obtain

$$x_d - x_{e,d} = (x_n - x_{e,d}) \cdot \exp\left(- Q_d \cdot t_d\right) \quad (6)$$

$$x_n - x_{e,n} = (x_d - x_{e,n}) \cdot \exp\left(- Q_n \cdot t_n\right) \quad (7)$$

where t_d and t_n denote the length of day and night respectively. (Since we only wish to demonstrate the principle, the second degree term is neglected. A similar procedure is used when $R \neq 0$.) These two equations are solved to find x_d and x_n. An iterative method is used, and convergence is obtained to determine the composition down to 80 km, and for H_2 and H_2O over the whole range 65–105 km. Daily mean values are used for the dissociation rates. For components having lifetimes shorter than a day, Equations (6) and (7) reduce to $x_d = x_{e,d}$ and $x_n = x_{e,n}$.

The results so obtained provide us with good initial data for the main program for the height range 80–105 km. This is very important, because the lifetimes of $O(^3P)$ and H are very long in this region. Therefore a step integration must be continued over many days in order to give convergence. Below 80 km the simplified program does not work since we have a considerable diurnal variation in $O(^3P)$ and H. However, at these levels the lifetimes of $O(^3P)$ and H are so short that convergence in the time-dependent model is obtained after integration (with $\Delta t = 60$ s) over two days. Initial values (except for H_2O and H_2) for the 65–80 km range may be computed on the assumption of photochemical equilibrium. Dissociation rates are computed at intervals of one hour except near sunrise and sunset, when new values are computed at intervals of 10 min.

At the lower boundary, 65 km, photochemical values are taken as boundary conditions, except for H_2, for which a mixing ratio of 5×10^{-7} is used. At the upper boundary, 105 km, a value of 0.4 is taken for the ratio $[O]/[O_2]$, in agreement with observations. It is further assumed that H_2O takes up $\frac{1}{300}$ and H_2 takes up $\frac{1}{25}$ of the available hydrogen at 105 km. These are low values, but the choice of boundary conditions for these components is of little significance below 100 km. For the other components photochemical values are assumed.

Two components have been subject to more interest than the others: atomic oxygen and ozone. Their diurnal variations have to some extent been disregarded in discussions. The diurnal variation of atomic oxygen is shown in Figure 7 for 45° latitude summer. Down to 85 km it is seen to be insignificant. Below 80 km, however, it is drastic: from very low night-time values a sudden increase starts at sunrise. A maximum value is found at noon, then a slow decrease in the afternoon until a drastic decrease takes place just after sunset.

Also ozone shows a remarkable diurnal variation (see Figure 8). Down to 85 km we may talk about a daytime and a night-time value, the variations occurring at sunset and sunrise. During the day and the night the variations are very small. At 80 km we still have a constant night-time value, but during the day we have an increase from sunrise till noon. Below 80 km the diurnal variation is much more complicated. We still have a nearly constant night-time value. At sunrise we find a sudden decrease by approximately an order of magnitude followed by a steady increase until a little before noon, when a decrease starts and goes on till sunset. Then we get a sudden increase to the almost constant night-time value.

The diurnal variation of atomic hydrogen, shown in Figure 9, is very much like that of atomic oxygen. Above 80 km it is insignificant. Below 75 km we have very low night-time values, a sudden increase at sunrise to daytime values of the order 10^7–10^8 cm^{-3}, followed by a sudden decrease at sunset.

The meridional distribution of daytime concentration of atomic oxygen is shown in Figure 10. For all latitudes the peak concentration is found at 92 km, the peak value

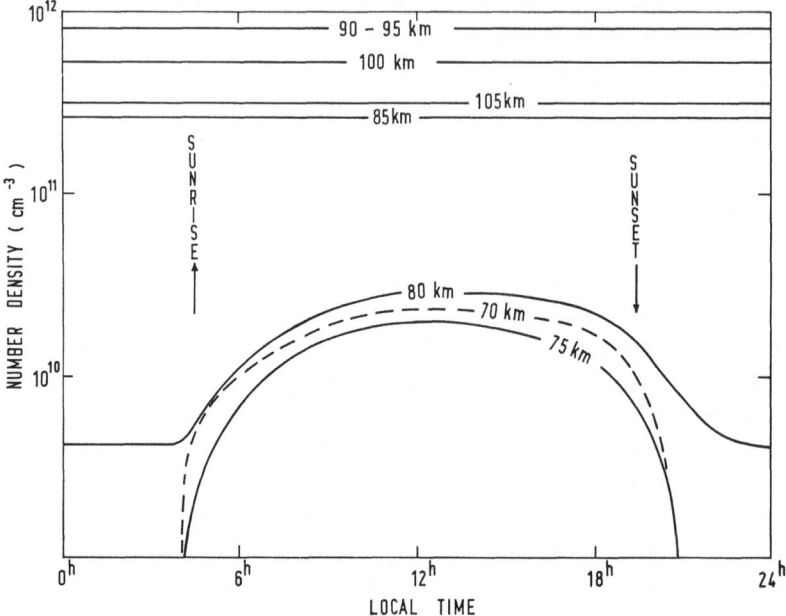

Fig. 7.   Diurnal variation of atomic oxygen (45° latitude, summer).

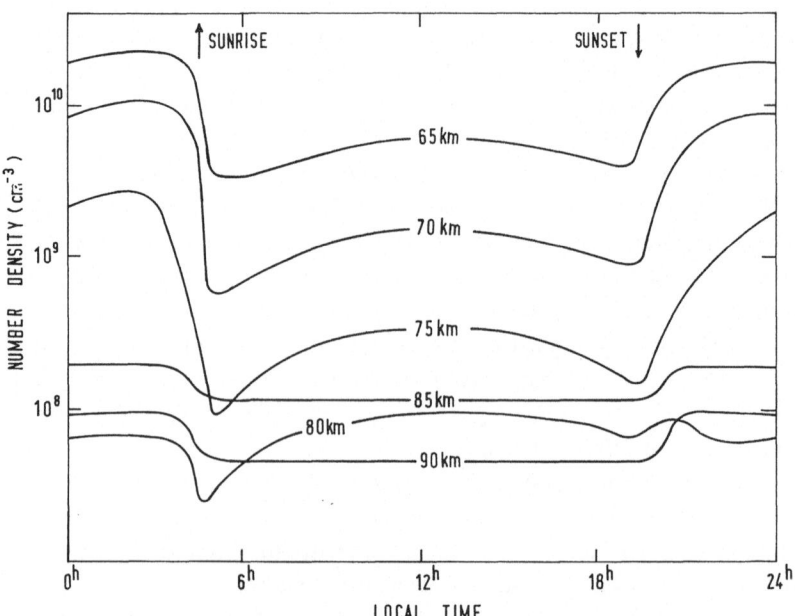

Fig. 8.   Diurnal variation of ozone (45° latitude, summer).

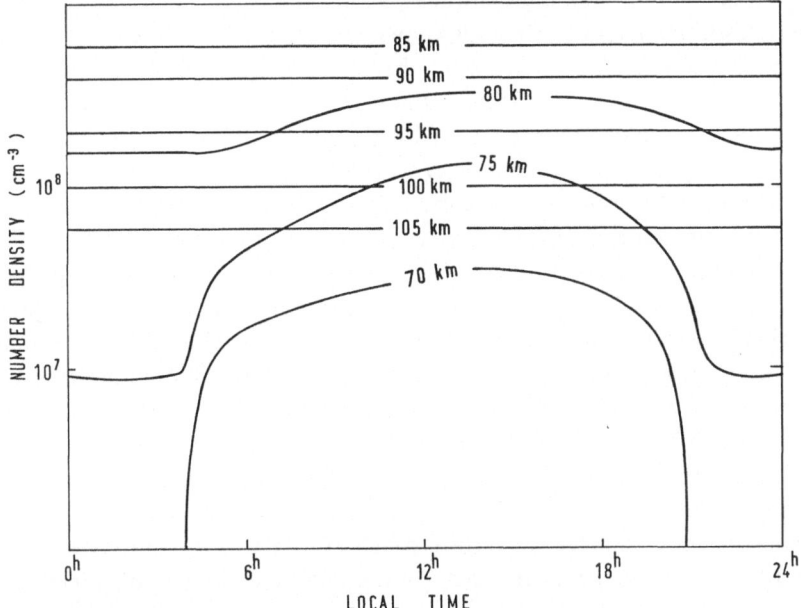

Fig. 9.   Diurnal variation of atomic hydrogen.

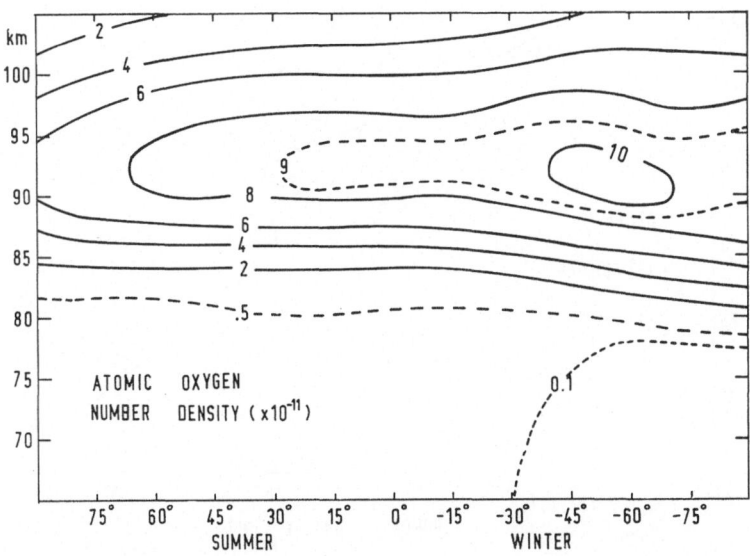

Fig. 10.   Meridional distribution of daytime atomic oxygen.

varying from 10^{12} cm^{-3} in high latitudes, winter, to 7×10^{11} cm^{-3} in high latitudes, summer. Also, the height of the well known cut-off at about 80 km varies very little with latitude and season. Consequently, the horizontal gradients are found to be very small. If this is the case also in the real atmosphere, horizontal transport should have a relatively small effect on the meridional distribution of atomic oxygen and ozone. This may, however, not be true for the high latitude winter atmosphere. The high values shown in Figure 10 result from the fixed upper boundary condition $[O]/[O_2] = 0.4$. Since atomic oxygen is constantly broken down in this dark part of the atmosphere, this ratio can be maintained only by horizontal transport from sunlit regions.

While the number density of ozone varies very little during the night, the daytime values (Figure 8) vary considerably below 80 km from sunrise to sunset, with a maximum a little before noon. Figure 11 shows the vertical distribution of ozone at nighttime and at noon for 45° latitude, summer. Experimental data are available up to about 65 km. The noon curve in Figure 11 may therefore be checked against observations. It turns out that the theoretical value at 65 km, 4×10^9 cm^{-3}, agrees fairly well with observations.

For the sake of completeness, vertical profiles for OH, HO$_2$, and H$_2$O$_2$ are given in Figure 12. These components are small constituents at all levels, with a mixing ratio of 10^{-8} or less. They therefore take up only an insignificant part of the oxygen and hydrogen, but are nevertheless important links in the chain of chemical reactions. $O(^1D)$ is present in very small quantities. It is in photochemical equilibrium during the day and is completely destroyed at night.

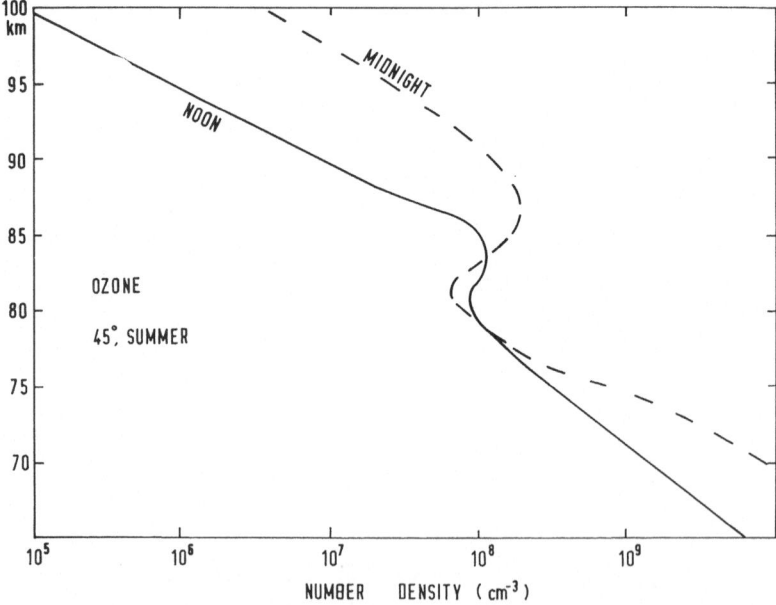

Fig. 11.   Vertical profile for ozone, 45° latitude, summer. The solid curve shows noon values, dashed curve shows midnight values.

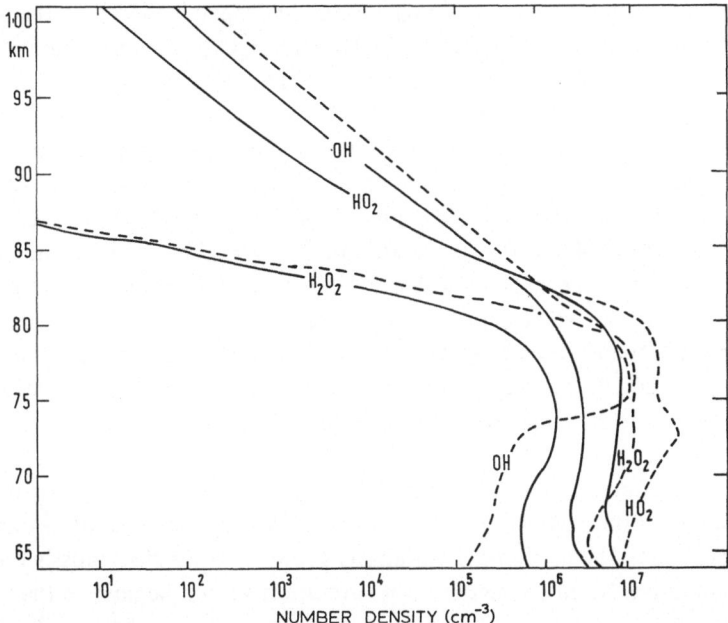

Fig. 12. Vertical distribution of OH, HO₂, and H₂O₂ (45° latitude, summer).

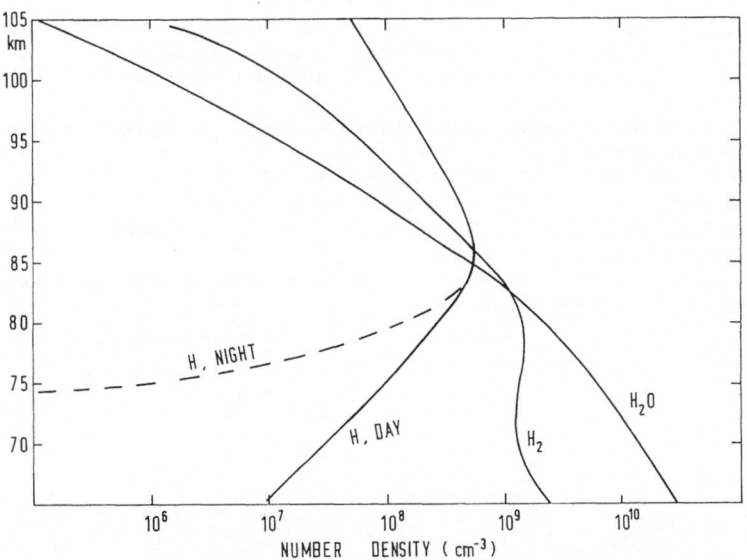

Fig. 13. Vertical distribution of H, H₂, and H₂O (45° latitude, summer).

Finally we shall see how the main part of the hydrogen is shared by the three major hydrogen components, H, H_2, and H_2O. H_2 and H_2O have very long lifetimes at all levels and their diurnal variation is insignificant. The same is true for atomic hydrogen above 80 km; below that level it is destroyed at night, as was shown in Figure 9. The vertical distribution of the three components is shown in Figure 13 for 45° latitude, summer. It will be seen that, in our model, water vapor is the most important hydrogen component up to the mesopause. The mixing ratios obtained are 6.3×10^{-6} at 65 km, 2.6×10^{-6} at 80 km, 4.8×10^{-7} at 90 km and 2.8×10^{-8} at 100 km. These figures depend strongly upon the profile chosen for K_z. Lower K_z-values would give a more rapid decrease in the water vapor mixing ratio.

Near the mesopause H_2 and H_2O seem to be of equal importance, while the importance of H gradually increases with height. But still at 100 km almost 25% of the hydrogen seems to be in the form of H_2, while H takes up about 75%.

The meridional model presented should be regarded as a first step towards a better understanding of the chemical composition of the mesosphere and lower thermosphere on a global scale. The meridional variations shown above reflect mostly the variation in solar radiation. In future models more attention should be paid to the variation in K_z with latitude and season. Furthermore horizontal transport has to be considered. However, it is believed that the main features of the results shown above are fairly representative and that future models will differ from the present one only in details.

References

Ackerman, M.: 1971, this volume, pp. 149–59.

Bates, D. R. and Nicolet, M.: 1950. *J. Geophys. Res.* **55**, 301–27.

Cole, A. E., Court, A., and Kantor, A. J.: 1965, in S. L. Valley (ed.), *Handbook of Geophysics and Space Environments*, Air Force Cambridge Research Laboratories, ch. 2.

Colegrove, D. F., Johnson, F. S., and Hanson, W. B.: 1966, *J. Geophys. Res.* **71**, 2227–36.

Hesstvedt, E.: 1968, *Geofys. Publik.* **27**, No. 4.

Huffman, R. E.: 1968, in *Symposium on Laboratory Measurements of Aeronomic Interest*, York University, Toronto, p. 96.

Johnson, F. S. and Wilkins, E. M.: 1965, *J. Geophys. Res.* **70**, 1281–4.

Kuhn, W. R. and London, J.: 1969, *J. Atmospheric Sci.* **26**, 189–204.

Schumb, W. C., Gatherfield, C. N., and Wentworth, R. L.: 1955, *Hydrogen Peroxide*, Reinhold Publishing Corporation, New York.

Watanabe, K.: 1958, in H. E. Landsberg and J. van Mieghem (eds), *Advances in Geophysics*, V, Academic Press, New York, p. 153.

OXYGEN, HYDROGEN AND NITROGEN CONSTITUENTS
IN THE MESOSPHERE AND IONIZATION PROCESSES

L. THOMAS

S.R.C. Radio and Space Research Station, Ditton Park, Slough, Bucks., England

1. Introduction

It has long been realized that minor neutral constituents have a controlling influence on the thermal balance of the mesosphere. Following the suggestion by Nicolet (1945) that solar Lyman-α radiation can penetrate to mesospheric heights and ionize nitric oxide, it has become evident that this and other minor constituents also dominate the ionization processes of the D region (Reid, 1967). Information on the height distributions of certain minor constituents, such as ozone and nitric oxide, has been provided by rocket-borne observations, but the data available for others are based on photochemical treatments. The important feature included in recent theoretical models has been the realization that certain constituents have long photochemical lifetimes and their distributions can be seriously affected by transport processes; particular attention has been paid to the effects of eddy and molecular diffusion (Hesstvedt, 1968, 1969; Bowman *et al.*, 1970; Shimazaki and Laird, 1970).

The aim of the present paper is to outline current ideas on the ionization processes operating in the D region at middle latitudes and to indicate the roles played by oxygen, hydrogen and nitrogen constituents.

In Section 2 the sources of ionization are discussed, particular attention being given to the photo-ionization of NO by Lyman-α radiation and of $O_2(^1\Delta_g)$ by radiation in the range 1027 Å to 1118 Å (Hunten and McElroy, 1968; Huffman *et al.*, 1971).

Mass-spectrometer observations of positive-ion composition (Narcisi and Bailey, 1965) are described in Section 3. The interpretation of the water cluster ions, $H^+ \cdot (H_2O)_n$, observed below the mesopause is considered in terms of the water vapour reaction scheme devised by Ferguson and Fehsenfeld (1969), and Good *et al.* (1970b). This scheme draws attention to the importance of the height distribution of H_2O at mesospheric heights, and further studies of reactions by Burke (1970) and Ferguson (private communication, 1970) suggest that the height variations of H and O might also be involved.

In Section 4 a negative-ion model based on laboratory measurements of Fehsenfeld *et al.* (1967, 1969a, b) and Fehsenfeld and Ferguson (1968) is presented. This involves the three-body attachment of electrons to O_2, electron detachment from O_2^- by reactions with $O_2(^1\Delta_g)$ and O, and a sequence of negative-ion changes involving reactions with O, O_3, CO_2, NO, and NO_2. It is found that the diurnal variations of these constituents control the day-to-night changes in negative-ion species.

At each stage the information available about the relevant minor constituents is presented. Finally, the uncertainties in D-region processes are mentioned, particularly

Fiocco (ed.), Mesospheric Models and Related Experiments, 65–77. All Rights Reserved.

those arising from lack of knowledge concerning the concentrations of these constituents in the mesosphere.

2. The Production of Ionization

The various sources of ionization operating during solar minimum conditions are
illustrated in Figure 1. The photo-ionization rates correspond to solar zenith angles of
about 60°, and the production rates for cosmic rays refer to a geomagnetic latitude of
50° (Webber, 1962).

The results for 2–8 Å and the total production curve for Lyman-β, extreme ultraviolet and X-rays of wavelength between 40 and 75 Å have been taken from Bourdeau
et al. (1966). The production rates for the ionization of $O_2(^1\Delta_g)$ by radiation in the
range 1027 to 1118 Å have been recently calculated by Huffman *et al.* (1971) and represent a revision of the results of Hunten and McElroy (1968), who first drew attention to the importance of this source of ionization. Huffman *et al.* have taken account

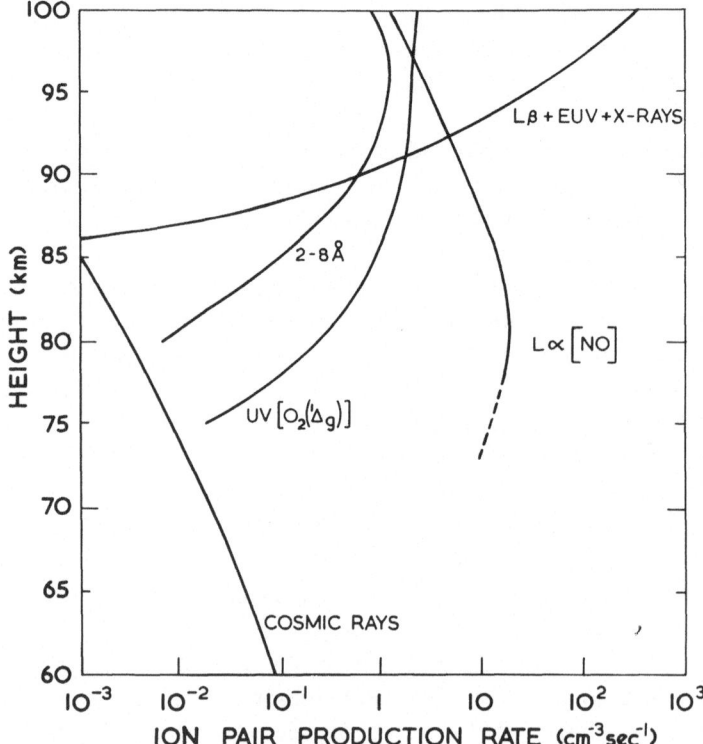

Fig. 1.   Ionization rates in the normal daytime *D* region and lower *E* region for solar minimum conditions and solar angles near 60°. The results for 2–8 Å, Lyman-β, EUV and soft X-rays are based on
calculations of Bourdeau *et al.* (1966); the results for the ionization of $O_2(^1\Delta_g)$ have been given by
Huffman *et al.* (1971); and those for the ionization of NO by Lyman-α radiation by Thomas (1971).
The ionization rates for galactic cosmic rays are those calculated by Webber (1962) for a geomagnetic
latitude of 50°.

of new photo-ionization cross-section data for $O_2(^1\Delta_g)$, recent solar-flux measurements, and absorption by atmospheric CO_2. The results shown for the ionization of nitric oxide by Lyman-α radiation have been derived by Thomas (1971).

It is evident that for these conditions the photo-ionization of the minor constituents NO and $O_2(^1\Delta_g)$ represents the main sources of ionization and it is of interest to consider the height distributions of these constituents. The distribution of NO adopted in the calculations shown in Figure 1 was based on a concentration of 3.9×10^7 cm^{-3} at 85 km and a height variation similar to that of the total atmospheric density of the U.S. Standard Atmosphere (Dubin et al., 1962). This distribution can be compared with that deduced by Barth (1966) from rocket-borne observations of the NO fluorescence in the dayglow or that derived by Mitra (1968) from ionospheric and photochemical data, shown in Figure 2. It can be seen that the distribution adopted is consistent with that of Barth but the concentrations are almost three orders of magnitude larger than those of Mitra. The $O_2(^1\Delta_g)$ distribution adopted in the calculation of Huffman et al. (1971) was derived from a rocket-borne observation of the dayglow emission of the 0–0 band of the infra-red atmospheric system $O_2(a^1\Delta_g - X^3\Sigma_g^-)$ at 1.27 μ (Evans et al., 1968). This distribution, as reported by Hunten and McElroy (1968, 1969), is also shown in Figure 2.

No information is available about the seasonal changes in production rates arising out of changes in height distributions of NO and $O_2(^1\Delta_g)$. However, ground-based observations of the twilight emission of the 0–1 band of the infra-red atmospheric system at 1.58 μ by Vallance Jones and Gattinger (1963) have indicated higher concentrations of $O_2(^1\Delta_g)$ in winter than summer.

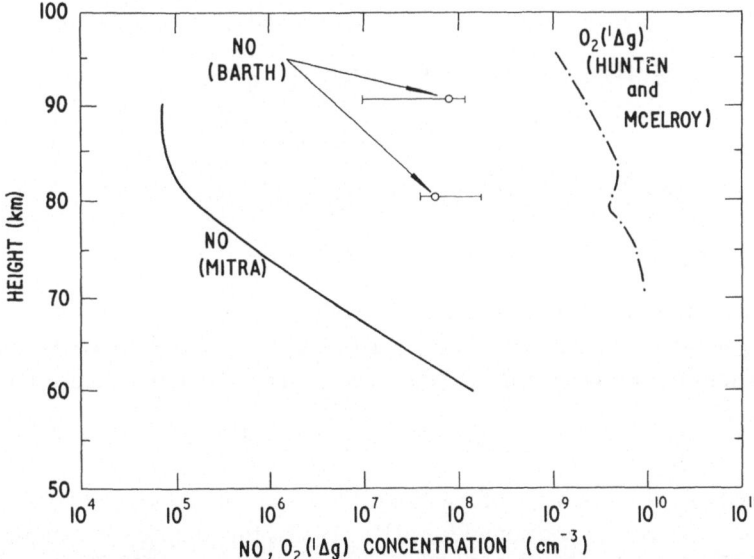

Fig. 2. Height distributions of NO concentration derived from rocket measurements of the dayglow (Barth, 1966) and from ionospheric and photochemical data (Mitra, 1968), and of $O_2(^1\Delta_g)$ deduced from rocket observations of the dayglow (Hunten and McElroy, 1968, 1969).

3. Positive-ion Reactions

On the basis of the D-region model proposed by Nicolet and Aikin (1960), it was expected that NO^+ and O_2^+ would be the major positive ions. Rocket-borne mass-spectrometer observations by Narcisi and Bailey (1965) confirmed that these ions were important at heights above about 82 km, together with layers of metal ions. However, at lower heights the major ions were of mass 37 and 19 μ, corresponding to $H^+ \cdot (H_2O)_2$ and H_3O^+, respectively. The concentrations of these ions decreased rapidly with height near 82 km and this decrease has been a feature of several observations made during a variety of geophysical conditions (Narcisi, private communication, 1969).

Laboratory measurements have shown that water cluster ions, $H^+ \cdot (H_2O)_n$, can arise from the reaction of H_2O^+ ions formed by photo-ionization with water molecules, but in the atmosphere these primary ions are lost instead by rapid charge exchange with O_2. It is therefore necessary to consider a reaction scheme to produce cluster ions which does not start with H_2O^+ ions and such a scheme has been devised by Ferguson and Fehsenfeld (1969), and Good et al. (1970b). In their model, hydrated O_2^+ or NO^+ ions are formed by reactions:

$$P^+ + P + M \rightarrow P \cdot P^+ + M \tag{a}$$

$$P^+ \cdot P + H_2O \rightarrow P^+ \cdot H_2O + P \tag{b}$$

or

$$P^+ + H_2O + M \rightarrow P^+ \cdot H_2O + M \tag{c}$$

where $P = O_2$ or NO.

For $P = O_2$, subsequent reactions with H_2O molecule lead to H_3O^+ or $H^+ \cdot (H_2O)_2$:

$$O_2^+ \cdot H_2O + H_2O \rightarrow H_3O^+ + OH + O_2 \tag{d}$$

or

$$\rightarrow H_3O^+ \cdot OH + O_2 \tag{e}$$

$$H_3O^+ + H_2O + M \rightarrow H^+ \cdot (H_2O)_2 + M \tag{f}$$

$$H_3O^+ \cdot OH + H_2O \rightarrow H^+ \cdot (H_2O)_2 + OH \tag{g}$$

and higher order ions can be similarly formed.

Owing to the smaller ionization potential of NO compared with O_2, the corresponding reactions do not occur for $P = NO$. Instead, the formation of cluster ions begins with

$$NO^+ \cdot (H_2O)_3 + H_2O \rightarrow H^+ \cdot (H_2O)_3 + HNO_2 \tag{h}$$

On the basis of this scheme of reactions, Ferguson and Fehsenfeld (1969) were able to compute the height variation of ions $H^+ \cdot (H_2O)_2$ for comparison with the observations of Narcisi and Bailey (1965). Their analysis revealed a difficulty in reproducing the observed rapid reduction in concentration near 82 km using height distributions of H_2O provided by recent photochemical treatments. The effect of including eddy diffusion in such treatments is to compensate for the loss of water vapour by photo-dissociation at heights above 75 km. This is illustrated in Figure 3 which shows the height distributions of H_2O derived by Bowman *et al.* (1970) using a mixing ratio of 5 parts per million and 10 parts per million at a height of 60 km and an eddy diffusion coefficient of 4.5×10^6 cm^2 s^{-1}; also shown are the corresponding results found for the larger H_2O concentration in the absence of diffusion.

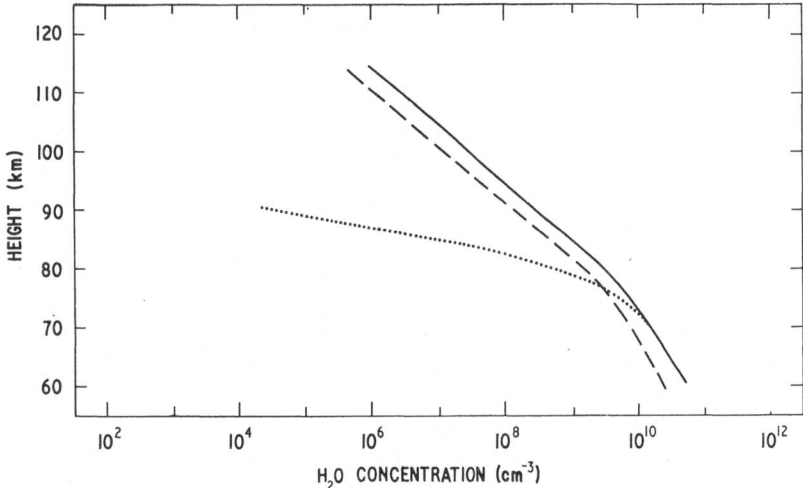

Fig. 3.   Height distributions of H_2O derived from theoretical calculations (Bowman *et al.*, 1970). The broken and continuous curves have been derived for an eddy diffusion coefficient of 4.5×10^6 cm^2 s^{-1} and correspond to water concentrations at 60 km of 5 parts per million and 10 parts per million, respectively. The dotted line was derived without including the effect of eddy diffusion.

A recent development reported by Ferguson (private communication, 1970) has been the discovery that the $O_2^+ \cdot O_2$ ions formed by reaction (a) with $P = O_2$ can be converted to O_2^+ by the reaction:

$$O_4^+ + O \rightarrow O_2^+ + O_3. \tag{i}$$

Although the rate coefficient for this reaction has not been determined accurately, Ferguson states that it is probably near 10^{-10} cm^3 s^{-1}. This reaction then tends to interrupt the sequence of reactions leading from O_2^+ to the water cluster ions.

The overall scheme beginning with O_2^+ and NO^+ ions is presented in Figure 4. This shows the lifetime of each ion species for an altitude of 80 km under the influence of the reactions listed in Appendix 1, together with the reactants involved. These lifetimes have been based on the following concentrations:

$$[O_2] = 8.7 \times 10^{13} \text{ cm}^{-3}, \qquad [N_2] = 3.3 \times 10^{14} \text{ cm}^{-3}, \qquad [NO] = 10^8 \text{ cm}^{-3},$$

$$[N] = 10^6 \text{ cm}^{-3}, \qquad [O] = 10^{11} \text{ cm}^{-3}, \qquad [H_2O] = 1.8 \times 10^9 \text{ cm}^{-3};$$

the value for NO corresponds to that reported by Barth (1966), that for O is as given by Hesstvedt (1969), and that for H_2O is taken from the broken line in Figure 3.

In Figure 4 a distinction has been drawn between very fast reactions corresponding to lifetimes less than 2 sec, represented by heavy lines, moderately fast reactions corresponding to lifetimes between 2 and 100 sec, represented by thin lines, and slow reactions corresponding to still longer lifetimes and represented by broken lines. It is to

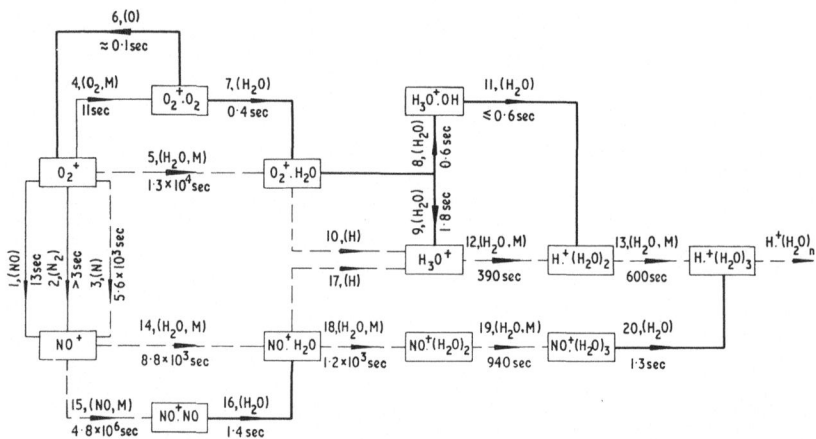

Fig. 4. Schematic representation of positive-ion reactions leading to water cluster ions. The lifetime of each ion under the influence of the reactions listed in Appendix 1 is shown for a height of 80 km, together with the reactants involved. Changes corresponding to lifetimes of less than 2 s, 2–100 s and greater than 100 s are represented by heavy lines, thin lines and broken lines respectively. No values are available for the rate coefficients of reactions 10 and 17.

be noted that the dissociative recombination of positive ions with electrons could compete with the ion-neutral reactions involved in Figure 4. Since the electron concentration at 80 km is normally 10^3 cm^{-3} or less, the recombination coefficients need to be greater than 10^{-5} cm^3 s^{-1} in order that recombination should compete with reactions corresponding to lifetimes of 100 sec or less.

Ferguson and Fehsenfeld (1969) pointed out that in spite of the larger production rate of NO^+, O_2^+ ions represent the main precursor of water cluster ions $H^+ \cdot (H_2O)_n$. This is evident from Figure 4. Burke (1970) has, however, drawn attention to reactions of the type:

$$P^+ \cdot (H_2O)_n + H \rightarrow H_3O^+ \cdot (H_2O)_{n-1} + P. \tag{j}$$

He pointed out that when $n=1$, the reaction is exothermic for $P=NO$ and O_2, and H_3O^+ can be formed from both primary NO^+ and O_2^+ ions. However, as seen from Figure 4, the formation of $NO^+ \cdot H_2O$ is very much slower than $O_2^+ \cdot H_2O$ because of the small concentration of NO compared with O_2. Furthermore, since the H concentration at mesospheric heights is about an order of magnitude less than that of H_2O

(Hesstvedt, 1969; Bowman *et al.*, 1970; Shimazaki and Laird, 1970) the reaction between $O_2^+ \cdot H_2O$ and H will need to have a rate coefficient greater than 10^{-10} cm^3 s^{-1} in order to compete with the reaction involving H_2O in the formation of H_3O^+. The two reactions corresponding to (j) are shown by broken lines in Figure 4.

It is to be noted that the reduction in O_2^+ production by the photo-ionization of $O_2(^1\Delta_g)$ reported by Huffman *et al.* (1971), Figure 1, implies a very much smaller production of water cluster ions below about 75 km. Furthermore, a comparison of the NO$^+$ production at 80 km, shown in Figure 1, with the loss rates corresponding to dissociative recombination with electron concentrations of 10^3 cm^{-3} and presently accepted values for the rate coefficient (Weller and Biondi, 1968), and to the ion-neutral reactions shown in Figure 4, shows a discrepancy of more than an order of magnitude. Since the Lyman-α flux is well established this would seem to imply that the NO concentration adopted in the photo-ionization calculations is too large by at least an order of magnitude.

4. Negative-ion Reactions

Early attempts to study the negative-ion chemistry of the D region were confined to the ion O_2^- known to be formed rapidly by three-body electron attachment (Chanin *et al.*, 1959):

$$e + O_2 + O_2 \rightarrow O_2^- + O_2. \tag{k}$$

These attempts considered that electrons could be released by collisional detachment and photodetachment, the latter being possible for all wavelengths up to the infra-red range since the electron affinity of the ion is only 0.43 eV (Pack and Phelps, 1966a).

More recent studies have shown that the changes in electron concentrations occurring near sunrise are associated with the arrival of ultra-violet radiation, this probably implying the presence of a negative ion of greater electron affinity. In addition, laboratory studies have indicated a complicated sequence of negative-ion changes involving reactions with minor constituents such as O, O_3, $O_2(^1\Delta_g)$, NO, NO_2, CO_2 and probably H_2O. An outline of the negative-ion changes based on the laboratory measurements of Fehsenfeld *et al.* (1967, 1969a, b) and Fehsenfeld and Ferguson (1968) is given in Figure 5 (Thomas, 1971). This shows the lifetimes of electrons and negative ions at a height of 65 km during day and night under the influence of reactions shown in Appendix 2, together with the reactants involved. The concentrations of neutral constituents adopted are shown in Appendix 3. A distinction is drawn between reactions corresponding to lifetimes less than 2 s, represented by thick lines, reasonably fast reactions corresponding to lifetimes of 2–100 sec, represented by thin lines, and slow reactions corresponding to longer lifetimes and represented by broken lines.

It is evident that the major differences between day and night arise from the day-to-night changes in concentrations of O, $O_2(^1\Delta_g)$, O_3, NO and NO_2. Some experimental data for such changes are available for O_3, as is shown in the observations of Johnson *et al.* (1952), Weeks and Smith (1968), Carver *et al.* (1966), and Reed (1968) presented in Figure 6. It appears that a ten-fold increase of concentration occurs between day

DAY

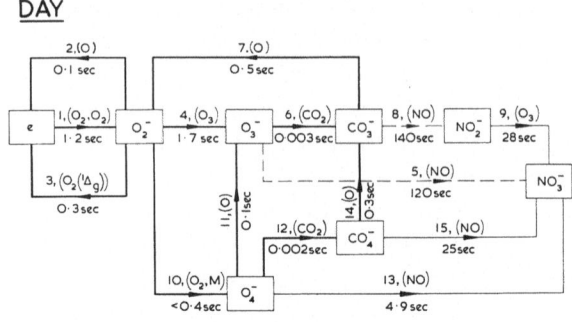

NIGHT

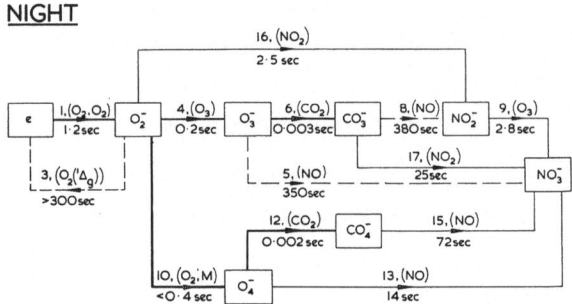

Fig. 5. Schematic representation of negative-ion changes occurring during day and night. The lifetimes of electrons and of each ion under the influence of reactions listed in Appendix 2 are shown for
a height of 65 km, together with the reactants involved. Changes corresponding to lifetimes of less
than 2 s, 2–100 s and greater than 100 s are represented by heavy lines, thin lines and
broken lines, respectively.

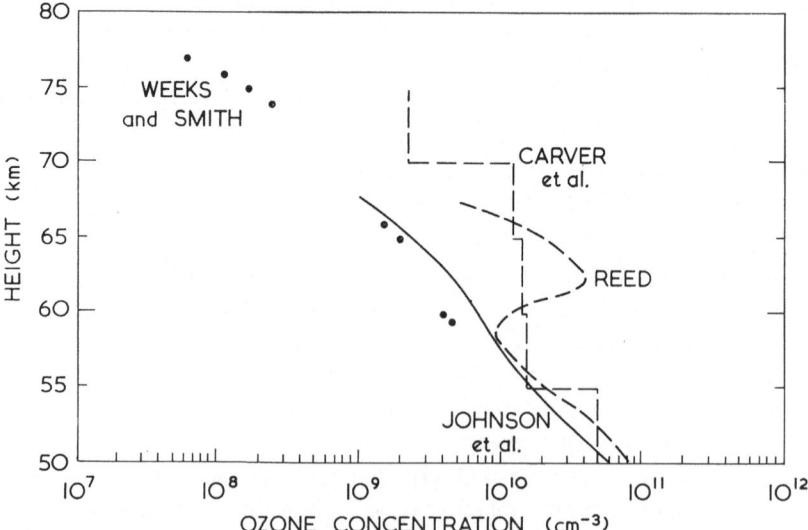

Fig. 6. Height distributions of ozone measured during the daytime (continuous lines and dots) by
Johnson *et al.* (1952), and Smith and Weeks (1968), and during the nighttime (broken lines) by Carver
et al. (1966) and Reed (1968).

and night for heights above about 65 km. For the other constituents, information on day-to-night changes in concentrations has been derived from photochemical treatments (Hunt, 1966; Hesstvedt, 1968, 1969; Shimazaki and Laird, 1970). Large discrepancies are found between the magnitudes of the diurnal change of O concentrations predicted by these treatments. These arise from uncertainties in concentrations of other constituents, particularly oxygen-hydrogen compounds known to have a large influence (Bates and Nicolet, 1950); lack of information about transport processes, particularly eddy diffusion; and uncertainties in rate coefficients. The results of a study carried out by Hesstvedt (1969) are shown in Figure 7. No information is available about the day-to-night change in height distribution of $O_2(^1\Delta_g)$ but it is to be expected from Gattinger (1969) that concentrations below 70 km at nighttime are more than three orders of magnitude smaller than the daytime values.

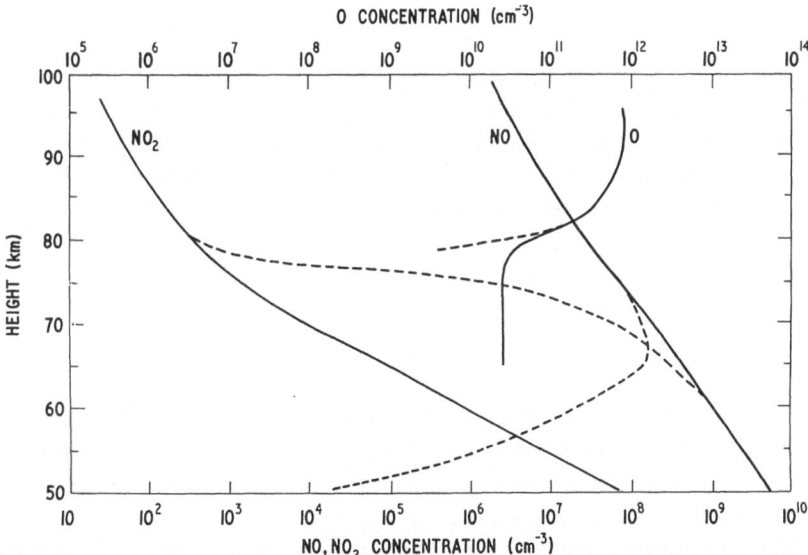

Fig. 7. Height distributions of O, NO and NO_2 during daytime (continuous curves) and nighttime (broken curves) derived from theoretical calculations which include the effects of vertical eddy diffusion (Hesstvedt, 1969).

In considering theoretically the height distributions of NO and NO_2 it is usually assumed that reactions between hydrogen and nitrogen compounds are unimportant and that reactions between oxygen and nitrogen components have negligible effects on the O and O_3 concentrations at mesospheric heights. These latter reactants are, however, involved in nitrogen oxide reactions. For example, the processes controlling the production and loss of NO_2 are believed to be:

$$NO + O + M \rightarrow NO_2 + M \tag{l}$$

$$NO_2 + O \rightarrow NO + O_2 \tag{m}$$

and at heights below 70 km:

$$NO + O_3 \rightarrow NO_2 + O_2 .$$ (n)

Theoretical studies have predicted marked diurnal changes in the NO and NO_2 concentrations at low mesospheric heights. The results of Hesstvedt (1969) for these two constituents are also shown in Figure 7. It is to be noted that the daytime curve for NO shows concentrations only slightly smaller than those of Barth (1966) in Figure 2.

It is seen from Figure 7 that the day-to-night changes in concentrations of minor constituents involved in the negative-ion chemistry are less marked at heights above 80 km. This implies that there is a smaller diurnal change in the predominant negative-ion species. In contrast to the results for 65 km shown in Figure 5, in which the predominant negative ions at nighttime would seem to be those of carbon and nitrogen oxides, it might be expected that above 80 km the continuing presence of O at night-time would mean that O_2^- is the main ion present as in daytime (Reid, 1970). The concentrations of this ion would, however, be small because of associative detachment by O atoms (reaction 2 of Appendix 2). It has been shown that because of their large dipole moments water molecules tend to cluster to negative ions (Pack and Phelps, 1966b), and it would be expected that negative ions which have relatively long life-times, such as NO_3^-, will be hydrated. Recent mass-spectrometer observations by Narcisi *et al.* (1969) have shown the presence of negative ions of mass 62 and 80 amu which would correspond to NO_3^- and the first hydrate.

5. Conclusions

Many of the uncertainties encountered in studies of the ionization processes operating in the D region of the ionosphere arise from our lack of knowledge concerning the height distribution of oxygen, hydrogen and nitrogen constituents in the mesosphere and how they change diurnally and seasonally.

The doubt about the concentration of NO is particularly serious since it is involved in the production of ionization (Figure 1), and in the positive- and negative-ion chemistry (Figures 4 and 5). There is a major discrepancy between the production of NO^+ ions, based on NO concentrations similar to those reported by Barth (1966), and loss by dissociative recombination and ion-neutral reactions.

Further measurements of $O_2(^1\Delta_g)$ concentrations are needed to understand its role in the production of ionization (Figure 1) and electron detachment from O_2^- (Figure 5).

Information about the detailed height distribution of H_2O is urgently required for the interpretation of water cluster ions $H^+ \cdot (H_2O)_n$, and probably for studies of hydrated negative ions. It seems likely that the reaction between $O_2^+ \cdot O_2$ ions and O also plays a major part in the positive-ion chemistry and the detailed height variation of O in daytime is also required.

The relative concentrations of O and O_3, and the NO_2 concentration, are of great importance in the negative-ion chemistry (Figure 5) and improved information is needed about the diurnal variations of the height distributions of these minor constituents.

Acknowledgements

The author is indebted to Drs Huffman, Paulsen, Larrabee and Cairns for their results concerning the photo-ionization of $O_2(^1\Delta_g)$ and to Dr Ferguson for recent rate-coefficient data, these being provided prior to publication. Acknowledgement of Figures 5 and 6 is made to the *Journal of Atmospheric and Terrestrial Physics*.

This paper is published by permission of the Director of the Radio and Space Research Station of the Science Research Council.

Appendix 1: Ion-Molecule Reactions Considered in Figure 4

.eaction	Rate Coefficient [a]	Reference
1. $O_2^+ + NO \rightarrow NO^+ + O_2$	8.0×10^{-10}	Ferguson et al. (1965)
2. $O_2^+ + N_2 \rightarrow NO^+ + NO$	$< 10^{-15}$	Ferguson et al. (1965)
3. $O_2^+ + N \rightarrow NO^+ + O$	1.8×10^{-10}	Ferguson et al. (1965)
4. $O_2^+ + O_2 + M \rightarrow O_2^+ \cdot O_2 + M$	2.4×10^{-30} [b]	Good et al. (1970b)
5. $O_2^+ + H_2O + M \rightarrow O_2^+ \cdot H_2O + M$	1.0×10^{-28} [b]	Good et al. (1970b)
5. $O_2^+ \cdot O_2 + O \rightarrow O_2^+ + O_3$	10^{-10}	Ferguson (private communication, 1970)
7. $O_2^+ \cdot O_2 + H_2O \rightarrow O_2^+ \cdot H_2O + O_2$	1.3×10^{-9}	Good et al. (1970b)
8. $O_2^+ \cdot H_2O + H_2O \rightarrow H_3O^+ \cdot OH + O_2$	9.0×10^{-10}	Good et al. (1970b)
9. or $\rightarrow H_3O^+ + OH + O_2$	3.0×10^{-10}	Good et al. (1970b)
0. $O_2^+ \cdot H_2O + H \rightarrow H_3O^+ + O_2$	No value	Burke (1970)
1. $H_3O^+ \cdot OH + H_2O \rightarrow H^+ \cdot (H_2O)_2 + OH$	$> 10^{-9}$	Good et al. (1970b)
2. $H_3O^+ + H_2O + M \rightarrow H^+ \cdot (H_2O)_2 + M$	$3.4 \times 10^{-27} (M = N_2)$ $3.7 \times 10^{-27} (M = O_2)$	Good et al. (1970a) Good et al. (1970b)
3. $H^+ \cdot (H_2O)_2 + H_2O + M \rightarrow H^+ \cdot (H_2O)_3 + M$	$2.3 \times 10^{-27} (M = N_2)$ $2.0 \times 10^{-27} (M = O_2)$	Good et al. (1970a) Good et al. (1970b)
4. $NO^+ + H_2O + M \rightarrow NO^+ \cdot H_2O + M$	1.5×10^{-28}	Lineburger and Puckett (1969)
5. $NO^+ + NO + M \rightarrow NO^+ \cdot NO + M$	5.0×10^{-30} [b]	Lineburger and Puckett (1969) Puckett et al. (1969)
6. $NO^+ \cdot NO + H_2O \rightarrow NO^+ \cdot H_2O + NO$	4.0×10^{-10}	Puckett et al. (1969)
7. $NO^+ \cdot H_2O + H \rightarrow H_3O^+ + NO$	No value	Burke (1970)
8. $NO^+ \cdot H_2O + H_2O + M \rightarrow NO^+ \cdot (H_2O)_2 + M$	1.1×10^{-27} [b]	Ferguson (private communication, 1970)
9. $NO^+ \cdot (H_2O)_2 + H_2O + M \rightarrow NO^+ \cdot (H_2O)_3 + M$	1.4×10^{-27} [b]	Ferguson (private communication, 1970)
0. $NO^+ \cdot (H_2O)_3 + H_2O \rightarrow H^+ \cdot (H_2O)_3 + HNO_2$	4.3×10^{-10}	Ferguson (private communication, 1970).

In $cm^3 s^{-1}$ for two-body and $cm^6 s^{-1}$ for three-body reactions.
The rate coefficients for reactions 4 and 5 were measured with O_2 as the third body, that of reaction 15 with NO
s the third body and those of reactions 18 and 19 with N_2 as the third body. In the calculation of lifetimes, the
.alues shown were adopted for $M = N_2$ and O_2.

Appendix 2: Electron Attachment, Detachment and Negative-ion Molecule Reactions Considered in Figure 5

Reaction	Rate coefficient[a]	Reference
1. $e + O_2 + O_2 \rightarrow O_2^- + O_2$	1.6×10^{-30}	Chanin et al. (1959)
2. $O_2^- + O \rightarrow O_3 + e$	3.3×10^{-10}	Fehsenfeld et al. (1967)
3. $O_2^- + O_2(^1\Delta_g) \rightarrow 2\,O_2 + e$	2.0×10^{-10}	Fehsenfeld et al. (1969a)
4. $O_2^- + O_3 \rightarrow O_3^- + O_2$	3.0×10^{-10}	Fehsenfeld et al. (1967)
5. $O_3^- + NO \rightarrow NO_3^- + O$	1.0×10^{-11}	Fehsenfeld et al. (1967)
6. $O_3^- + CO_2 \rightarrow CO_3^- + O_2$	4.0×10^{-10}	Fehsenfeld et al. (1967)
7. $CO_3^- + O \rightarrow O_2^- + CO_2$	8.0×10^{-11}	Fehsenfeld et al. (1967)
8. $CO_3^- + NO \rightarrow NO_2^- + CO_2$	9.0×10^{-12}	Fehsenfeld et al. (1967)
9. $NO_2^- + O_3 \rightarrow NO_3^- + O_2$	1.8×10^{-11}	Fehsenfeld and Ferguson (1968)
10. $O_2^- + O_2 + M \rightarrow O_4^- + M$	$> 10^{-30}$	Fehsenfeld et al. (1969b)
11. $O_4^- + O \rightarrow O_3^- + O_2$	4.0×10^{-10}	Fehsenfeld et al. (1969b)
12. $O_4^- + CO_2 \rightarrow CO_4^- + O_2$	4.3×10^{-10}	Fehsenfeld et al. (1969b)
13. $O_4^- + NO \rightarrow NO_3^- + O_2$	2.5×10^{-10}	Fehsenfeld et al. (1969b)
14. $CO_4^- + O \rightarrow CO_3^- + O_2$	1.5×10^{-10}	Fehsenfeld et al. (1969b)
15. $CO_4^- + NO \rightarrow NO_3^- + CO_2$	4.8×10^{-11}	Fehsenfeld et al. (1969b)
16. $O_2^- + NO_2 \rightarrow NO_2^- + O_2$	8.0×10^{-10}	Fehsenfeld and Ferguson (1968)
17. $CO_3^- + NO_2 \rightarrow NO_3^- + CO_2$	8.0×10^{-11}	Ferguson (1969)

[a] In $cm^3\ s^{-1}$ for two-body and $cm^6\ s^{-1}$ for three-body reactions.

Appendix 3: Concentrations at 65 km Adopted in Deriving Lifetimes in Figure 5

	O	O_2	$O_2(^1\Delta_g)$	O_3	CO_2	N_2	NO	NO_2
Day	2.5(10)	7.3(14)	1.6(10)	2.0(9)	1.0(12)	2.7(15)	8.2(8)	2.0(5)
Night	1.0(4)	7.3(14)	< 1.6(7)	2.0(10)	1.0(12)	2.7(15)	2.9(8)	5.1(8)

These concentrations are given in cm^{-3}, 2.5(10) signifying 2.5×10^{10}.

In this table the values for O_2 and N_2 correspond to the U.S. Standard Atmosphere model (Dubin et al., 1962). The daytime value for O has been taken from the results of the theoretical study by Hesstvedt (1969) shown in Figure 7, and the nighttime values are known to be very much smaller (Hunt, 1966), the value $10^4\ cm^{-3}$ being taken for illustrative purposes. The daytime value for $O_2(^1\Delta_g)$ represents an extrapolation of the curve in Figure 2 and the nighttime value is assumed to be more than three orders of magnitude smaller. The values for CO_2 correspond to a mixing ratio of 3×10^{-4}, equal to that found in the lower atmosphere. The daytime value for NO has been derived using the value of $3.9 \times 10^7\ cm^{-3}$ at 85 km reported by Barth (1966) and a completely mixed distribution lower down, and the nighttime value and the two values for NO_2 have been derived from the theoretical values of Hesstvedt (1969) after including an increase by a factor of 2 necessary to reconcile his daytime value with that deduced from Barth (1966).

References

Barth, C. A.: 1966, *Ann. Geophys.* **22** 198–207.
Bates, D. R. and Nicolet, M.: 1950, *J. Geophys. Res.* **55**, 301–27.
Bourdeau, R. E., Aikin, A. C., and Donley, J. L.: 1966, *J. Geophys. Res.* **71**, 727–40.
Bowman, M. R., Thomas, L., and Geisler, J. E.: 1970, *J. Atmospheric Terrest. Phys.* **32**, 1661–74.
Burke, R. R.: 1970, *J. Geophys. Res.* **75**, 1345–7.
Carver, J. H., Horton, B. H., and Burger, F. G.: 1966, *J. Geophys. Res.* **71**, 4189–91.
Chanin, L. M., Phelps, A. V., and Biondi, M. A.: 1959, *Phys. Rev. Letters* **2**, 344–6.
Dubin, M., Sissenwine, N., and Wexler, H.: 1962, *U.S. Standard Atmosphere*, U.S. Government Printing Office, Washington D.C.
Evans, W. F., Hunten, D. M., Llewellyn, E. J., and Vallance Jones, A.: 1968, *J. Geophys. Res.* **73**, 2885–96.
Fehsenfeld, F. C. and Ferguson, E. E.: 1968, *Planetary Space Sci.* **16**, 701–2.
Fehsenfeld, F. C., Albritton, D. L., Burt, J. A., and Shiff, H. I.: 1969a, *Can. J. Chem.* **47**, 1793–5.
Fehsenfeld, F. C., Ferguson, E. E., and Bohme, D. K.: 1969b, *Planetary Space Sci.* **17**, 1759–62.
Fehsenfeld, F. C., Schmeltekopf, A. L., Schiff, H. I., and Ferguson, E. E.: 1967, *Planetary Space Sci.* **15**, 373–9.
Ferguson, E. E.: 1969, *Can. J. Chem.* **47**, 1815–20.
Ferguson. E. E. and Fehsenfeld, F. C.: 1969, *J. Geophys. Res.* **74**, 5743–51.
Ferguson, E. E., Fehsenfeld, F. C., Goldan, P. D., and Schmeltekopf, A. L. 1965, *J. Geophys. Res.* **70**, 4323–9.
Gattinger, R. L.: 1969, *Ann. Geophys.* **25**, 825–30.
Good, A., Durden, D. A., and Kebarle, P.: 1970a, *J. Chem. Phys.* **52**, 212–21.
Good, A., Durden, D. A., and Kebarle, P.: 1970b, *J. Chem. Phys.* **52**, 222–9.
Hesstvedt, E.: 1968, *Geophys. Publik.* **27**, 1–35.
Hesstvedt, E.: 1969, *Rept. Inst. Geophys., University of Oslo*, December 1969.
Huffman, R. E., Paulsen, D. E., Larrabee, J. C., and Cairns, R. B.: 1971, *J. Geophys. Res.* **76**, 1028–38.
Hunt, B. G.: 1966, *J. Geophys. Res.* **71**, 1385–98.
Hunten, D. M. and McElroy, M. B.: 1968, *J. Geophys. Res.* **73**, 2421–8.
Hunten, D. M. and McElroy, M. B.: 1969, *J. Geophys. Res.* **74**, 3067.
Johnson, F. S., Purcell, J. D., Tousey, R., and Watanabe, K.: 1952, *J. Geophys. Res.* **57**, 157–76.
Lineburger, W. C. and Puckett, L. J.: 1969, *Phys. Rev.* **187**, 286–91.
Mitra, A. P.: 1968, *J. Atmospheric Terrest. Phys.* **30**, 1065–1114.
Narcisi, R. S. and Bailey, A. D.: 1965, *J. Geophys. Res.* **70**, 3687–700.
Narcisi, R. S., Bailey, A. D., Della Lucca, L. E., and Sherman, C.: *Trans. Amer. Geophys. Union* 1969, **50**, 654.
Nicolet, M.: 1945, *Mem. Inst. Roy. Meteorol. Belg.* **19**, 83.
Nicolet, M. and Aikin, A. C.: 1960, *J. Geophys. Res.* **65**, 1469–83.
Pack, J. L. and Phelps, A. V.: 1966a, *J. Chem. Phys.* **44**, 1870–83.
Pack, J. L. and Phelps, A. V.: 1966b, *J. Chem. Phys.* **45**, 4316–29.
Puckett, L. J., Teague, M. W., and Kregel, M. D.: 1969, Presented at DASA Symposium on Physics and Chemistry of the Upper Atmosphere, Stanford Res. Inst., California, June 1969.
Reed, E. I.: 1968, *J. Geophys. Res.* **73**, 2951–57.
Reid, G. C.: 1967, in *Space Res.* **7**, Vol. I (ed. by R. L. Smith-Rose) North-Holland Publ. Co., Amsterdam, 197–211.
Reid, G. C.: 1970, *J. Geophys. Res.* **75**, 2551–62.
Shimazaki, T. and Laird, A. R.: 1970, *J. Geophys. Res.* **75**, 3221–35.
Thomas, L.: 1971, *J. Atmospheric Terrest. Phys.* **33**, 157–95.
Vallance Jones, A. and Gattinger, R. L.: 1963, *Planetary Space Sci.* **11**, 961–74.
Webber, W.: 1962, *J. Geophys. Res.* **67**, 5091–106.
Weeks, L. G. and Smith, L. G.: 1968, *Planetary Space Sci.* **16**, 1189–95.
Weller, C. S. and Biondi, M. A.: 1968, *Phys. Rev.* **172**, 198–206.

ENERGY CONVERSIONS AND MEAN VERTICAL MOTIONS IN THE HIGH LATITUDE SUMMER MESOSPHERE AND LOWER THERMOSPHERE

PAUL J. CRUTZEN*

Clarendon Laboratory, Oxford University, Oxford, England

As is well known, an appreciable part of the solar energy initially absorbed in the upper atmosphere is not immediately converted to kinetic energy, but appears in chemical form (dissociation energy or energy of excited species). It can remain as such for very long periods before being converted to other forms of energy. A non-negligible part of the initially available chemical energy is radiated away, the remaining part heating the atmosphere at locations that may be remote from those where the chemical energy was created. This is accomplished by atmospheric motions of different types, from molecular diffusion to mean large scale motions.

The chemical species which carry most of the initially stored energy, e.g. atomic oxygen in the ground state and the first electronically excited state $O(^1D)$, initiate a large variety of chemical processes in which many vibrationally and electronically excited species participate. Houghton (1969) has discussed the last stages in the decay of the chemical energy, the relaxation of vibrationally excited molecular oxygen and nitrogen and the role played by the radiating gases carbon-dioxide and water vapour.

In this paper the ideas in Houghton's article will be somewhat extended and their importance illustrated by numerical results. Assuming that the mean vertical motion is the prime factor in advecting heat and chemical properties at high latitudes in the summer season, the vertical motions and the chemical state of the upper atmosphere are derived which are simultaneously consistent with the mean thermal state.

Reaction Scheme

(1a)	$O_2 + h\nu \to O + O(^1S)$	$\lambda < 1340$ Å
(1b)	$\to O + O(^1D)$	$\lambda < 1759$ Å
(1c)	$\to 2O$	$\lambda < 2454$ Å
(2a)	$O_3 + h\nu \to O(^1D) + O_2(^1\Delta_g)$	$\lambda < 3030$ Å
(2b)	$\to O + O_2(^1\Delta_g)$	$\lambda < 3500$ Å
(2c)	$\to O + O_2$	$\lambda < 11\,400$ Å
(3a)	$H_2O + h\nu \to H + OH$	$\lambda = 1215.7$ Å (75%)
(3b)	$\to H_2 + O$	$\lambda = 1215.7$ Å (25%)
(3c)	$\to H + OH$	$\lambda < 2420$ Å

* ESRO post-doctoral fellow on leave from the Meteorological Institute, Stockholm University, Sweden.

Fiocco (ed.), Mesospheric Models and Related Experiments, 78–88. All Rights Reserved.
Copyright © 1971 by D. Reidel Publishing Company, Dordrecht-Holland.

(4)	$O(^1D) + H_2O \rightarrow 2OH$	$k_4 = 10^{-10}$
(5)	$O(^1S) + H_2O \rightarrow 2OH$	$k_5 = 4 \times 10^{-10}$
(6a)	$O(^1D) + O_2 \rightarrow O_2(^1\Sigma_g^+) + O$	$k_{6a} = 2 \times 10^{-11}$
(6b)	$\rightarrow O_2^* + O$	$k_{6b} = 3 \times 10^{-11}$ (text)
(7)	$O(^1D) + N_2 \rightarrow O + N_2$	$k_7 = 8 \times 10^{-11}$
(8)	$O(^1S) + O_2 \rightarrow O + O_2$	$k_8 = 2 \times 10^{-13}$
(9)	$O(^1S) \rightarrow O(^1D) + h\nu$	$\tau_9 = 1.3$ sec
(10)	$O_2(^1\Delta_g) + O_2 \rightarrow 2O_2$	$k_{10} = 2.4 \times 10^{-18}$
(11)	$O_2(^1\Delta_g) \rightarrow O_2 + h\nu$	$\tau_{11} = 64$ min
(12)	$O_2(^1\Sigma_g^+) + M \rightarrow O_2 + M$	$k_{12} = 2 \times 10^{-15}$
(13)	$O_2(^1\Sigma_g^+) \rightarrow O_2 + h\nu$	$\tau_{13} = 12$ sec
(14)	$O + O + M \rightarrow O_2 + M$	$k_{14} = 2.7 \times 10^{-33}$
(15)	$O + O_2 + M \rightarrow O_3 + M$	$k_{15} = 1.16 \times 10^{-35} \exp(1050/T)$
(16)	$O + O + O \rightarrow O(^1S) + O_2$	$k_{16} = 2 \times 10^{-34}$
(17)	$O + OH \rightarrow O_2 + H$	$k_{17} = 5 \times 10^{-11}$
(17a)	$O_3 + OH \rightarrow OH_2 + O_2$	$k_{17a} = 10^{-13}$
(18)	$O + HO_2 \rightarrow OH + O_2$	$k_{18} = 2 \times 10^{-11}$
(19)	$O_3 + H \rightarrow OH^* + O_2$	$k_{19} = 2.6 \times 10^{-11}$
(20)	$H + O_2 + M \rightarrow HO_2 + M$	$k_{20} = 5 \times 10^{-32}$
(21)	$OH + HO_2 \rightarrow H_2O + O_2$	$k_{21} = 10^{-11}$
(22)	$OH + OH \rightarrow H_2O + O$	$k_{22} = 3 \times 10^{-12}$
(23)	$HO_2 + HO_2 \rightarrow H_2O_2 + O_2$	$k_{23} = 3 \times 10^{-12}$
(24)	$H + HO_2 \rightarrow H_2 + O_2$	$k_{24} = 4 \times 10^{-13}$ (text)
(25)	$H + H + M \rightarrow H_2 + M$	$k_{25} = 2.6 \times 10^{-32}$
(26)	$H_2 + O(^1D) \rightarrow H + OH$	$k_{26} = 10^{-10}$
(27)	$H_2O_2 + h\nu \rightarrow 2OH$	$J_{27} = 1.2 \times 10^{-4} s^{-1}$

All reaction rate coefficients are expressed in cm molecule sec units.

Most of the listed values for the rate coefficients are from review articles (Zipf, 1969; Schiff, 1969; Kaufman, 1969; Wayne, 1969; Schofield, 1967). The reaction scheme deviates little from the one originally proposed by Bates and Nicolet (1950). Several coefficients are still not well known, such as the important temperature dependence of k_{15} (Schiff, 1969). The production of $O(^1S)$ in the photolysis of O_2 was detected by Filseth and Welge (1969). The assumption that all dissociation at the Ly-α wavelength yields $O(^1S)$ leads to lifetimes for H_2O around the mesopause two times shorter than would be the case if only reaction (3) were considered. It appears, however, from observations of the green line dayglow (Wallace and McElroy, 1966; Dandekar, 1969) that the quantum efficiency for $O(^1S)$ formation cannot be unity at 1216 Å. Anticipating the results of the present model, the dayglow at noon would be 30 kR. According to Dandekar (1969) the upper limit for the emission due to photodissociation is 5 kR, indicating a quantum yield of not more than 16%.

For information about the products in the photolysis of ozone, reference is made to the article by Clark and Wayne (1970) and to Wayne's paper (this volume, p. 240).

The dissociation of water vapour and the production of H_2 at the Ly-α wavelength has been discussed by McNesby and Okabe (1964), and reactions (6a) and (6b) will be discussed in more detail later.

Solar intensity data and absorption data for molecular oxygen and ozone are from Watanabe (1958); Detwiler (1961); Johnson (1954); Ditchburn and Young (1962); Thompson et al. (1963); and Vigroux (1953). The penetration of ultraviolet light between 1750 and 2000 Å was simulated using data of Hudson et al. (1969).

The total mixing ratio of hydrogen atoms in the form of H_2O, H, OH, HO_2 and H_2O_2 was assumed to be 2.5×10^{-6}. That for H_2 was assumed to be equal to 5×10^{-7} over the entire altitude range. Some justification for the later assumption is that the photochemical lifetime for H_2 is of the order of years. Furthermore, in reactions (24) and (25) vibrationally excited H_2 may be formed, which, if not adequately deactivated, may react with ground state atomic oxygen $H_2(v \geqslant 1) + O \rightarrow H + OH$. The energy present in the vibration is sufficiently large to overcome the activation energy (10.2 kcal/mole).

Atomic oxygen in the ground state and in the first electronically excited state carries most of the initially produced chemical energy. Ground state atomic oxygen participates in the following important exothermic reactions:

(28) $O + O + M \rightarrow O_2 + M + 117$ kcal/mole

(29) $O + OH \rightarrow O_2 + H + 16.6$ kcal/mole

(30) $O + HO_2 \rightarrow O_2 + OH + 55$ kcal/mole

(31) $O + O_2 + M \rightarrow O_3 + M + 24$ kcal/mole

(32) $O_3 + H \rightarrow OH + O_2 + 77$ kcal/mole

(33) $H + O_2 + M \rightarrow HO_2 + M + 46$ kcal/mole

Not all the released chemical energy will be converted to kinetic energy. An appreciable part will appear as vibrational energy in O_2, N_2 and OH.

Electronically excited species could be formed in reactions (28), (30) and (32), but this appears unlikely from airglow observations.

Vibrationally excited OH* is produced mainly in the highest accessible levels $v = 9$, 8 and 7 (Anlauf et al., 1968; Charters et al., 1970). The following deactivation process was proposed as being of importance by Anlauf et al. (1968):

(34) $OH(v = i) + O_2 \rightarrow OH(v = i - 5) + O_2(^1\Sigma_g^+)$

Other resonant processes which may be significant are:

(35) $OH(v = 9) + N_2 \rightarrow OH(v = 7) + N_2(v = 2) + 18$ cm^{-1}

(36) $\quad \text{OH}(v=7) + \text{N}_2 \rightarrow \text{OH}(v=3) + \text{N}_2(v=5) - 37 \text{ cm}^{-1}$

(37) $\quad \text{OH}(v=7) + \text{O}_2 \rightarrow \text{OH}(v=2) + \text{O}_2(v=10) + 31 \text{ cm}^{-1}$

(38) $\quad \text{OH}(v=4) + \text{O}_2 \rightarrow \text{OH}(v=3) + \text{O}_2(v=2) - 11 \text{ cm}^{-1}$

(39) $\quad \text{OH}(v=2) + \text{N}_2 \rightarrow \text{OH}(v=0) + \text{N}_2(v=3) + 69 \text{ cm}^{-1}$

(40) $\quad \text{OH}(v=8) + \text{N}_2 \rightarrow \text{OH}(v=7) + \text{N}_2(v=1) + 83 \text{ cm}^{-1}$.

Although reactions (35)–(39) are multiple quantum transfer processes and therefore less probable than single quantum processes, the presence of the H atom does enchance relaxation due to the 'hydrogen effect' (Lambert, 1967).

As the radiation lifetime of OH has recently been reported by Potter *et al.* (1970) to be 6×10^{-2} s, deactivation of OH* may be of importance already at the mesopause level (communication with Dr Lambert).

We chose the deactivation rate constant to be $3 \times 10^{-15} \text{ cm}^3 \text{ molecule}^{-1} \text{ s}^{-1}$.

For heat balance considerations it is of some importance to know what products are formed in the deactivation of $\text{O}(^1D)$ by O_2. In reaction (5b) O_2^* denotes vibrationally excited molecular oxygen. The choice of the rate constants k_{5a} and k_{5b} was based upon a comparison with the study by Wallace and Hunten (1968).

Deactivation of $\text{O}(^1D)$ by N_2 may lead to vibrationally excited N_2

(41) $\quad \text{O}(^1D) + \text{N}_2 \rightarrow \text{O} + \text{N}^*(v \leqslant 7)$.

Nothing is known about how much of the energy initially present in $\text{O}(^1D)$, 43 kcal/mole, is converted into kinetic energy. For the calculations we have assumed that no kinetic energy is created, thereby again underestimating the heating rates.

It now remains to find out what happens to the vibrationally excited O_2 and N_2. We make the reasonable assumption that all vibrational quanta originally produced in N_2 degrade to the first vibrational level, since processes of the type

(42) $\quad \text{N}_2(v=n) + \text{N}_2(v=0) \rightarrow \text{N}_2(v=n-i) + \text{N}_2(v=i)$

are relatively fast (Zipf, 1969; Walker, 1968). The degradation of vibrational quanta in O_2 has attracted attention in connection with Krassovsky's proposal (1951) that vibrational excited O_2 might contribute to the OH airglow by the reaction

(43) $\quad \text{O}_2^*(v \leqslant 27) + \text{H} \rightarrow \text{O} + \text{OH}^*$.

Bates and Moiseiwitsch (1956) rejected this hypothesis because atomic oxygen would be very fast in deactivating O_2 either by atom-atom interchange reactions (especially for high vibrational levels)

(44a) $\quad \text{O}_2(v=n) + \text{O} \rightarrow \text{O} + \text{O}_2(v=n-i) + \text{kinetic energy}$

or by chemical interaction

(44b) $\quad \text{O}_2(v=n) + \text{O} \rightarrow \text{O}_2(v=n-i) + \text{O} + \text{kinetic energy}$.

The process corresponding to (42)

(45) $\quad \text{O}_2(v=n) + \text{O}_2 \rightarrow \text{O}_2(v=n-i) + \text{O}_2(i)$

may, however, not be neglected, especially in the lower regions of the thermosphere and in the mesosphere, where the atomic oxygen mixing ratio is lower and production of excited O_2 occurs mainly up to the second level by reaction (29). For these regions let us consider that most of the excited oxygen quanta will degrade to the first excited level.

It is an interesting fact that O_2 and N_2 undergo resonant transfer of vibrational energy with H_2O and CO_2 respectively (Houghton, 1969). The energy present in the O_2 and N_2 vibration therefore can be radiated away.

We consider the energy systems:

$$O_2(v=1) - H_2O(v_2) \quad \text{and} \quad N_2(v=1) - CO_2(v_3).$$

Flux of energy into the system $O_2(v=1) - H_2O(v_2)$ takes place mainly by the reassociation reactions (14), (17) and (18).

The following processes are of importance:

(46) $H_2O(v_2) \xrightarrow{6.3\mu} H_2O + hv$ $\tau_{46} = 52 \times 10^{-3}$ s

(47) $H_2O(v_2) + O_2 \leftrightarrow H_2O + O_2(v=1) + 41 \text{ cm}^{-1}$

                                                                $k_{47} = 4 \times 10^{-12}$

(48) $O_2(v=1) + M \rightarrow O_2 + M$ $k_{48} = 2 \times 10^{-19}(210 \text{ K})$

(48a) $O_2(v=1) + CO_2 \rightarrow CO_2(v_1, 2v_2) + O_2 + 209 \text{ cm}^{-1}$

                                                                $k_{48a} = 3 \times 10^{-15}(210 \text{ K})$

(49) $O_2(v=1) + O \rightarrow O_2(v=0) + O$ $k_{49} = 10^{-12}$?

The rate coefficients k_{48} and k_{48a} have been taken from Houghton (1969) and apply at 210 K. At higher temperatures the values are larger. No measurements of k_{49} have been made in the atmospheric temperature range. At high temperatures (>1500 K) Kiefer and Lutz (1967) found atomic oxygen to be very fast in the relaxation of O_2, obtaining $k_{49} \approx 10^{-12}$. The temperature dependence was found to be small. The fast relaxation was ascribed to a chemical effect, transient formation of O_3 being involved. We have assumed here that k_{49} is large even in the atmospheric temperature range and that the value 10^{-12} may be used although experimental values of k_{49} at lower temperatures are needed. With the given value for k_{49}, processes (48) and (48a) may be neglected above approximately 70 km. Below 70 km all energy going into v_2 vibrations will anyhow be converted to kinetic energy (Houghton, 1969). Denoting the influx of quanta into the system by P_1 and equating P_1 to the flux of quanta out of the system by (49) we obtain

(50) $P_1 = A_{46}(H_2O^*) - k_{49}(O)(O_2^*),$

where $A_{46} = \tau_{46}^{-1}$.

Equilibrium for excited water vapour molecules yields

(51) $(H_2O^*)[A_{46} + k_{47}(O_2)] = k_{-47}(H_2O)(O_2^*).$

The ratio between that part of the system energy going into kinetic energy and that radiated out as 6.3 μ radiation by the water vapour can now easily be written down:

$$(52) \quad \frac{\text{K.E.}}{\text{R.E.}} = \frac{k_{49}(O)(O_2^*)}{A_{46}(H_2O^*)}$$

and therefore with the aid of (51)

$$(53) \quad \frac{\text{K.E.}}{\text{R.E.}} = \frac{k_{49}(O)}{A_{46}} \frac{A_{46} + k_{47}(O_2)}{k_{-47}(H_2O)}.$$

As the atomic oxygen concentration is much larger than the water vapour concentration above about 70 km it follows that almost all energy flowing into the $O_2(v=1)-$ $-H_2O(v_2)$ energy system is converted to kinetic energy. An appreciable proportion of the system energy will, however, be radiated away by either $H_2O(v_2)$ or $CO_2(v_2)$ if atomic oxygen is much slower in the relaxation of O_2 than assumed here (Houghton, 1969). The distribution of water vapour and atomic oxygen may therefore be important factors also for this type of heat budget considerations.

Flux of energy into the system $N_2(v=1)-CO_2(v_3)$ occurs by absorption by CO_2 of 4.3 μ and 2.7 μ solar radiation (Houghton, 1969; Williams, this volume, p. 177). We will deal with the possible flux of energy into the system by reaction (41). The following processes were considered by Houghton (1969):

$$(54) \quad N_2(v=1) + CO_2 \leftrightarrow CO_2(v_3) + N_2 - 18 \text{ cm}^{-1},$$
$$k_{54} = 4 \times 10^{-13}$$

$$(55) \quad CO_2(v_3) \underset{4.3\mu}{\overset{\longleftrightarrow}{\longrightarrow}} CO_2 + h\nu, \qquad \tau_{55} = 2.4 \times 10^{-3} \text{ s}$$

$$(56) \quad N_2(v=1) + O_2 \rightarrow N_2 + O_2(v=1) + 777 \text{ cm}^{-1},$$
$$k_{56} = 2 \times 10^{-18} \,(210 \text{ K})$$

$$(57) \quad CO_2(v_3) + O_2 \rightarrow CO_2(v_2) + O_2(v=1) + 128 \text{ cm}^{-1},$$
$$k_{57} = 4 \times 10^{-15} \,(210 \text{ K}).$$

The mean free path of a photon emitted in the 4.3 μ band, before reabsorption, is only about 200 meters at 80 km. This has led to the suggestion that, provided most of the original $O(^1D)$ energy goes into nitrogen vibrations, measurements of 4.3 μ fluxes at low solar zenith angles may be used for almost in situ determinations of ozone concentrations between 55 and 90 km (Crutzen, 1970).

For the computations it has been assumed that reations (54)–(57) are adequate. Thus the proportion α of the available energy in N_2 going to kinetic energy can be estimated from Houghton (1969). Above the mesopause little heating can take place.

Remembering the probably very strong effect of atomic oxygen on the relaxation of molecular oxygen, it may be of interest to inquire about a similar effect with molecular nitrogen:

$$(58) \quad N_2(v=n) + O \rightarrow N_2(v=n-i) + O + \text{kinetic energy}.$$

Breshears and Birg (1968) measured k_{58} (for $v=1$) around 3500 K and obtained 5×10^{-14} cm^3 mole^{-1} s^{-1}; at lower temperatures k_{58} should be much smaller. The relaxation by atomic oxygen is about ten times faster than by molecular oxygen. The importance of process (59) is very much dependent on the ratio $k_{58}(O)/k_{54}(CO_2)$ which increases strongly with height above the mesopause and may possibly reach a significant value.

The equations describing the mean vertical motion and the chemical state of the upper atmosphere, which are consistent with the mean thermal state, are:

$$(59) \qquad c_p \varrho w \left[\frac{\partial T}{\partial z} + \frac{g}{c_p} \right] = Q_n,$$

$$(60) \qquad w(M) \frac{\partial}{\partial z} \left[\frac{(X)}{(M)} \right] = P(X).$$

c_p = thermal heat capacity per unit mass of air at constant pressure,
T = temperature in K,
g = gravitational acceleration,
ϱ = air density,
z = height,
w = dz/dt, the vertical wind velocity,
(X) = concentration of chemical species X,
$P(X)$ = production rate of X by photochemical processes,
Q_n = net heating rate.

The following terms contribute to Q_n:
(i) Absorption of solar UV radiation by O_2 and O_3, and 4.3 μ radiation by CO_2. Only that part of the absorbed energy which is immediately converted to kinetic energy is considered.
(ii) Conversion of chemical energy to kinetic energy by reactions.
(iii) Thermal cooling by CO_2 at 4.3 and 15 μ and O_3 at 9.6 μ.
The sum of the contributions to the energy budget by thermal cooling processes and by absorption of solar radiation by CO_2 is shown in column 7 of Table I.

The thermal cooling rates adopted in this study are those calculated by Kuhn and London (1969) for high latitude summer conditions. The vibrational relaxation time they used was 2×10^{-5}. This important parameter for the calculation of heating rates above 70 km is still not well known. Furthermore, enhanced vibrational relaxation by atomic oxygen is an interesting possibility (Crutzen, 1970).

The lowest level considered in this study was at 50 km. Photochemical equilibrium for all constituents except H_2 was assumed at the lower boundary.

The most striking feature in Figure 1, which shows the calculated distribution of the main chemical species, is the sudden decline of the water vapour mixing ratio

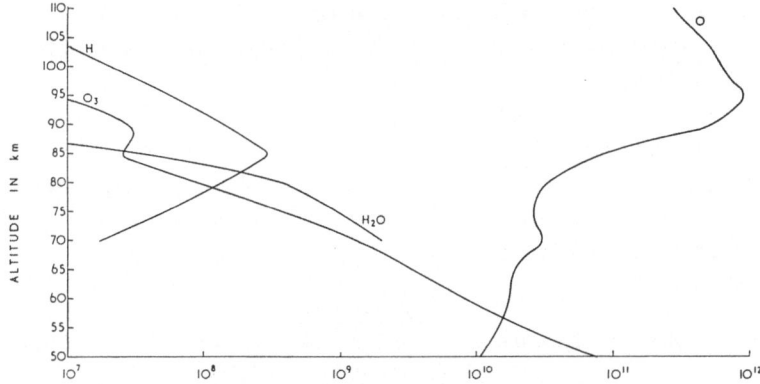

Fig. 1.   Concentrations at noon of some atmospheric gases.

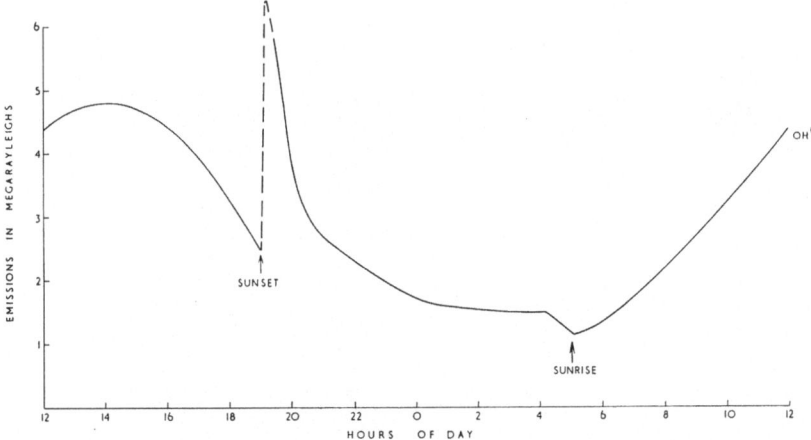

Fig. 2.   Total OH-airglow emission above 50 km as a function of time.

above 80 km. This result is quite different from that obtained by Hesstvedt (1968) in a model in which the effect of eddy diffusion was considered.

It agrees better, however, with earlier results (Hesstvedt, 1965) from a model in which only the effect of mean vertical motions was taken into account. Another finding is that also in this model the ozone concentrations show a maximum between 85 and 90 km, which is very pronounced during the morning and reflects the large diurnal variations obtained for the odd oxygen concentrations between 70 and 85 km (Table III). It is due mainly to the very effective removal of the odd oxygen content during the short period of darkness and its restoration during the morning hours. It is also reflected in the diurnal variation of the OH emission.

The main results of the energy budget calculations are shown in Table I. The most striking features are:

(i) The creation of large amounts of chemical energy (column 4). Conversion of this energy to kinetic energy can lead to substantial heating in atmospheric regions which are far away from the summer pole.

(ii) The possible emission of large amounts of energy by non-thermal processes, mainly by CO_2 and OH (columns 3, 5, 6).

(iii) The relatively small amounts of energy associated with atmospheric UV heating rates as compared with the quantities which are converted to chemical energy or are emitted as airglow above 85 km (column 2).

TABLE I

Heating rates, airglow emissions, theoretical vertical wind velocities and chemical energy creation in high latitude summer regions

Altitude in km	Heating °C/d	Total air glow emission °C/d	Chemical energy creation °C/d	$CO_2(\nu_3)$ emission °C/d	OH emission °C/d	Assumed infrared cooling °C/d	W mm/s
110	40	48	180	45	0	+2	31
105	20	26	110	23	0	+3	15
100	14	8.4	41	7	0.3	+4	9
95	9.3	4.7	13	2.4	1.4	+6	3
90	7.7	4.3	1.7	1	2.7	+6	2
85	3.8	1.6	0.6	0.2	1.3	−0.6	5
80	2.6	0.7	0.2	0.1	0.5	−3.8	12
75	2.6	0.2	–	–	0.16	−4	11
70	5.0	0.2	–	–	0.08	−1	10
65	5.6	0.1	–	–	0.03	+2	7
60	7.6	–	–	–	–	+5	4
55	10	–	–	–	–	+6.5	4
50	13	–	–	–	–	+6	8

TABLE II

Height in km	R	Q_1 (in %)	Q_2 (in %)
110	1.34	0	0
105	0.92	0	0
100	0.67	0	0
95	0.37	0	0
90	0.20	0	0
85	0.10	23	4
80	0.14	80	20
75	0.24	100	50
70	0.33	100	80
65	0.36	100	100
60	0.36	100	100

R = ratio between that part of the absorbed energy going into $O(^1D)$ excitation and that going into kinetic energy, according to the results of this model.

Q_1 = Minimum proportion of N_2^* energy being converted to kinetic energy according to Houghton (1969).

Q_2 as Q_1 but according to Williams (1970).

TABLE III

Calculated concentrations of ozone and atomic oxygen at different times of the day

Height (km)	Just before sunset (O_3)	(O)	Just after sunrise (O_3)	(O)	At noon (O_3)	(O)
95	7.7(6)	8.8(11)	7.6(6)	8.6(11)	7.8(6)	9(11)
90	3(7)	5(11)	2.8(7)	4.6(11)	2.9(7)	4.9(11)
85	9.0(7)	1.1(11)	2.5(7)	6.5(10)	3.3(7)	8.7(10)
80	2.0(8)	6.0(10)	3.9(6)	1.1(9)	1.2(8)	3.5(10)
75	6.5(8)	4.7(10)	3.7(7)	2.3(9)	5.5(8)	3.9(10)
70	1.2(9)	2.3(10)	7.7(8)	8.5(9)	1.8(9)	3.4(10)

The numbers shown in Table I illustrate how important it is to know to what extent the assumptions about the decay of excited particles are valid. This is the main conclusion of this paper.

Acknowledgements

I am indebted to Dr J. T. Houghton, Dr J. D. Lambert, Dr R. P. Wayne and Mr J. Cramp for helpful discussions, and to Dr A. B. Callear for correspondence. The work was carried out while the author was at Oxford University as a post-doctoral ESRO fellow.

References

Anlauf, K. G., MacDonald, R. G., and Polanyi, J. C.: 1968, *Chem. Phys. Letters* **1**, 619–22.
Bates, D. R. and Moiseiwitsch, B. L.: 1956, *J. Atmospheric Terrest. Phys.* XX, 305–8.
Bates, D. R. and Nicolet, M.: 1950, *J. Geophys. Res.* **55**, 301–27.
Breshears, W. D. and Bird, P. F.: 1968, *J. Chem. Phys.* **48**, 4768–473.
Charters, P. E., MacDonald, R. G., and Polanyi, J. C.: 1970, 'Formation of Vibrationally Excited OH by the Reaction $H + O_3$', to be published.
Crutzen, P. J.: 1970, 'Comments on J. T. Houghton's paper (1969)', *Quart. J. Roy. Meteor. Soc.*, to be published.
Dalgarno, A.: 1963, *Planetary Space Sci.* **10**, 19–28.
Dandekar, B. S.: 1969, *Planetary Space Sci.* **17**, 1609–18.
Detwiler, C. R., Garrett, D. L., Purcell, J. D., and Tousey, R.: 1961, *Ann. Geophys.* **17**, 263–72.
Ditchburn, R. W. and Young, P. A.: 1962, *J. Atmospheric Terrest. Phys.* **24**, 127–139.
Filseth, S. V. and Welge, K. H.: 1969, *J. Chem. Phys.* **51**. 839.
Hesstvedt, E. 1965, *Tellus* **17**, 341–9.
Hesstvedt, E.: 1968, *Geofys. Publik.* **27**, 1–35.
Houghton, J. T.: 1969, *Quart. J. Roy. Meteor. Soc.* **95**, 1–20.
Hudson, R. D., Carter, V. L., and Breig, E. L.: 1969, *J. Geophys. Res.* **74**, 4079–86.
Johnson, F. S.: 1954, *J. Meteorol.* **11**, 431–9.
Jones, I. T. N. and Wayne, R. P.: 1970, *Proc. Roy. Soc. London* (to be published).
Kaufman, F.: 1969, *Can. J. Chem.* **47**, 1917–27.
Kiefer, J. H. and Lutz, R. W.: 1967, *XIth Symposium on Combustion*, The Combustion Institute, Pittsburg, Pennsylvania, 67–76.
Krassovsky, V. I. and Lukashenia, V. T.: 1951, *Dokl. Akad. Nauk S.S.S.R.* **80** 735–9.
Kuhn, W. R. and London, J.: 1969, *J. Atmospheric Sci.* **26**, 189–204.
Lambert, J. D.: 1967, *Quart. Rev. Chem. Soc., London* **21**, 67–78.
McNesby, J. R. and Okabe, H.: 1963, in W. A. Noyes, Jr., G. S. Hammond, and J. N. Pitts, Jr. (eds.), *Advances in Photochemistry*, Vol. 3, John Wiley and Sons, Inc., 157–240.

Potter, A. E., Jr., Coltharp, R. N., and Worley, S. D.: to be published.
Schiff, H. I.: 1969, *Can. J. Chem.* **47**, 1903–16.
Schofield, K.: 1967, *Planetary Space Sci.* **15**, 643–70.
Thompson, B. A., Harteck, P. and Reeves, R. R., Jr.: 1963, *J. Geophys. Res.* **68**, 6431–6.
Vigroux, E.: 1953, *Ann. Phys.* **8**, 709–63.
Walker, J. C. G.: 1968, *Planetary Space Sci.* **16**, 321–7.
Wallace, L. and McElroy, M. B.: 1966, *Planetary Space Sci.* **14**, 677–708.
Wallace, L. and Hunten, D. M.: 1968, *J. Geophys. Res.* **73**, 4813–34.
Watanabe, K.: 1958, *Adv. Geophys.* **5**, 153–221.
Wayne, R. P.: 1969, in W. A. Noyes, Jr., G. S. Hammond, and J. N. Pitts, Jr. (eds.), *Advances in Photochemistry*, Vol. 9, John Wiley and Sons, Inc., pp. 311–71.
Williams, A. P.: 1971, this volume, pp. 177–87.
Zipf, E. C.: 1969, *Can. J. Chem.* **47**, 1863–70.

PHOTOCHEMISTRY AND THE ESCAPE EFFICIENCY
OF TERRESTRIAL HYDROGEN

R. T. BRINKMANN*

Molecular Oxygen in the Earth's Atmosphere

Molecular oxygen is important in atmospheric models because (1) it is the only major constituent which participates appreciably in chemical reactions with other species, including hydrogen-containing species, (2) it shields all molecules below a certain level from the solar ultraviolet and thereby determines the dissociation rates of the various species, and (3) it is an important source of O atoms through normal dissociation and predissociation.

Absorption Data. We are primarily concerned with the absorption by O_2 of solar photons in the range 1200–2400 Å. The number of incident photons at shorter wavelengths is small (see Table 1) and the absorbing power of O_2 at longer wavelengths is negligible. O_2 is rather transparent to Lyman-α radiation ($k \approx 0.3$ cm^{-1}; Ogawa, 1970): Since Ly-α is strong (about 3×10^{11} photons cm^{-2} s^{-1} at 1 AU), there is a source of O atoms and dissociating ultraviolet down to about 70 km. While appreciable, Ly-α is not a dominant source. The broad, strong absorption feature from 1300 Å to 1750 Å is the Schumann-Runge continuum, an important source of O atoms above 80 km though not dominant over this entire region: the solar flux (Table I) is several times less in the region of the S-R continuum than in the region of the S-R bands. No important pressure or path length dependences of the effective absorption coefficient have been observed in the S-R continuum.

The Herzberg continuum extends from about 2400 Å to shorter wavelengths. It is a weak absorption system but, for range 2000–2400 Å, is an important source of O atoms low in the atmosphere.

The Schumann-Runge band system (1750–2000 Å) has been frequently studied (see, for example, Blake *et al.*, 1965, 1966), but a gap has existed between those who study its absorption in the laboratory and those who apply the laboratory results to atmospheric calculations. Each of the peaks appearing in the early data is a vibrational band consisting of unresolved rotational lines. In more recent work (Hudson and Carter, 1968), obtained at much higher resolution, the individual rotational lines are resolved, in most cases, down to their baselines. Thus it is possible to measure the true line widths (apart from instrumental broadening); above $v' = 2$ the widths correspond to lifetimes of the order of 10^{-11} s or less. Since optically allowed transitions are not expected to be faster than about 10^{-9} s, this is convincing evidence for strong

* Visiting Scientist, ESRIN; present address Lunar Science Institute, Houston, Texas, U.S.A.

Fiocco (ed.), Mesospheric Models and Related Experiments, 89–102. All Rights Reserved.
Copyright © 1971 by D. Reidel Publishing Company, Dordrecht-Holland.

TABLE I
Incident solar flux at 1 AU

Wavelength range (Å)	Flux (photons cm^{-2} s^{-1})
< 1200	6 $\times 10^{10}$
1200–1230	30 $\times 10^{10}$
1230–1300	1.4 $\times 10^{10}$
1300–1350	1.7 $\times 10^{10}$
1350–1400	2.1 $\times 10^{10}$
1400–1450	3.0 $\times 10^{10}$
1450–1500	5.5 $\times 10^{10}$
1500–1550	8.4 $\times 10^{10}$
1550–1600	12.9 $\times 10^{10}$
1600–1650	19.7 $\times 10^{10}$
1650–1700	30 $\times 10^{10}$
1650–1700	30 $\times 10^{10}$
1700–1750	47 $\times 10^{10}$
1750–1800	72 $\times 10^{10}$
1800–1850	110 $\times 10^{10}$
1850–1900	178 $\times 10^{10}$
1900–1950	303 $\times 10^{10}$
1950–2000	481 $\times 10^{10}$
2000–2050	725 $\times 10^{10}$
2050–2100	1100 $\times 10^{10}$
2100–2200	4800 $\times 10^{10}$
2200–2300	7300 $\times 10^{10}$
2300–2400	7200 $\times 10^{10}$
2400–2500	9000 $\times 10^{10}$
2500–2600	13000 $\times 10^{10}$
2600–2700	30000 $\times 10^{10}$
2700–2800	31000 $\times 10^{10}$
2800–2900	46000 $\times 10^{10}$
2900–3000	88000 $\times 10^{10}$
3000–3500	600000 $\times 10^{10}$
3500–4000	1100000 $\times 10^{10}$

predissociation, which had been inferred earlier. The existence of strong predissociation makes the S-R bands an important source of O atoms and explains the absence of detectable fluorescence in the upper atmosphere.

Since the natural width will be greater than the Doppler width the lines can be regarded as Lorentzian. The effective absorption coefficient for a single Lorentz line, when observed with a resolution much poorer than the half-width of the line and with sufficient path thickness for the line centers to be optically thick, varies as the inverse square root of the path thickness; i.e., the transmission is given by $\exp(-k'x^{1/2})$, where k' is the effective absorption coefficient for unity absorber thickness.

What is the situation when we have a system of lines and absorption is important in the regions where more than one line makes an appreciable contribution to the absorption coefficient? Two classes of band models have been theoretically investigated: the regular (Elsasser) model and the random (Goody) model. While numerous variations of each model exist, the basic features are well preserved in all of them (Goody, 1964). We consider the simplest formulation: uniformly wide, uniformly strong Lorentz lines spaced uniformly (regular model) or randomly but with the same mean line density (random model). Transmissions at any wavelength calculated for a finite random system of lines or at a fixed wavelength for an infinite system of lines are expectation values only.

For the random model, it can be readily shown that the effective absorption coefficient obeys (in the mean) the same law as for the single line; i.e., that

$$T_{ran}(x) = \exp(-k'x^{1/2})$$

if the line centers are strongly absorbed. The regular model is not so amenable to analytic representation and numerical calculations of the transmission (Elsasser function) have been tabulated for different values of the relevant parameters:

$$T_{reg}(x) = \exp(-k_{ef}(x)\,x) = El(x).$$

The difference at large path lengths is that k_{ef} in the regular model tends to be more nearly path length independent. Figure 1 shows the transmission of the two models for similar lines, equal mean line density and a line-width: line-spacing ratio of 1:100. Important differences arise before one gets to large overall absorptions. We might expect that the S-R bands give transmissions intermediate between the regular and random models, since the lines are not evenly spaced but 'spectral gappiness' is clearly limited by quantum mechanical selection rules. Enough information is now available on the S-R bands (see Kockarts, this volume) to permit straightforward calculation of this behavior. However, we rely here on experimental measurements to justify a simplified, approximate approach.

Blake et al. (1965, 1966) have obtained measurements of the effective absorption coefficient in the S-R bands: even for modest optical thicknesses the behavior seems to be intermediate between the random and regular cases. Nevertheless, the departures from random model behavior, up to an optical thickness of about 3, are not major and we approximate the effective absorption coefficient as a function of (O_2) path thickness in the S-R bands as follows:

$$k_{ef}(x) = k_{cont} + (k' - k_{cont})/x^{1/2}.$$

x is the path thickness of O_2, k_{cont} is the Herzberg absorption coefficient and k' is the total absorption coefficient measured at unity path thickness.

Temperature effects are not important at normal atmospheric temperatures.

O *Atom Production*. One of the effects of the predissociation and the path length dependence of the absorption coefficient is a strong source of atomic oxygen around 80 km. Earlier model studies have not included this source, which has important implications around 80 km where large mixing ratio gradients appear in the minor species. Hudson *et al.* (1969) have calculated the atomic oxygen production rate as a function of altitude resulting from photodissociation of O_2 both with and without the predissociation mechanism. They also calculated the O atom production via photodissociation of O_3, using Hunt's (1966) non-equilibrium O_3 profile and an unshielded photodissociation rate of 1.02×10^{-2} s^{-1}. We have recomputed the latter O atom production curve using the O_3 profile of Shimazaki and Laird (1970) (the noon curve

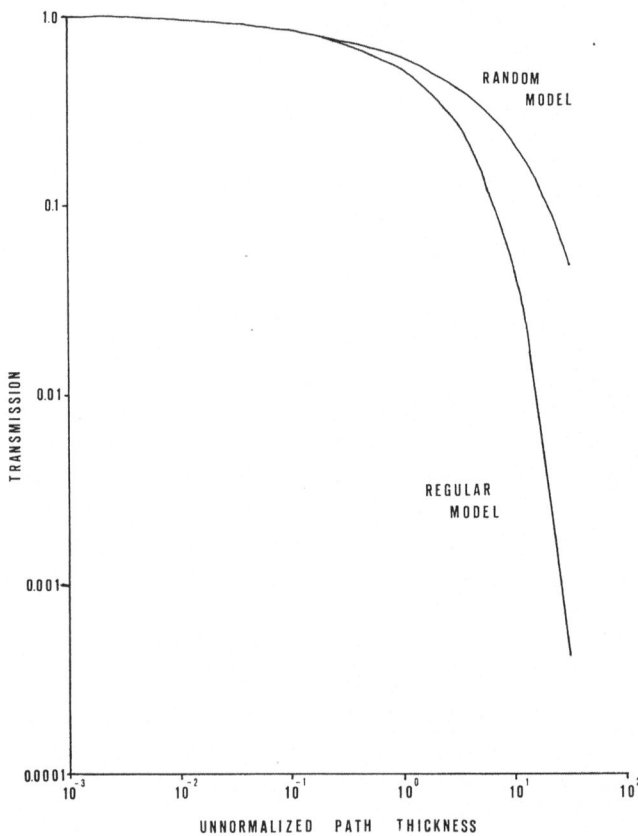

Fig. 1.   Transmission in the two basic band models for uniform Lorentz lines with line-width: line-spacing $= 0.01$. Even at modest absorptions significant differences exist.

from their Figure 6a for three different eddy diffusion coefficient profiles) and an unshielded photodissociation rate of 0.635×10^{-2} s^{-1}. The results presented in Figure 2, with the calculations of Hudson *et al.* (1969), are for an overhead Sun; the ratios should be typical of diurnal averages. Taking predissociation in the S-R bands into account lowers the altitude at which O_2 and O_3 photodissociation are equal sources of O atoms from (for $K_{ed} = 10^6$) about 90 km to 74 km. Between 70 and

80 km the importance of the effect seems to be strongly model dependent but below 70 km O_3 dissociation is the more important source. There may well be strong downward diffusion of O atoms which could push the transition altitude (at which O_3 becomes a more important O atom source than O_2) down by another 5 km or so. The upper altitude limit at which O atom production by predissociation becomes important is about 100 km.

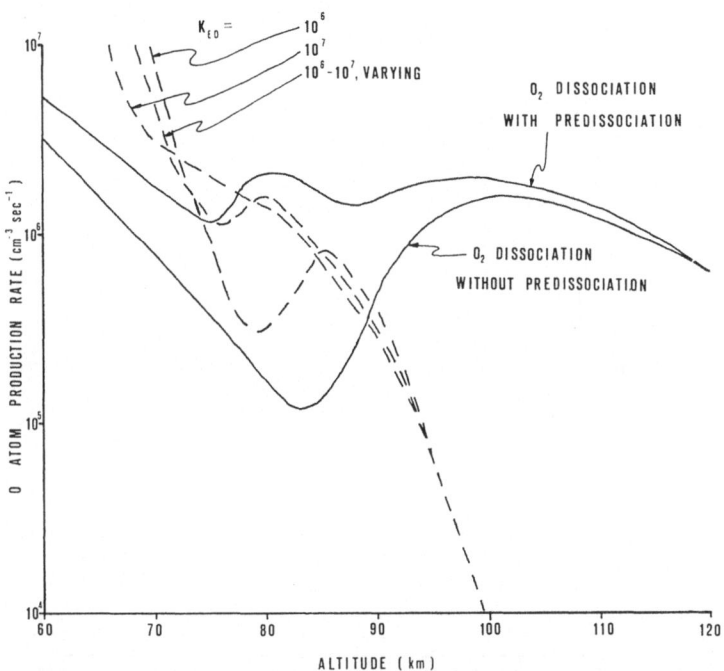

Fig. 2. O atom production rates by O_2 photodissociation with and without predissociation in the S-R bands (after Hudson *et al.*, 1969) and by O_3 photodissociation and upward diffusion (see text for explanation).

H_2O *Photodissociation.* Another effect of taking into account the path length dependence of the absorption coefficient is that water vapor will be dissociated at lower altitudes. If the water vapor mixing ratio changes with altitude, a different total H_2O dissociation, or H atom production, rate will be obtained. If not, the densities of the minor O–H species and the escape efficiency of hydrogen will still be affected. Figure 3 shows the photodissociation rate of H_2O in the Earth's atmosphere under full Sun vs. wavelength, assuming a uniform H_2O mixing ratio of 5 ppm by volume (3 ppm by weight). Not shown is the effect of Ly-α which, if the H_2O mixing ratio can be maintained up to 80 km, can make a significant contribution to the O production rate at these high altitudes. The mixing ratio of water vapor at these altitudes is uncertain. Figure 4 shows the dissociation rate of H_2O for the same conditions as Figure 3 but as a function of altitude, integrated over wavelength. The volume dissociation rate is roughly constant from 40 km to 85 km and drops off fairly rapidly below 40 km. Since the H_2O density in this model increases by a factor of about 200 between 80 km and 40 km the lifetime of H_2O molecules against photodissociation increases rapidly with decreasing altitude in this range.

Fig. 3.   Dissociation rate of H_2O under full Sun versus wavelength for a particular model atmosphere. About half the dissociations occur in the region 1800 ± 40 Å.

Hydrogen Atoms in the Earth's Atmosphere

Hydrogen atoms are produced primarily by the photodissociation of water vapor. Chemical reactions contribute in a limited way. Water vapor can be reformed chemically, in which case the photon's energy has been wasted as regards altering the chemical composition of the atmosphere directly. Some liberated hydrogen atoms will escape out the top of the atmosphere, thus altering the chemical composition of the atmosphere. In the first step of the escape process, a hydrogen atom liberated either directly in a photodissociation, or as a result of a chemical reaction, must survive chemical entrapment back to H_2O long enough for diffusion to raise it to the homopause at 100–120 km. It is widely believed that only a small fraction of the chemical reactions into which an H atom might stray will trap it in a water molecule. These reaction cycles make it difficult to calculate the escape efficiency of hydrogen atoms in a straightforward manner. In the second step, hydrogen atoms which have risen to the homopause are swept through the thermosphere to the exobase, where they rattle

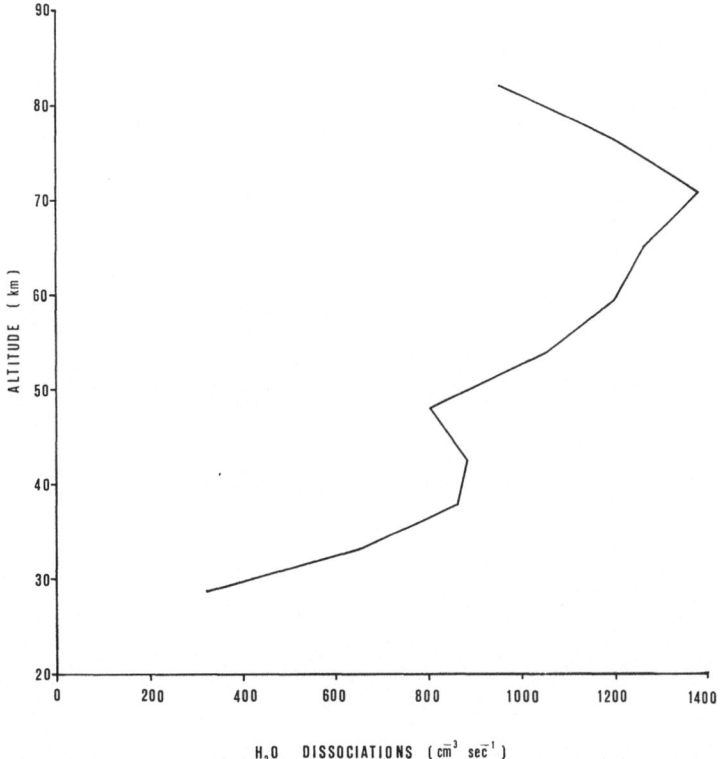

Fig. 4. Dissociation rate of H₂O under full Sun versus altitude for a particular model atmosphere. The volume dissociation rate is roughly constant from 40 to 85 km.

around until they have sufficient energy, the proper direction and a suitable altitude for escape. The journey from the homopause is uneventful because number densities are so low that chemical reactions are ineffective and because of the rather high thermal gradient, leading to a strong thermal term in the molecular diffusion equation.

We now define the *escape efficiency*, ε, of hydrogen atoms to be the ratio of hydrogen atoms which *can* escape (i.e., twice the photodissociation rate of H_2O, in the context of our model) to those which *do* escape. The escape efficiency has been only infrequently studied quantitatively (Donahue, 1969).

A model must include mixing if its ε is not to be zero. Below the homopause eddy diffusion and mean motion are the dominant mixing mechanisms. Mean motion is difficult to treat, but perhaps eddy diffusion alone gives a reasonable picture. Horizontal eddy diffusion can be neglected. Hesstvedt (1968) first dealt with diffusion and photochemistry in an oxygen-hydrogen atmosphere. Shimazaki and Laird (1970) have published results of similar calculations including nitrogen and molecular diffusion. Neither study includes the predissociation or path length dependence of the effective

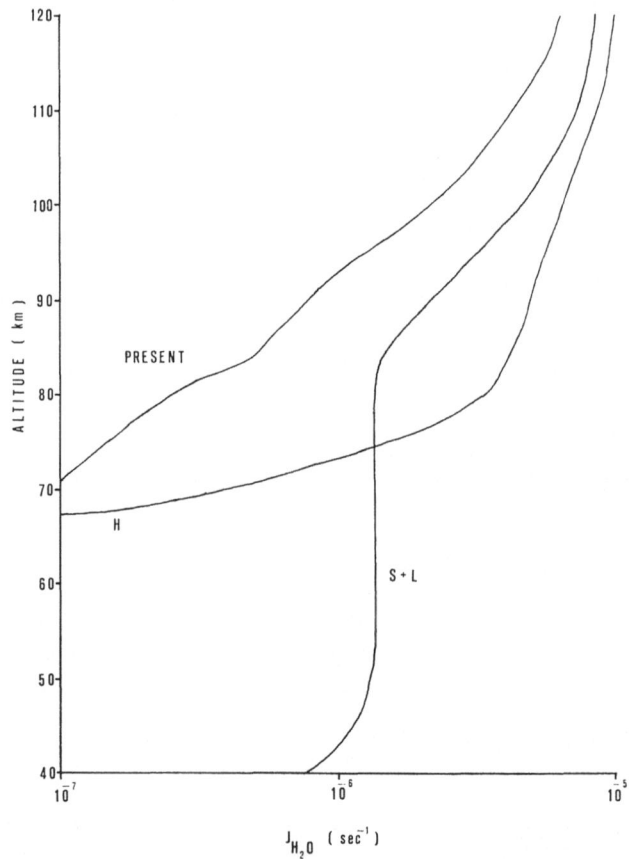

Fig. 5.   Values of the photodissociation rate constant, J_{H_2O}, versus altitude for the three sets of calculations considered: H = Hesstvedt, S + L = Shimazaki and Laird.

absorption coefficient in the S-R bands. Hesstvedt's use of photochemical values as boundary conditions at 100 km seems to invalidate his H_2O density above 80 km or so. Shimazaki and Laird (1970) use an unrealistic value for J_{H_2O} (see Figure 5). Although this quantity must decrease several orders of magnitude between about 85 and 45 km, Shimazaki and Laird's J_{H_2O} values are essentially constant over this height range. The basic solar fluxes, absorption data, rate constants, etc., used in the two calculations are similar, and it is of interest to compare the results. At 70, 80, 90 and 100 km the densities for the species $O(^3P)$, H_2O, OH, H_2O_2, HO_2, H, H_2 and O_3 vary by factors as high as 10^5, with a mean disagreement of a factor of ~ 20, Hesstvedt usually being higher on the minor species. On the other hand, the overall, qualitative behavior of the two sets of results is similar. If we use these calculations to estimate the hydrogen escape efficiency we must hope that ε will not be strongly model dependent. If a correction is made for the J_{H_2O} values of Shimazaki and Laird between 45 and 85 km this may indeed be the case.

Calculations were carried out by the author to study the effect of the S-R band O atom source and the path length dependence of the effective absorption coefficient. The model was similar to Hesstvedt's (1968) with regard to reaction scheme and rates. Mean atmospheric number density and temperature profiles for the Earth were obtained from standard atmospheres. Solar fluxes used were those of Wilson *et al.*, 1954; Malitson *et al.*, 1960; Purcell *et al.*, 1960; and Detwiler *et al.*, 1961, digitized by Brinkmann *et al.* (1966), except for the Lyman-α line (apparently the non-linear response of the film was a problem) for which a flux of 5 erg cm^{-2} s^{-1} was adopted. Other sources were used for the solar flux to longer and shorter wavelengths and for the absorption cross sections. The calculations are intended to be illustrative only and not definitive; the approach was similar to that in prior work. The explicit method of solution was used, which is less stable than the implicit method of Shimazaki and Laird, but perhaps simpler to program. Quasi-stability was enforced, where necessary, by assuming approximately steady-state values when the characteristic time for a species was less than the time step and exponential change when this was not the case, and by arbitrarily restricting the fractional change in any species at any altitude during a time step. In some altitude ranges the method resulted in a superposition of 'noise' on the curve. The noise has been averaged out. This procedure is justified since (1) noisy segments between smooth segments (where, presumably, stability has not had to be enforced) do not distort the shape of the curve, and (2) reasonable results are obtained. Calculations are presented here for steady-state, overhead Sun conditions only. Time steps were very small initially ($\sim 10^{-7}$ s) and were increased by a fixed percentage each step. To guarantee that near-equilibrium was reached calculations were extended to more than 10^6 s for all species. For H_2 times of 10^8 to 10^{10} s were used since its characteristic time is so long. Since our method does not strictly conserve mass these longer times were used only to obtain the H_2 distribution; after 10^6 s conservation of mass is still an excellent approximation. In spite of these precautions none of the treatments available thus far generates much confidence in its numerical accuracy and reliability. Nevertheless, for illustrative purposes what we have here should be useful. The upper boundary condition in this calculation is as follows: all hydrogen diffusing through the top, regardless of form, is presumed to escape in the strong thermal gradient above 125 km; all oxygen diffusing through the top, regardless of form, is presumed to be returned as oxygen atoms.

Figure 6 shows results from the three sets of calculations for atomic oxygen, $O(^3P)$. The effect of the O atom production by the S-R bands is obvious. The irregularity at 80 km in both the calculations of Hesstvedt and of Shimazaki and Laird is removed. Outside the region 65–90 km the agreement is quite good. As expected, the O and O_2

densities are equal at about 120 km. Figure 7 shows corresponding results for H_2O. The agreement with the calculations of Shimazaki and Laird is quite good. That Hesstvedt's H_2O distribution lies considerably higher than the others below 90 km is not as important as the change of slope above 80 km, since it could simply represent a higher initial mixing ratio. This dropoff is apparently due to the choice of boundary conditions at 100 km, and the H_2O distribution above about 80 km should not be taken seriously. We conclude that diffusion is extremely effective in maintaining the H_2O mixing ratio, even somewhat above 100 km. Shimazaki and Laird, who included molecular diffusion in their calculations, found that the H_2O mixing ratio increases at higher altitudes.

We now consider the problem of deducing the hydrogen escape efficiencies. The escape efficiency will be the atomic hydrogen flow rate (atoms cm^{-2} s^{-1}, taken at an altitude above the photochemistry; we use 100 km since that is as high as Hesstvedt's calculations extend) divided by twice the mean H_2O dissociation rate. The flow rate is given by

$$\text{flux} = - K_{ed} \left\{ \frac{d\varrho_H}{dz} + \frac{\varrho_H}{H_{mix}} + \frac{\varrho_H}{T} \frac{dT}{dz} \right\}.$$

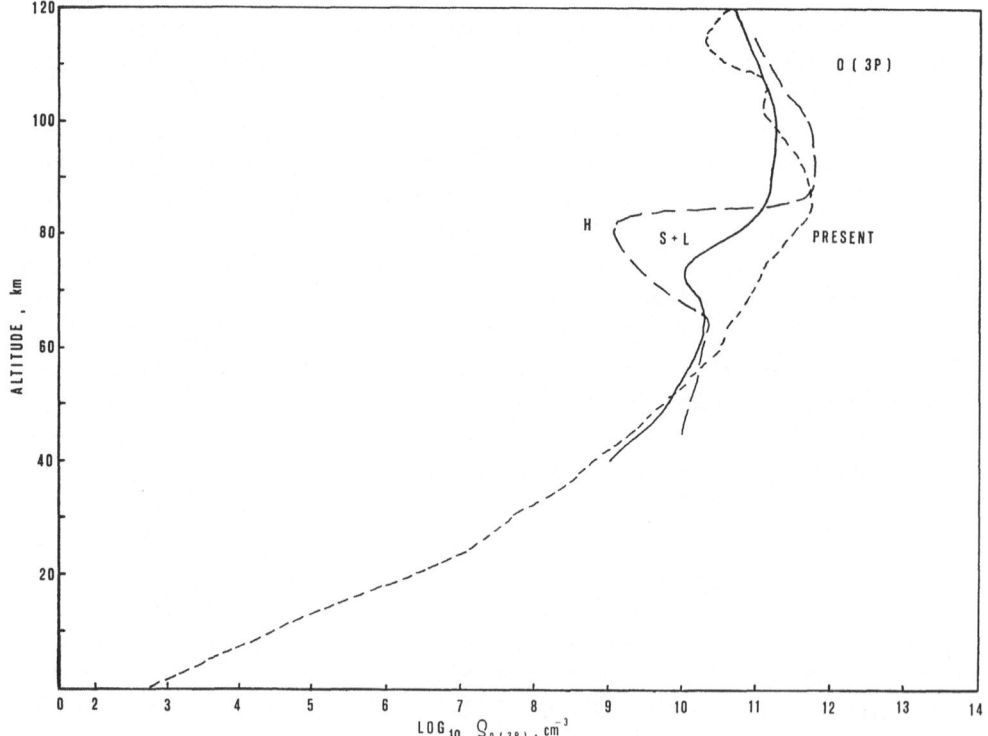

Fig. 6.   Number densities of $O(^3P)$ calculated for full Sun as a function of altitude.

K_{ed} is the eddy diffusion coefficient, ϱ_H the density of hydrogen and H_{mix} the scale height of the mixed atmosphere. The thermal term was not included in the present calculations except for the results of Shimazaki and Laird. The column H_2O dissociation rate can be obtained by integrating over altitude the product of H_2O density, $\varrho_{H_2O}(z)$, and photodissociation rate, $J_{H_2O}(z)$. Both densities and J-values are available for all three sets of calculations. Table II illustrates the rough calculations. The model of Shimazaki and Laird, as mentioned earlier, produces an unreasonable amount of dissociation of H_2O at lower altitudes because of the use of a bad J_{H_2O} profile. The probability of escape from these low levels may be quite small. Using Hesstvedt's model as a guide (the two should be similar for this purpose), we thus disregard all photodissociation of H_2O occurring below 65 km in Shimazaki and Laird's model. This gives a much smaller but more realistic dissociation rate. In the present calculations, in which shielding by O_2 is not as effective because the path length dependence of the absorption coefficient was considered, H_2O dissociation is important at lower altitudes. Even in this case, however, the maximum photodissociation rate occurs at about 70 km, in contrast to the calculations of Shimazaki and Laird, in which the maximum occurs at or below 40 km (see Table II). These rates refer to full Sun: for a diurnal average we use the approximation of dividing by four.

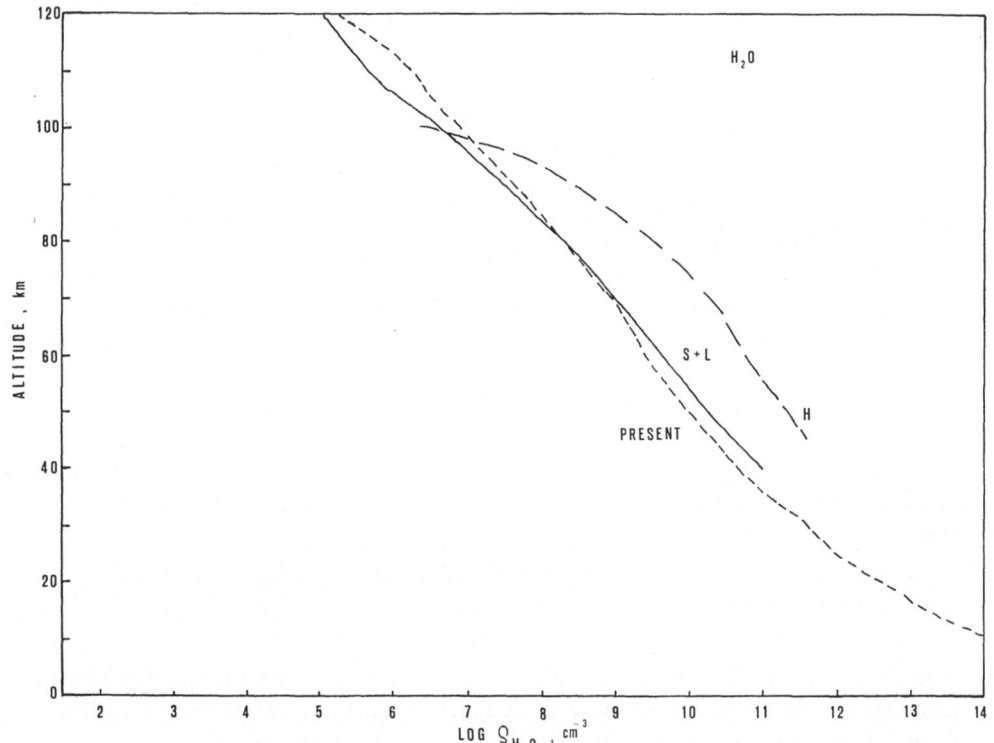

Fig. 7.   Number densities of H_2O calculated for full Sun as a function of altitude.

R. T. BRINKMANN

TABLE II
H$_2$O photodissociation rates under full Sun

	ϱ_{H_2O}	×	J_{H_2O}	×	height range	=	dissn. rate
Hesstvedt							
97.5–100 km	2.5×10^6 cm^{-3}		6.3×10^{-6} s^{-1}		2.5×10^5 cm		0.4×10^7 cm^{-2} s^{-1}
92.5–97.5 km	3.0×10^7 cm^{-3}		5.4×10^{-6} s^{-1}		5.0×10^5 cm		8.1×10^7 cm^{-2} s^{-1}
87.5–92.5 km	2.0×10^8 cm^{-3}		4.8×10^{-6} s^{-1}		5.0×10^5 cm		48.0×10^7 cm^{-2} s^{-1}
82.5–87.5 km	1.0×10^9 cm^{-3}		4.1×10^{-6} s^{-1}		5.0×10^5 cm		205.0×10^7 cm^{-2} s^{-1}
77.5–82.5 km	4.0×10^9 cm^{-3}		3.1×10^{-6} s^{-1}		5.0×10^5 cm		620.0×10^7 cm^{-2} s^{-1}
72.5–77.5 km	1.0×10^{10} cm^{-3}		1.6×10^{-6} s^{-1}		5.0×10^5 cm		800.0×10^7 cm^{-2} s^{-1}
67.5–72.5 km	2.0×10^{10} cm^{-3}		4.0×10^{-7} s^{-1}		5.0×10^5 cm		400.0×10^7 cm^{-2} s^{-1}
62.5–67.5 km	3.0×10^{10} cm^{-3}		5.0×10^{-8} s^{-1}		5.0×10^5 cm		75.0×10^7 cm^{-2} s^{-1}
							2157×10^7 cm^{-2} s^{-1}
Shimazaki and Laird							
95–100 km	7.0×10^6 cm^{-3}		6.0×10^{-6} s^{-1}		5×10^5 cm		2.1×10^7 cm^{-2} s^{-1}
85–95 km	3.0×10^7 cm^{-3}		2.0×10^{-6} s^{-1}		1×10^6 cm		6.0×10^7 cm^{-2} s^{-1}
75–85 km	2.0×10^8 cm^{-3}		1.4×10^{-6} s^{-1}		1×10^6 cm		28.0×10^7 cm^{-2} s^{-1}
65–75 km	1.0×10^9 cm^{-3}		1.0×10^{-6} s^{-1}		1×10^6 cm		100.0×10^7 cm^{-2} s^{-1}
55–65 km	5.0×10^9 cm^{-3}		1.0×10^{-6} s^{-1}		1×10^6 cm		500.0×10^7 cm^{-2} s^{-1}
45–55 km	1.5×10^{10} cm^{-3}		1.0×10^{-6} s^{-1}		1×10^6 cm		1500.0×10^7 cm^{-2} s^{-1}
35–45 km	1.0×10^{11} cm^{-3}		6.0×10^{-7} s^{-1}		1×10^6 cm		6000.0×10^7 cm^{-2} s^{-1}
							8136×10^7 cm^{-2} s^{-1}
						136×10^7	
Present calculation							
95–100 km	1.0×10^7 cm^{-3}		1.7×10^{-6} s^{-1}		5×10^5 cm		0.9×10^7 cm^{-2} s^{-1}
85–95 km	4.0×10^7 cm^{-3}		8.3×10^{-7} s^{-1}		1×10^6 cm		3.3×10^7 cm^{-2} s^{-1}
75–85 km	2.0×10^8 cm^{-3}		2.5×10^{-7} s^{-1}		1×10^6 cm		5.0×10^7 cm^{-2} s^{-1}
65–75 km	9.0×10^8 cm^{-3}		8.9×10^{-8} s^{-1}		1×10^6 cm		8.0×10^7 cm^{-2} s^{-1}
55–65 km	2.4×10^9 cm^{-3}		2.2×10^{-8} s^{-1}		1×10^6 cm		5.3×10^7 cm^{-2} s^{-1}
45–55 km	1.0×10^{10} cm^{-3}		3.3×10^{-9} s^{-1}		1×10^6 cm		3.3×10^7 cm^{-2} s^{-1}
35–45 km	4.0×10^{10} cm^{-3}		6.3×10^{-10} s^{-1}		1×10^6 cm		2.5×10^7 cm^{-2} s^{-1}
25–35 km	4.0×10^{11} cm^{-3}		3.4×10^{-11} s^{-1}		1×10^6 cm		1.4×10^7 cm^{-2} s^{-1}
15–25 km	3.5×10^{12} cm^{-3}		2.1×10^{-13} s^{-1}		1×10^6 cm		0.1×10^7 cm^{-2} s^{-1}
							30×10^7 cm^{-2} s^{-1}

Next we determine the fluxes of H atoms at 100 km. The scale height of the mixed atmosphere is taken to be 6 km. The flow rates were as follows: 7.3×10^8 cm^{-2} s^{-1}, Hesstvedt; 4.4×10^7 cm^{-2} s^{-1}, Shimazaki and Laird; 1.6×10^7 cm^{-2} s^{-1}, this author. Little significance should be attached to these numbers as they are strongly model dependent. The escape efficiencies for the three models are as follows: Hesstvedt, $\varepsilon = 0.066$; Shimazaki and Laird, $\varepsilon = 0.065$; present author, $\varepsilon = 0.107$. While the closeness of the agreement between the results of Hesstvedt and of Shimazaki and Laird is coincidental, the results of all three approaches are of the same order of magnitude, lending substance to our hope that ε might be model insensitive.

Additional preliminary calculations have been made by the author to ascertain the effect on ε of modifying the model. For instance, when values of the eddy diffusion coefficient were decreased by a factor of 10, H densities increased and ε was altered by less than a factor of 2. Large decreases in the O_2 mixing ratio (of possible importance in the study of the evolution of the Earth's atmosphere) had little effect on the hydrogen escape efficiency. Thus the preliminary indications are that ε is fairly independent of the particular model chosen.

Perhaps the most important implication concerns the effort that has gone into obtaining measurements of Ly-α, Ly-β and H-α emissions from hydrogen in the Earth's outer atmosphere and inverting the equation of radiative transfer to obtain hydrogen densities and escape rates. The consensus was that the mean escape rate cannot be much more than about 2×10^8 cm^{-2} s^{-1}. However, the mean H production rate (twice the H_2O photodissociation rate) is considerably greater, 3.2×10^9 cm^{-2} s^{-1} (Brinkmann, 1969). When the escape efficiency was believed to be large (0.5) a discrepancy of nearly an order of magnitude existed, borne out by *in situ* measurements from satellite and rocket mass spectrometers and other instrumentation; these gave higher values than the Ly-α people were willing to accept (e.g., Brace *et al.*, 1967; Hoffman, 1967; Reber *et al.*, 1967). With a lowered value of ε (0.1 to 0.07), however, the disagreement all but disappears. We must await future *in situ* measurements to see whether there is some difficulty with the recent experiments.

Acknowledgments

The calculations reported in this paper were performed primarily at the University of Southern California Computing Center in Los Angeles, and partly at the Aerospace Corporation, El Segundo, California. Thanks are expressed to Robert Carlson at U.S.C. and David Cartwright and David Elliot at Aerospace. The work was not supported in any other way. Conversations with William De More at the Jet Propulsion Laboratory were helpful in bringing the reaction scheme up to date.

References

Blake, A. J., Carver, J. H., and Haddad, G. N.: 1965, 'Molecular Oxygen Photo-Absorption Cross Sections Between 1050 Å and 2350 Å', Internal Report, Dept. of Physics, University of Adelaide.
Blake, A. J., Carver, J. H., and Haddad, G. N.: 1966, *J. Quant. Spectr. Radiative Transfer*, **6**, 451–459.
Brace, L. H., Reddy, B. M., and Mayr, H. G.: 1967, *J. Geophys. Res.* **72**, 265.
Brinkmann, R. T., Green, A. E. S., and Barth, C. A.: 1966, 'A Digitalized Solar Ultraviolet Spectrum', JPL Tech. Rep. No. 32-951, Jet Propulsion Lab., Pasadena, California.
Detwiler, C. R., Garrett, D. L., Purcell, J. D., and Tousey, R.: 1961, *Ann. Geophys.* **17**, 263.
Donahue, T. M.: 1969, *J. Geophys. Res.* **74**, 1128.
Goody, R. M.: 1964, *Atmospheric Radiation, I. Theoretical Basis*, Clarendon Press, Oxford, pp. 97–171.
Hesstvedt, E.: 1968, *Geofys. Publik.* **27**, 1.
Hoffman, J. H.: 1967, *J. Geophys. Res.* **72**, 1883.
Hudson, R. D. and Carter, V. L.: 1968, *J. Opt. Soc. Am.* **58**, 1621.
Hudson, R. D., Carter, V. L., and Breig, E. L.: 1969, *J. Geophys. Res.* **74**, 4079.
Hunt, B. G.: 1966, *J. Geophys. Res.* **71**, 1385.
Malitson, H. H., Purcell, J. D., Tousey, R., and Moore, C. E.: 1960, *Astrophys. J.* **132**, 746.

Ogawa, M. and Yamawaki, K. R.: 1970, *Appl. Opt.* **9**, 1709.

Purcell, J. D., Packer, D. M., and Tousey, R.: 1960, 'The Ultraviolet Spectrum of the Sun', in *Space Res.* **1**, North-Holland Publ. Co., Amsterdam, pp. 581–589.

Reber, C. A., Cooley, J. E., and Harpold, D. N.: 1967, *Trans. Am. Geophys. Union* **48**, 75.

Shimazaki, T. and Laird, A. R.: 1970, *J. Geophys. Res.* **75**, 3221.

Wilson, N., Tousey, R., Purcell, J. D., Johnson, F. S., and Moore, C. E.: 1954, *Astrophys. J.* **119**, 590.

ON THE INTRODUCTION OF TRANSPORT PROCESSES
INTO MESOSPHERIC MODELS

A. D. CHRISTIE

Meteorological Service of Canada

Abstract. There is now substantial evidence that waves of planetary and smaller scale are present and contribute, by the process of slantwise convection, to transport of atmospheric constituents of aeronomic interest, and to their distributions in mesosphere and lower thermosphere.

A critical evaluation of current simple one and two dimensional models has shown that they may only be considered as providing extremely crude simulation of the process involved.

A complete and self-consistent set of equations exists specifying the interaction of the dynamical and photochemical processes. The major barrier preventing integration of the complete set of equations is related to current computational limitations.

A discussion of current knowledge of the meteorology of the region bounding the mesopause is presented and, consistent with this insight given into the synoptic patterns and the energetics, a proposed set of limited models is outlined.

Fiocco (ed.), Mesospheric Models and Related Experiments, 103. All Rights Reserved.
Copyright © 1971 by D. Reidel Publishing Company, Dordrecht-Holland.

DYNAMICAL MODELLING OF THE STRATOSPHERE
AND MESOSPHERE

R. J. MURGATROYD

Meteorological Office, Bracknell, Berks., England

1. The Thermal Structure and General Circulation of the Upper Atmosphere

The upper atmosphere is assumed in this paper to start at the tropopause and extend up to the lower thermosphere. Higher levels are of course of considerable interest but our knowledge of their composition, heating fields and dynamics is rather sparse. The absorption of the solar beam between about 1300 and 2000 Å by molecular oxygen leads to considerable heating in the lower thermosphere around 100 km. By about 80 km these wavelengths are heavily depleted and the heating rate becomes comparatively small. Absorption in the near infra-red by carbon dioxide is also of importance at this altitude. Other wavelengths that penetrate to this level (e.g. in the X-ray, Lyman-α, and near ultra-violet and visible regions) are not important in producing heating here. Major heating occurs between about 75 and 35 km where the absorption by ozone in the 2000–3000 Å wavelength range becomes the principal contributor. From 35 to 20 km absorption by ozone in the near ultra-violet and visible regions produces the main heating and in the lower stratosphere small heating rates result from absorption by carbon dioxide, water vapour, and other minor constituents in the near infra-red.

The vertical temperature profile and also its horizontal distribution broadly follow the distribution of the solar heating, i.e. the mesopause and lower stratosphere regions are comparatively cold while the stratopause and lower thermosphere regions have high temperatures. Average cross-sections of temperature are given for different seasons in Figure 1. These may be compared with similar solar heating cross-sections (shown e.g. by Murgatroyd and Goody, 1958).

The effect of the solar heating is to some extent balanced by the infra-red emission by carbon dioxide (15 μ band) and ozone (9.6 μ band) which tends to produce maximum cooling in the high temperature regions. On the other hand in low temperature (summer mesopause and equatorial tropopause) regions, where the emission is comparatively small, infra-red convergence may take place and heating will occur as the atmospheric radiation tends to smooth out the vertical temperature profile. A cross-section of the pattern of the net radiation field (solar heating minus atmospheric cooling) in the stratosphere, mesosphere and lower thermosphere and of surface transfer and latent heat sources in the troposphere is given in Figure 2. The main importance of this diagram is that it delineates the sources and sinks of heating which provide the drive for the general circulation. Energetically the circulation may be regarded as a means of heat transport from sources to sinks, which indicates that in the lower stratosphere (and the troposphere) its function is to transfer heat from equa-

Fiocco (ed.), Mesospheric Models and Related Experiments, 104–121. All Rights Reserved.
Copyright © 1971 by D. Reidel Publishing Company, Dordrecht-Holland.

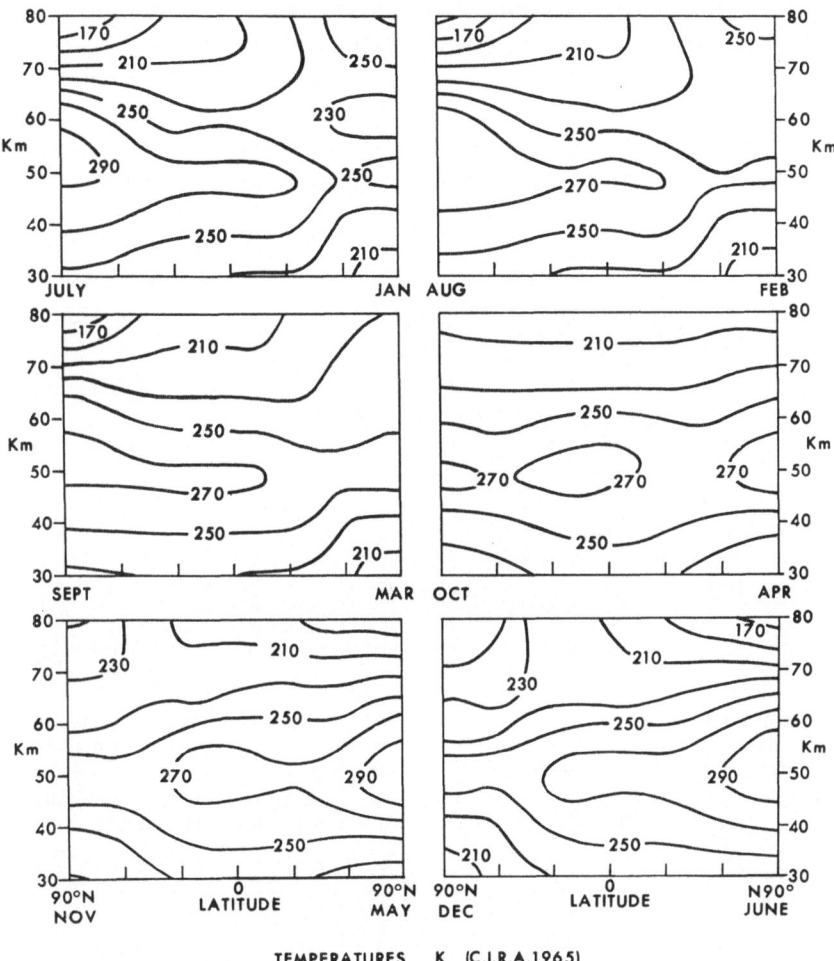

TEMPERATURES   K   (C.I.R.A.1965)

Fig. 1.   Mean cross-section of temperature K in different months (after CIRA, 1965).

tor to pole while in the upper stratosphere, mesosphere and probably also lower thermosphere the transport must effectively be from summer high latitudes to winter high latitudes at the solstices (and equator to pole at the equinoxes). It should also be noted that whereas the net heating is positive in the highest temperature regions of the upper stratosphere (and the troposphere) it is negative in the highest temperature regions of the lower stratosphere and mesopause regions. In the former case it is trying to produce further unbalance between latitudes (i.e. create available potential energy) to drive the circulation, whereas in the latter it is tending to smooth out the latitudinal temperature differences, i.e. damp down the circulation. The circulation in the troposphere, upper stratosphere and lower thermosphere hence acts in the manner of a heat engine whereas in the lower stratosphere and mesopause regions its action is

analogous to that of a refrigerator, i.e. it is driven or forced almost certainly by the
circulations in the layers below (see Murgatroyd, 1966, 1970).

The mean winds may be related to the temperatures at extra-tropical latitudes
through the geostrophic equation and this been useful in preparing more detailed
cross-sections as both the wind and the temperature data are limited in upper parts of
this region. Figure 3 gives a mean cross-section of zonal (west-east) wind components

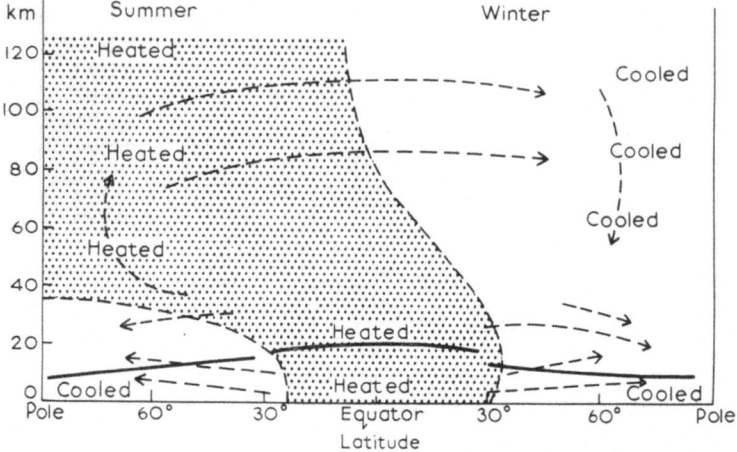

Fig. 2.   Heat transfers (dashed lines) due to the net radiation field (solar heating minus atmospheric
cooling) and other heat sources at the solstices. Regions of heating are shown stippled. (After Murgatroyd, ESRO SP-30, Fig. 15b, September 1968.) Latent heat and surface transfer are important in
the troposphere. Conduction and absorption of upwards propagating wave energy may well invalidate
the distribution shown in the thermosphere, where it is very tentative.

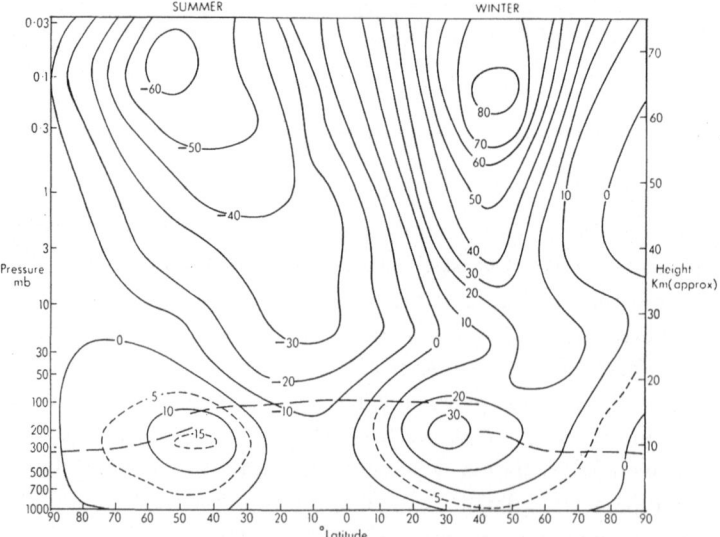

Fig. 3.   Mean pole-to-pole cross-section of zonal wind m s^{-1} at the solstices. (After Murgatroyd, 1970.)

for the solstices. It will be noted that the wind throughout most of the stratosphere and mesosphere is in the mean westerly in winter and easterly in summer with maxima around 60–70 km, and changeovers take place around the equinoxes. The questions of the magnitudes and patterns of the mean meridional (south-north) and vertical wind components which are ageostrophic (and very small and difficult to measure) will be discussed below.

It is most important, however, to realise that the wind and temperature fields are very variable in space and time, as illustrated in Figure 4. The nature of the general circulation is determined by the requirements of its heat and momentum balance and their most efficient transfer between latitudes by the mean circulations and the eddies. It is well known that synoptic systems (depressions, anticyclones, etc.) exist in the troposphere and their quasi-horizontal motions and asymmetric structure provide the principal means of transport rather than the mean meridional (axially symmetric cells), at least in middle and high latitudes. In the winter stratosphere and meso-sphere very large scale (wave numbers 1 to 4 mainly) systems exist and provide the

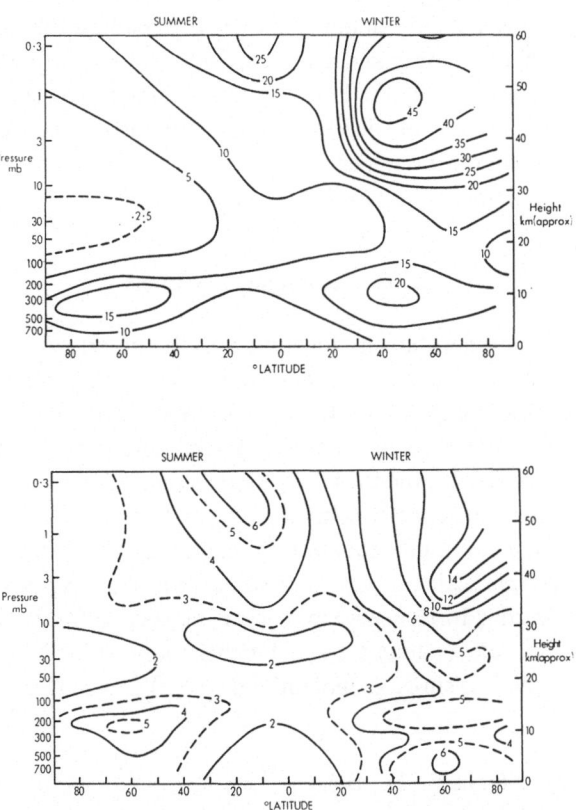

Fig. 4.   Mean monthly standard deviations of temperature, °C (lower diagram), and vector mean wind, m s⁻¹, at the solstices. (After Murgatroyd, 1970.)

main transport mechanisms, except probably at low latitudes where mean circulation cells are again important. At all latitudes, however, both mean and eddy motions of different scales are necessarily involved in the transport processes although it is often difficult to diagnose in detail their respective roles. Sequences of synoptic charts in the stratosphere show that whereas in summer conditions are rather steady, these large scale systems are dominant in winter. At this time of the year they move and develop leading to phenomena such as 'sudden warmings' when rapid and large changes of temperature and ozone amounts are observed at different locations. These systems extend throughout the stratosphere and mesosphere and appear to be linked with correlations between 'D' layer anomalies and stratospheric temperatures. At higher levels, i.e. near the mesopause, they probably decrease in intensity as the horizontal temperature gradient reverses, in a manner similar to the decrease of intensity of tropospheric systems in the lower stratosphere. In the lower thermosphere the motions are increasingly dominated by gravity waves and tides. As Dr Theon is discussing the large scale synoptic systems of the mesosphere and Prof. Lindzen gravity and tidal wave phenomena in other papers, they will not be described further here. There are in addition other phenomena contributing to the variability, e.g. semi-annual and quasi-biennial oscillations particularly in low latitudes. If possible all these wave-motions as well as the major synoptic or planetary waves should be incorporated in a realistic dynamical model. Energetically the latter should include the thermal drive due to the net radiation and the other heat sources as well as forcing between layers, particularly that of the troposphere on the upper atmosphere. These problems and also the possibilities of realistic simplifications will now be considered.

2. The Governing Equations

The state of the atmosphere at any location may be described by the six variables pressure p, density ϱ, temperature T, zonal wind component u, meridional wind component v, vertical wind component w (or in pressure co-ordinates $\omega = \mathrm{d}p/\mathrm{d}t$ where t is time). There are correspondingly available three equations of motion (Newton's 2nd law on a rotating Earth), the thermodynamic equation, the equation of state, and the continuity equation. If separate constituents, e.g. tracers, ionised constituents, etc., are to be considered a continuity equation for each may be included. It is permissible when dealing with large scale motions to assume hydrostatic equilibrium, i.e. neglect vertical accelerations $\mathrm{d}w/\mathrm{d}t$. Hence we may write, neglecting friction and using spherical co-ordinates, latitude ϕ, longitude λ, Earth's radius E and angular velocity Ω, Coriolis parameter $f = 2\,\Omega \sin \phi$, geopotential $\Phi\,(=gz)$ and putting

$$\frac{\mathrm{d}}{\mathrm{d}t} = \frac{\partial}{\partial t} + \frac{u}{E \cos \phi} \frac{\partial}{\partial \lambda} + \frac{v}{E} \frac{\partial}{\partial \phi} + \omega \frac{\partial}{\partial p} \tag{1}$$

$$\frac{\mathrm{d}u}{\mathrm{d}t} - \left(f + \frac{u \tan \phi}{E}\right) v + \frac{1}{E \cos \phi} \frac{\partial \Phi}{\partial \lambda} = 0 \quad \text{(zonal)} \tag{2}$$

$$\frac{dv}{dt} + \left(f + \frac{u \tan \phi}{E} \right) u + \frac{1}{E} \frac{\partial \Phi}{\partial \phi} = 0 \quad \text{(meridional)} \tag{3}$$

$$\frac{\partial \Phi}{\partial p} + \frac{RT}{p} = 0 \quad \text{(hydrostatic)} \tag{4}$$

$$\frac{\partial \omega}{\partial p} + \frac{1}{E \cos \phi} \left(\frac{\partial u}{\partial \lambda} + \frac{\partial}{\partial \phi} v \cos \phi \right) = 0 \quad \text{(continuity)} \tag{5}$$

$$\frac{dT}{dt} - \frac{R}{c_p} \frac{T}{p} \omega - \frac{\dot{q}}{c_p} = 0 \quad \text{(thermodynamic)} \tag{6}$$

$$\frac{dr}{dt} - S = 0 \quad \text{(individual constituent)} \tag{7}$$

In the upper atmosphere below 80 km the diabatic heating rates $\dot{q}/c_p$ may be taken to be the net radiational heating rates. The source term S in the continuity Equation (7) for the mass mixing ratio r of an individual constituent is the net rate of change produced by physical processes, e.g. photochemistry, recombinations, attachments, radioactive decay, etc., depending on the constituent considered. Usually S is temperature dependent and $\dot{q}$ depends on the composition at all levels so that Equations (6) and (7) are interdependent.

By combining Equation (5) with Equations (2), (3) (6) and (7) respectively these equations can be rewritten in their 'flux forms'. Further, if we now denote any parameter by its time mean (barred) value plus a deviation (primed) value, e.g. write

$$u = \bar{u} + u'$$
$$v = \bar{v} + v'$$
$$\omega = \bar{\omega} + \omega' \tag{8}$$
$$T = \bar{T} + T'$$
$$r = \bar{r} + r'$$

we obtain equations of the form:

$$\left[\frac{\partial \bar{T}}{\partial t} + \frac{\bar{u}}{E \cos \phi} \frac{\partial \bar{T}}{\partial \lambda} + \frac{\bar{v}}{E} \frac{\partial \bar{T}}{\partial \phi} + \bar{\omega} \frac{\partial \bar{T}}{\partial p} - \frac{R}{c_p} \frac{\bar{\omega} \bar{T}}{p} \right] + \left[\frac{1}{E \cos \phi} \frac{\partial}{\partial \lambda} \overline{u'T'} + \right.$$
$$\left. + \frac{1}{E \cos \phi} \frac{\partial}{\partial \phi} \overline{v'T'} \cos \phi + \frac{\partial}{\partial p} \overline{\omega'T'} - \frac{R}{c_p p} \overline{\omega'T'} \right] - \frac{\dot{q}}{c_p} = 0 \tag{6a}$$

with similar types of equations containing $\partial \bar{u}/\partial t$, $\partial \bar{v}/\partial t$ and $\partial \bar{r}/\partial t$ etc. The first term in brackets contains only the average values and the second products of deviations. The type of average chosen may be in time, space or space-time when we would have for example

$$\overline{vT} = \bar{v}\bar{T} + \overline{v'T'} \tag{9}$$

$$[\overline{vT}] = [\bar{v}][\bar{T}] + [\bar{v}^*\bar{T}^*] + [\overline{v'T'}]. \tag{10}$$

In Equation (9) the time average of the product vT is simply the product of the individual time means (barred values) plus the time mean of the products of the individual deviations. In Equation (10) the space-time mean (with the square brackets denoting means round latitude circles) may be resolved into three terms. The first, which is the product of the individual space-time means, is the contribution by the mean circulation. The second, which involves deviations (denoted by asterisks) of the time means from the space-time means, is the contribution due to standing eddies, i.e. those fixed in location and generally tied to topographical features. The last term is the space-averaged contribution of the moving systems – the transient eddies. Products of the type $\overline{v'T'}$, e.g. covariance between the motion fields and the elements they advect, are the variables in the flux terms. We can now consider the possible approaches to modelling based on equations of the type of Equation 6(a) and the effects of attempting to simplify them.

3. The Representation of the Eddy Fluxes

If an element θ is advected by a wind component v it is possible, given a complete series of observations of v and θ in space and time, to calculate simply the terms $[\bar{v}][\bar{\theta}]$, $[\bar{v}^* \bar{\theta}^*]$, and $[\overline{v'\theta'}]$ and assign relative contributions to the mean, standing and transient eddy terms. Even when observations of u and v are readily obtainable those of w have to be estimated indirectly using equations derived from Equations (2) to (7) and usually containing approximations (e.g. the 'adiabatic' thermodynamic equation, the 'ω' equation, the simplified vorticity equation, etc.). w for the large scale systems is usually in the order of mm s^{-1}. If detailed data are not available a theoretically or even empirically based relationship between the covariance terms and some function of the better known mean fields must be sought. Presumably the physical mechanism particularly for the standing eddies can be regarded as a spectrum of waves of θ interacting with a spectrum of waves of v, with the products determined by their various amplitudes and phases, but the necessary data to express the fluxes in this form do not exist. Hence it appears that the relationship to be sought must rest on some statistical basis. If we write

$$\overline{v'\theta'} = a'\sigma_\theta\sigma_V \tag{11}$$

where a' is a correlation coefficient this does not help because a' is not usually known and the standard deviations σ_θ, σ_V are also poorly established. The next common approach is to use a mixing length theory or at least a flux-gradient relationship of the form

$$[\overline{v'\theta'}] = - K \frac{\partial[\bar{\theta}]}{\partial y} \tag{12}$$

for all eddy transports. K here is an eddy diffusivity and is analogous to a molecular diffusivity. In this case it is taken to represent the effect of all scales of motion. In the stratosphere and mesosphere the value of K will be determined by the planetary waves for horizontal transport at least, although smaller scales may also be of importance for vertical transport. At higher levels the effect of gravity waves and probably tides as transport mechanisms must also be incorporated. There are, however, several other important considerations regarding the applicability of this formulation and the possible magnitudes of this coefficient. It is implicit in the mixing length theory that K should be a positive quantity, i.e. that the flux should always be directed down the mean gradient. Observations show, however, that this apparently does not always take place when only horizontal or vertical components of transfer are considered separately. For example large scale vertical heat fluxes in the troposphere are upwards while the mean gradient of potential temperature is directed downwards, large scale fluxes of heat and ozone in the lower stratosphere are generally polewards while the average potential temperature and ozone mixing ratio gradients are generally equatorwards, etc. This does not imply, however, that when the total flux and gradient vectors are considered the flux is counter-gradient but merely that separate consideration of their resolved components can give this effect. The observations are in fact consistent with the phenomena of 'slant convection' (Sheppard, 1963) and it is possible to reformulate the relationship by treating the diffusivity as a tensor instead of a scalar quantity. For a two-dimensional (y, z) model for example, Reed, and German (1965) have shown that consistent results may be obtained using relationships of the form

$$
\begin{aligned}
-\overline{[v'\theta']} &= K_{yy} \frac{\partial [\bar{\theta}]}{\partial y} + K_{yz} \frac{\partial [\bar{\theta}]}{\partial z} \\
-\overline{[w'\theta']} &= K_{zy} \frac{\partial [\bar{\theta}]}{\partial y} + K_{zz} \frac{\partial [\bar{\theta}]}{\partial z}.
\end{aligned}
\tag{13}
$$

The theory requires that the advected quantity θ be conservative during the motions considered. Hence it may well succeed with quantities such as radioactive substances, potential vorticity, ozone in the lower stratosphere, but not for example with angular momentum. It is difficult to assign values to the K's at all levels. Measurements based on the dispersion of particle trajectories suggest that the order of K_{yy} is 10^6, $|K_{yz}|$ 10^3–10^4, and K_{zz} 1–10 m^2 sec^{-1} for the large scale systems (see e.g. Murgatroyd, 1969), and models of the dispersion of radioactive tracers in the stratosphere, for example, based on this approach give results in good agreement with observations. Theoretical considerations also suggest that

$$
K_{yy} = \tau \overline{[v'^2]}
\tag{14}
$$

where τ is an integral time scale and $\overline{[v'^2]}$ is the variance of the meridional wind component, and there is a similar type of expression for K_{zz}. τ depends effectively on the dominant scale of the systems involved and the wind variance is highly correlated with synoptic activity and wind speed. Thus in the troposphere and lower stratosphere K_{yy} is largest at mid-latitudes and in the upper troposphere. Figure 4 gives some guidance regarding the values to be expected in the upper stratosphere and lower mesosphere. Since the stratospheric systems have very large scales τ will probably be larger than in the troposphere. In winter $\overline{[v'^2]}$ is also much larger than at lower levels. Accordingly we might expect the corresponding values of K to be perhaps an order larger than at the lower levels. In summer, however, the expectation is that the effective values of K in the upper stratosphere and mesosphere will be very small since the wind variances are small. In both seasons there are decreases of the mean wind speed above about 65 km and probably also of the variances due to the large scale systems. At higher levels, however, with the growing importance and large amplitudes of gravity waves and tides it seems that diffusivity due to the large-scale motions will no longer be the major factor. Some authors have used values for K_{zz} as high as $10^2 - 10^3$ m^2 s^{-1} in the upper mesosphere and it is not clear from the above considerations how these can arise, particularly in the summer.

4. Mean Meridional and Vertical Motions in the Stratosphere and Mesosphere

It is very difficult to obtain fields of $[\bar{v}]$ and $[\bar{w}]$ from observations. $[\bar{v}]$ has characteristic values of order cm s^{-1} and is obtained as a mean round a latitude circle of much larger values of v (usually m s^{-1}). Hence a very large number of observations and a good network are required to obtain reliable values. Once $[\bar{v}]$ is known $[\bar{\omega}]$ may be determined from the zonally averaged continuity equations

$$\frac{\partial [\bar{\omega}]}{\partial p} + \frac{1}{E \cos \phi} \frac{\partial}{\partial \phi} [\bar{v}] \cos \phi = 0 \tag{15}$$

and the values of $[\bar{w}] = -(1/g\varrho)[\bar{\omega}]$ are usually two or three orders less than those of $[\bar{v}]$.

It is possible also to deduce values of $[\bar{v}]$ and $[\bar{\omega}]$ by using Equation (15) in conjunction with the zonally averaged thermodynamic or zonal wind equations, i.e.

$$\left[\frac{\partial [\bar{T}]}{\partial t} + \frac{[\bar{v}]}{E} \frac{\partial [\bar{T}]}{\partial \phi} + [\bar{\omega}] \left\{ \frac{\partial [\bar{T}]}{\partial p} - \frac{R}{c_p p} [\bar{T}] \right\} \right] +$$

$$+ \left[\frac{1}{E \cos \phi} \frac{\partial}{\partial \phi} [\overline{v'T'}] \cos \phi + \frac{\partial}{\partial p} [\overline{\omega'T'}] - \frac{R}{c_p p} [\overline{\omega'T'}] \right] = \frac{[\bar{q}]}{c_p} \tag{16}$$

$$\left[\frac{\partial [\bar{u}]}{\partial t} + \frac{[\bar{v}]}{E} \frac{\partial [\bar{u}]}{\partial \phi} + [\bar{\omega}] \frac{\partial [\bar{u}]}{\partial p} - \left(f + \frac{\tan \phi}{E} [\bar{u}] \right) [\bar{v}] \right] +$$

$$+ \left[\frac{1}{E \cos \phi} \frac{\partial}{\partial \phi} [\overline{u'v'}] \cos \phi - \frac{\tan \phi}{E} [\overline{u'v'}] + \frac{\partial}{\partial p} [\overline{u'\omega'}] \right] = 0. \tag{17}$$

The eddy terms here include both transient and standing components. Studies for the troposphere and stratosphere have shown that $[\overline{u'\omega'}]$ is a comparatively small term and hence, if $[\overline{u'v'}]$ and $[\bar{u}]$ are known, solutions may be obtained for $[\bar{v}]$ and $[\bar{\omega}]$. These have demonstrated that the tropospheric mean meridional circulation has three cells which extend into the lower stratosphere with intensities progressively decreasing with height. This approach is not possible at higher levels since global values of $[\overline{u'v'}]$ are not known (although Meteorological Rocket Network data are available over North America). Equation (16) could be used in a broadly similar way if values of $[\overline{v'T'}]$, $[\overline{\omega'T'}]$ were available as well as a field of $[\overline{T}]$ and $[\overline{q}]$. Murgatroyd and Singleton (1961) performed the calculation of $[\overline{v}]$ and $[\overline{\omega}]$ for the stratosphere and mesosphere for the case where the eddy terms (2nd term in large brackets in Equation (16)) were neglected. Their calculations indicated motions generally upwards in the summer hemisphere and downwards in the winter hemisphere of about 1 cm s^{-1} magnitudes, with flow in the upper mesosphere from summer to winter hemisphere. Their values, although considered by the author plausible as to general directions, are likely to be over-estimates in magnitude, particularly the winter means. The present suggestions are:

(1) Since the eddy activity is low in summer it is possible that a slow upwards and mainly equatorwards mean meridional circulation does in fact exist in the upper stratosphere and mesosphere at that time of the year.

(2) As the eddy activity is high in winter we might expect mean conditions somewhat analogous to the winter troposphere. Slant convection will then occur but there will also be a marked region of descent in mid-latitudes. (See also Newell, 1968.) This is also in accordance in the middle stratosphere with observations of the movement of radioactive debris (List and Telegedas, 1969). A schematic diagram of possible circulations is given in Figure 5.

5. The Possibilities of Dynamic Modelling

If the modelling is restricted to one dimension, i.e. some spatially averaged variation with height, the zonal and meridional terms disappear from Equations (2) to (7) and we are left with only the vertical motion terms. Even the mean vertical term of the form $[\overline{\omega}] (\partial[\overline{\theta}]/\partial p)$ should be taken as zero in a global average, but over limited regions it may be possible to discuss the respective roles of mean motion and diffusion. This type of approach was used for example by Brewer (1949) in discussing a steady state condition for water vapour in the lower stratosphere in the presence of upwards diffusion opposed by mean subsidence.

If a one-dimensional model, including diffusion expressed by a flux-gradient relationship of the form of Equation (12), is used it is not clear how it should be physically interpreted or what value should be chosen for K at the different levels in the mesosphere and above.

In the case of two-dimensional (y, z) cross-sections we have to deal with a set of equations of the type of Equations (16) and (17). Treating the problem as diagnostic and assuming that the mean fields are known, each equation contains an unknown $[\overline{v}]$ and a set of unknown flux divergences. ($[\overline{\omega}]$ is determinable from the continuity equation if $[\overline{v}]$ is known.) The central difficulty is again the expression of the eddy fluxes in terms of known fields, and if a K_{yy}, K_{yz}, K_{zz} formulation is used values must be assumed or some attempt made to obtain solutions between the data for a number of equations for different constituents. As remarked above the momentum equation is not suitable for this treatment although equations for potential vorticity, ozone, etc., may possibly be used. If suitable values for the K's are chosen, time integrations of these equations appear to present no special difficulties and models of this type have been constructed for example to clarify the nature of the annual ozone variations in the lower stratosphere (e.g. Berkovsky, 1967).

In general, however, it appears that the use of two-dimensional equations is likely to be largely based on empirically chosen parameters and, as in the case of single-dimension models, their magnitudes will be selected mainly on the results they give rather than on an independent basis. This is the approach made by several workers modelling ozone and/or radioactive distributions in the lower stratosphere by means

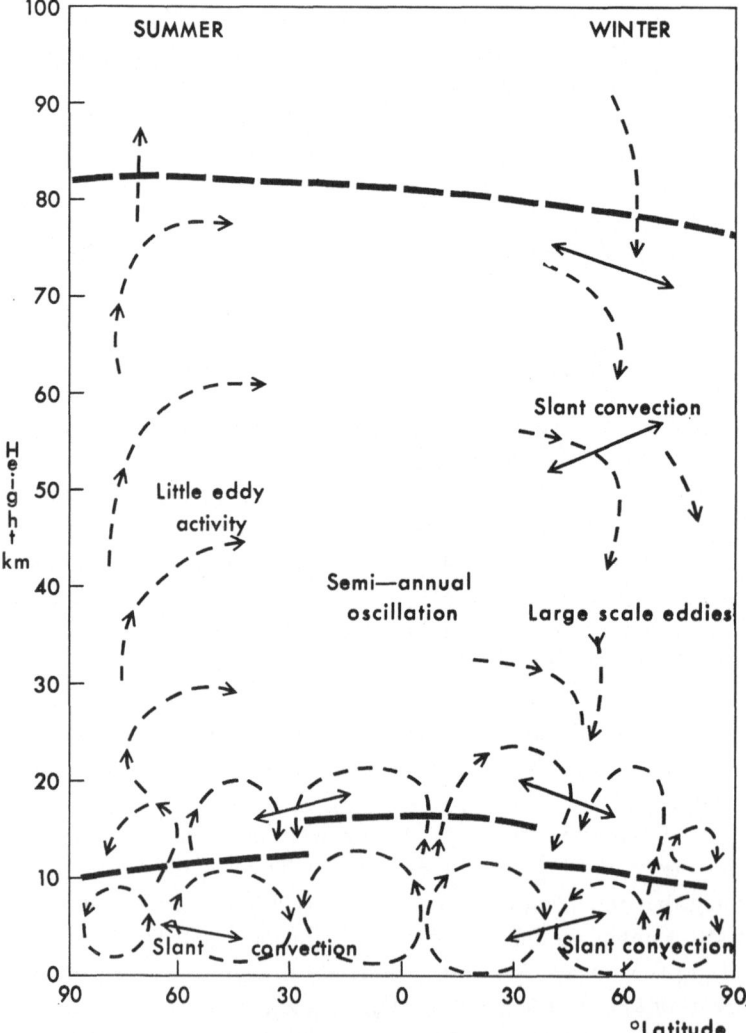

Fig. 5. Possible mean circulations in the stratosphere and mesosphere (dashed lines) and large scale eddy transfers (full lines with arrows). Dashed heavy lines denote mesopause and tropopause. (In the troposphere cells are shown globally for the equinoxes.)

of grid point or box models (e.g. Prabhakara, 1963; Davidson, Friend and Seitz, 1966; Pierson and Cambray, 1967) and these have achieved a reasonable degree of success particularly in establishing relative rates of transfer between different layers in the vertical and between hemispheres. This type of approach, however, requires a knowledge of the distributions of the constituents. For its application to the upper mesosphere and lower thermosphere a more detailed knowledge of the vertical and latitudinal distribution of a selection of the minor constituents seems to be essential.

The most fundamental approach is to form a full three-dimensional model based on time integrations of Equations (1) to (7) over a few months and containing all the relevant physical processes. The hope is then that the model predicts or at least suggests the behaviour of this part of the atmosphere. The principal limitations are those of computational effort and computer size, but it should be noted that tropospheric models have already been extended to about 37 km (4 mb level), i.e. including the lower and middle stratosphere (see Manabe and Hunt, 1968: Hunt and Manabe, 1968; Hunt, 1969). Models of this type, starting from an observed initial field of wind and temperature, show as the integration proceeds the nature of the evolution of these basic fields in accordance with Equations (1) to (7) and the changes of the heat sources and sinks. Their realism depends on the success in representing the radiative mechanisms (solar and terrestrial), the composition changes, the dynamics on all scales and the important feedbacks involved. So far, except for the rather successful 'parameterising' of the hydrologic cycle (including the effect of latent heat), feedback processes such as those involving ozone formation, radiation, and motion have not been fully incorporated. Nevertheless the agreement with observations and the pointers to the relevant atmospheric mechanisms which are important in several problems are noteworthy.

(a) Since the main features of their horizontal fields of temperature and wind are in good agreement with direct observations and diagnostic studies it appears likely that the fields of vertical motion and hence vertical transports, also shown by the model but difficult to measure or diagnose, are also realistic. This allows both horizontal and vertical components of the fluxes and energy conversion terms to be evaluated. These confirm the broad conclusions that the circulation of the lower stratosphere is mainly driven by that of the troposphere and that the middle stratosphere to some extent at least is driven by its own heat sources and sinks. Moreover in the models it is clear that the cold equatorial tropopause is formed as a result of dynamical rather than radiative processes.

(b) The movement of tracers such as water vapour, radioactivity, and ozone is determined in the lower stratosphere by both the mean circulations and the eddy motions, the former being important in low latitudes and the latter dominating the transfer in middle and high latitudes. In all cases the total transfer of these substances is down the concentration gradient.

The humidity of the stratosphere is low in the model calculations primarily due to the 'cold trap' effect at the equatorial tropopause as originally postulated by Brewer (1949). The ozone and radioactivity maxima at high latitudes in spring appear to occur mainly as a result of polewards transfer to middle latitudes by the mean cells (which as also found diagnostically are vertical extensions of the tropospheric cells), followed by primarily eddy transport to higher latitudes. There is considerable recycling within the stratosphere before descent to the troposphere through the tropopause gap can take place.

(c) If photochemical processes are included the ozone distribution and fluxes are considerably modified compared with models in which only advection is considered since replacement takes place in the upper stratosphere when there is transfer downwards by the motions. The detailed mechanism by which the eddy (synoptic) motions produce latitudinal transfer can also be followed in the model whereas very expensive sampling programmes would be required to investigate this observationally.

(d) The onset of large disturbances in the stratosphere circulation, particularly 'sudden warmings', has been shown diagnostically to be due to enhanced vertical flux of geopotential from the troposphere. This effect has also been demonstrated in some of the model calculations although a detailed representation has not yet been made entirely successfully. (See e.g. Miyakoda, Strickler and Hembree, 1970.)

The progress achieved by the models in the stratosphere strongly suggests that efforts should be made to extend this work to the mesosphere. In particular a great deal will be learned from the study of the large-scale fluxes within such models. Since the motions at grid points within the model are treated explicitly the values there at each time step will allow covariances and hence fluxes to be calculated directly. The only values of diffusivities 'K' within the model are those which may be introduced to ensure computational stability or to represent the contributions of smaller (sub-grid) scale motions than those dealt with explicitly, i.e. variations within the volume represented by the grid point and also within the time step intervals. In either case these diffusivities will be at least an order smaller than the values appropriate to the larger-scale eddies, so that the prospects of representing realistically the main features of the transports of constituents in the upper stratosphere and mesosphere in terms of mean motions and large scale eddies appear to be good. At higher levels, i.e. in the thermosphere where transport by tides, gravity waves, and molecular diffusion become important, there are also other difficulties, e.g. in determining the effective sources and sinks in the thermodynamic equation, adding other terms due to electrical forces, molecular viscosity, etc., to the equations of motion, but these are outside the scope of this paper. Meanwhile the broad features of the design of a model for the stratosphere and mesosphere will now be outlined.

6. Design of a Stratosphere-Mesosphere Dynamic Model

The size of the model will be determined by the computing facilities, but it should extend from at least tropopause to mesopause and 8 or 10 levels are desirable. (See also Murgatroyd, 1969a.) For ease of computation a pressure p co-ordinate is desirable vertically, although as vertical spacing in height is more convenient in upper air studies the equations can probably best be dealt with in terms of equal intervals of $\log p$.

Lateral grid point spacing will probably need to be in the region of 5° in latitude and

$10°$ in longitude. In general the closer the spacing the better the representation and the prospects of simulating baroclinic instability, but this has to be balanced against computing facilities and time. Since also the model should run over an integration time equivalent to 3–6 months to simulate seasonal changes, the machine running time should not exceed about one hour per day of integration.

Preferably the co-ordinates should be on a sphere and the coverage global in order to have any prospect of representing inter-hemispherical transfer and polar phenomena. However, the use of spherical co-ordinates introduces problems due to the convergence of the meridians towards the poles. On a regular grid the zonal distance between grid points becomes small and this tends to lead to roughnesses in the integrations and necessitates small time steps to avoid computational instability at high latitudes (defined by the Courant-Friedrichs-Lewy criterion). More efficient time integrations are possible using schemes in which the number of grid points is decreased with latitude, the integration time step varies with latitude, the highest latitudes are replaced by a 'polar cap', etc., but so far none of these has received general acceptance. If the problem at high latitudes is avoided by using a map projection serious difficulties arise in getting a satisfactory representation over both hemispheres.

On the other hand the model can be confined effectively to one hemisphere and a lateral boundary inserted near the equator with some sacrifice of generality as the fluxes (or zero transfer) across it will have to be specified. The problem of boundaries also arises in the vertical. Frequently the practice for the upper boundary is to specify zero vertical velocity there. Accordingly the vertical velocities at lower levels can then be found by integrating Equation (5) downwards. Probably the vertical velocity within the large-scale circulations found in this way becomes realistic within two or three levels down from the top of the model, and from this point of view it may be desirable to extend the model to a somewhat higher level than that at which the vertical velocities and fluxes are to be studied in detail.

The specification of the lower boundary is of basic importance since, as discussed above, the lower stratosphere's circulation is mainly driven by that of the troposphere. The geopotential Φ at each level is found by integrating Equation (4) upwards. A realistic evolution of the bottom boundary fields of Φ and T must therefore be specified (probably it is best to interpolate from sequences of actual observed fields, and of course experiments can be made to study the effect of varying these bottom boundary values and investigating such phenomena as 'sudden warmings'). It is also desirable to check the overall energy balance of the model by evaluating throughout the integration the rate of change of energy within the model, i.e. with $p_T \approx 0$,

$$\frac{\mathrm{d}}{\mathrm{d}t}\int_{p_B}^{p_T} \overline{\left(\frac{u^2+v^2}{2} + c_v T + \Phi\right)} \mathrm{d}p =$$

$$= \int_{p_B}^{p_T} \overline{\dot{q}}\,\mathrm{d}p + \omega_B \overline{\left[\frac{u^2+v^2}{2} + c_p T + \Phi\right]}_B - p_B \overline{\frac{\partial \Phi_B}{\partial t}} \cdot (18)$$

Equation (18) states that the rate of change of the total kinetic and potential energy in the whole mass of the model (the curly over-bar here represents an areal average) is equal to its rate of gain from diabatic heating, flux of total energy through its base and rate of change of the geopotential height $(\Phi_B = g_0 H_B)$ of its base. The suffixes T, B denote the top and bottom of the model. The kinetic energy terms are comparatively small and geostrophic values of u and v can be used in the lower boundary fluxes and in forming the $\partial u/\partial z$, $\partial v/\partial z$ terms for the lowest level in the model. Equation (18) is valid for the continuous equations and must also therefore be satisfied by the difference equation analogues chosen for use in the model.

As regards the radiative flux there is no special difficulty in calculating at suitable intervals and interpolating for the absorption of solar radiation at each level and time step with given zenith angles and concentrations of the constituents (e.g. O_2, O_3, CO_2, H_2O etc.), provided that the incident radiation at the top of the model is also specified. This could be taken as the extra-terrestrial intensity decreased in accordance with climatological values of the constituents above the top of the model. Again it may be necessary to discount the results for the model's highest level although the values will reduce quickly at lower levels to those determined by the model itself. If the absorption is not realised as heat immediately, i.e. there is chemical storage and later release, further problems arise but these have so far received little study. The solar absorption calculations here will generally be made at the same time as any photochemical calculations of the value S in Equation (7). The atmospheric infra-red radiation on the other hand presents difficulties in the specification of the upwards flux from the Earth's surface and troposphere through the lower boundary. This will vary with unknown factors such as cloud amounts, and it appears therefore that the best approach is to use linear cooling as the means of representing the infra-red fluxes in the model. In this case

$$\frac{\dot{q}}{c_p} = \left(\frac{dT}{dt}\right)_{SR} + (mT + c) \tag{19}$$

in Equation (6) where $(dT/dt)_{SR}$ is the calculated heating rate due to absorption of the solar beam, and the infra-red cooling at any level is given by $(mT+c)$ where the constants m and c are obtained empirically from the type of data given in Figure 6. The absorption of solar radiation back-scattered from the troposphere is likely to be small compared with the direct beam absorption and may be neglected or incorporated semi-empirically.

Equations (1) to (7) may be used directly in the so-called 'primitive equation' models. In this case Equations (2), (3), (6) and (7) are most conveniently combined with Equation (5) into their flux forms and then expressed as suitable finite difference equation analogues. It is of course necessary to ensure in the adopted scheme that the component fluxes of heat, momentum and tracer substances laterally and vertically out of the sides of each grid box are numerically equal in the finite difference forms to the values taken for inflow at the interfaces into the sides of adjacent boxes. The initial fields should be chosen such that they do not lead to large gravity waves (which are

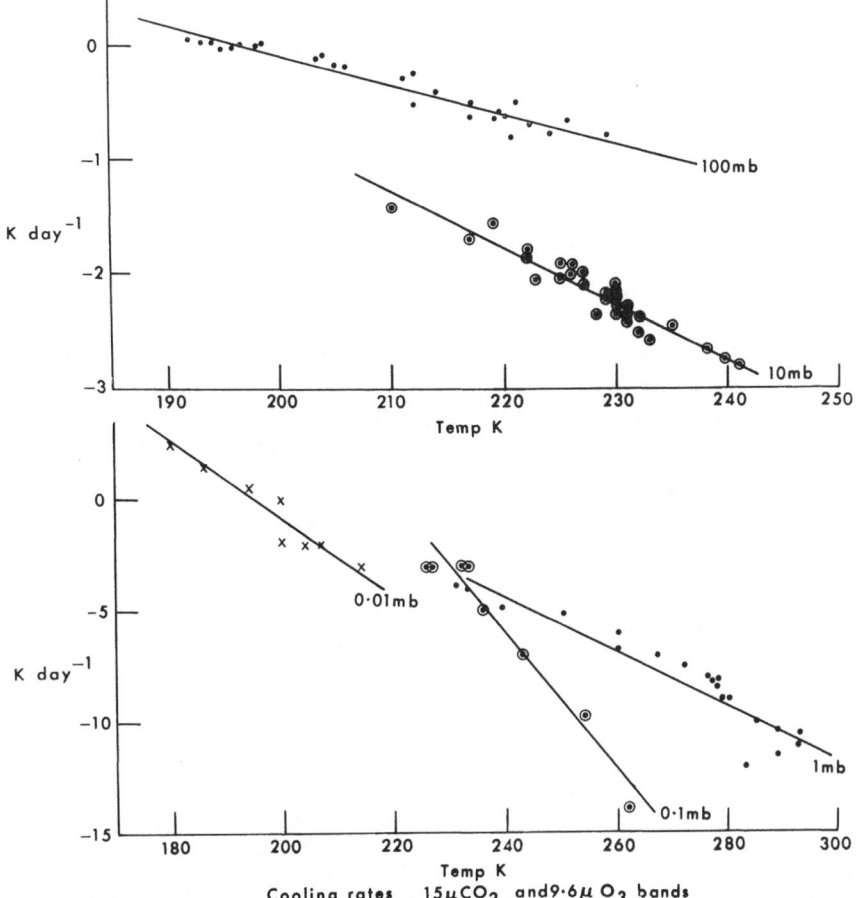

Fig. 6. Approximate linear relationships (cooling to space) between infra-red cooling rates and temperatures in the stratosphere and mesosphere. (Based on results of Murgatroyd and Goody (1958) and Rodgers (1967).)

also solutions of Equations (1)–(6)) in the integration, preferably by adjusting them beforehand to satisfy the 'balance equation' – a condition which ensures that the total time derivative of the divergence is zero at the beginning of the integration (see e.g. Thompson (1961)). In the author's experience models of the stratosphere and mesosphere designed on these lines are very sensitive to any discontinuities in the procedure which tend to produce gravity waves. Perhaps to some extent this reflects the behaviour of the atmosphere in which gravity waves increase markedly in amplitude with height in the mesosphere, but preliminary experiments have shown that unbalanced initial fields and discontinuities in the lower boundary data quickly produce large gravity waves which obscure the effects of planetary wave development within the model. In other respects the chosen initial values are not critical and could be climatologically based with the expectation that the fields will ultimately adjust to the energy inputs.

More simple models, e.g. spectral models using only a few of the low wave numbers and vorticity equation models, have also been devised for use in the stratosphere and mesosphere but these also have disadvantages and limitations. It appears that although promising results have been obtained by some workers (e.g. Byron-Scott (1967); Clark (1970)), the work on them so far has not been very extensive and has been largely concerned with the mechanisms involved in the development of sudden warmings.

Based on the above considerations a model of the form shown in Figure 7 is now being developed and further details have been given by Murgatroyd (1969a). As remarked above the principal difficulty encountered in its preliminary testing is in obtaining stable integrations basically undisturbed by large gravity waves near the top of the model, and it appears necessary to include considerable damping at these levels.

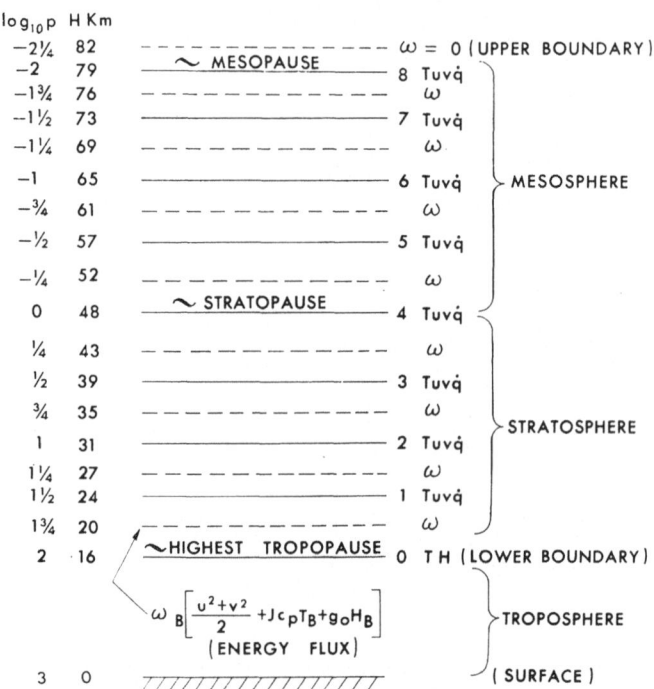

Fig. 7.   Arrangement of levels in a primitive equation model of the stratosphere and mesosphere. (See also Murgatroyd, 1969a.)

Preliminary indications suggest that the strength of the mean meridional circulations may well be considerably less than found in Murgatroyd and Singleton's (1961) paper. This would not be unexpected as subsequent work which included eddy contributions in the lower stratosphere (e.g. Vincent, 1968; Murgatroyd, 1969c) led to similar conclusions for that region. Details of the mean circulations and the other important transfer mechanisms by the eddies will, it is hoped, emerge from this work in the future.

Acknowledgments

This paper is published by permission of the Director-General of the Meteorological Office. (Copyright, Controller H.M.S.O.)

References

Berkovsky, L.: 1967, 'Numerical Models of the Atmosphere incorporating Upper Atmosphere Effects'. Dept. Meteor., Hebrew Univ. of Jerusalem. Final Report. Contract No. 05–3485.

Brewer, A. W.: 1949, *Quart. J. Roy. Meteorol. Soc.* **75**, 351–63.

Byron-Scott, R.: 1967, 'A Stratospheric General Circulation Experiment incorporating Diabatic Heating and Ozone Photochemistry', McGill Univ. Publ. in Meteorology, No. 87, Contract No. AF 19 (628)–4955, Sci. Rep., No. 4.

Clark, J. H. E.: 1970, *Mon. Weath. Rev.* **98**, 443–61.

COSPAR, W. G. IV: 1965, *COSPAR International Reference Atmosphere*, North-Holland Publ. Co., Amsterdam.

Davidson, B., Friend, J. P., and Seitz, H.: 1966, *Tellus* **18**, 301–15.

Hunt, B. G.: 1969, *Mon. Weath. Rev.* **97**, 287–306.

Hunt, B. G. and Manabe, S.: 1968, *Mon. Weath. Rev.* **96**, 503–39.

List, J. and Telegedas, K.: 1969, *J. Atmospheric Sci.* **26**, 1128–36.

Manabe, S. and Hunt, B. G.: 1968, *Mon. Weath. Rev.* **96**, 477–502.

Miyakoda, K., Strickler, R. F., and Hembree, G. D.: 1970, *J. Atmospheric Sci.* **27**, 139–54.

Murgatroyd, R. J.: 1966, 'Radiation Sources and Sinks in the Stratosphere and Mesosphere' in *Les Problèmes Météorologiques de la Stratosphère et de la Mésosphère*, Publications du Centre National d'Études Spatiales, Presses Universitaires de France, Paris, pp. 243–70.

Murgatroyd, R. J.: 1969a, 'Models of the Stratosphere and the Mesosphere', in: Aeronomy Report No. 32 'Meteorological and Chemical Factors in D-region Aeronomy – Record of the Third Aeronomy Conference', University of Illinois, Urbana, pp. 23–37.

Murgatroyd, R. J.: 1969b, *Phil. Trans. Roy. Soc. London* A **265**, 273–94.

Murgatroyd, R. J.: 1969c, *Quart. J. Roy. Meteorol. Soc.* **95**, 194–202.

Murgatroyd, R. J.: 1970, in G. A. Corby (ed.), *The Global Circulation of the Atmosphere*, Proceedings of the Conference held in London, Aug. 1969. Royal Meteorological Society, London.

Murgatroyd, R. J. and Goody, R. M.: 1958, *Quart. J. Roy. Meteorol. Soc.* **84**, 225–34.

Murgatroyd, R. J. and Singleton, F.: 1961, *Quart. J. Roy. Meteorol. Soc.* **87**, 125–36.

Newell, R. E.: 1968, *Meteorol. Monographs* **9**, 31, *Meteorological Investigations of the Upper Atmosphere*, Amer. Meteor. Soc. Boston, Mass., pp. 98–113.

Pierson, D. H. and Cambray, R. S.: 1967, *Nature* **216**, 755–8.

Prabhakara, C.: 1963, *Mon. Weath. Rev.* **91**, 411–31.

Reed, R. J. and German, K. E.: 1965, *Mon. Weath. Rev.* **93**, 313–21.

Rodgers, C.: 1967, 'The Radiative Heat Budget of the Troposphere and Lower Stratosphere', Report No. A2. Planetary Circulations Project. Dept. Meteor., Massachusetts Institute of Technology.

Sheppard, P. A.: 1963, *Rep. Prog. Phys.* **26**, 213–67.

Thompson, P. D.: 1961, *Numerical Weather Analysis and Prediction*, Macmillan Co., New York.

Vincent, D. G.: 1968, *Quart. J. Roy. Meteorol. Soc.* **94**, 333–49.

TIDES AND GRAVITY WAVES IN THE UPPER ATMOSPHERE

RICHARD S. LINDZEN*

Dept. of the Geophysical Sciences, University of Chicago, Chicago, Ill., U.S.A.

The subject of this review is amply dealt with in existing papers and monographs. Data on atmospheric tides below 120 km as well as theoretical developments through about 1967 are described in detail in Chapman and Lindzen (1970). Some more recent results are given in a sequence of papers (Lindzen, 1970a, b; Lindzen and Blake, 1970) dealing with the effects of viscosity and conductivity on tides – especially in the thermosphere. The subject of internal gravity waves in the upper atmosphere is well covered by Hines (1960, 1963). Some more recent developments are to be found in Pitteway and Hines (1963) and Booker and Bretherton (1967). A comprehensive review of the subject by Walter Jones will soon appear in *Handbuch der Physik*. The two subjects are by no means unrelated. Atmospheric tides are essentially internal gravity waves (modified by the earth's sphericity and rotation) for which the periods (integral fractions of a day) are known precisely and for which the excitations, while not perfectly known, are better known than they are for most other gravity waves.

I wish to avoid repeating in this paper what is already adequately handled in the literature. Instead I shall dwell on two topics that may prove particularly useful for aeronomers who are not well versed in dynamics. First I will give a particularly simple physical description of the properties of internal gravity waves. This description will, I hope, serve as an adequate introduction to the more sophisticated reviews cited above. Second, I will discuss the transport properties of internal gravity waves.

Any true wave must be associated with some restoring mechanism. For acoustic waves the restoring force arises from the compressibility of the air; for internal gravity waves, it is the buoyancy force exerted on a displaced fluid element in a stably stratified fluid.

Consider an element of fluid at some level z_0 in a fluid where density ϱ is decreasing with height at a rate $-d\varrho/dz$. The situation is depicted in Figure 1a. The mass of the fluid element is

$$\delta m = \varrho(z_0) \, \delta v$$

where δv = volume of fluid element.

If we displace δm a small distance δs it will be subject to a buoyancy force

$$- g \left(\varrho(z_0) - (\varrho(z_0) + (d\varrho(z_0)/dz) \, \delta s) \right) \delta v$$

acting to return δm to z_0; g is the acceleration of gravity. Variations of δv due to compressibility have, for the moment, been neglected. The equation of motion for δm is

$$\varrho(z_0) \, \delta v \, \frac{d^2 \, \delta s}{dt^2} = + g \, \frac{d\varrho}{dz} \, \delta s \, \delta v$$

* Alfred P. Sloan Foundation Fellow.

Fiocco (ed.), Mesospheric Models and Related Experiments, 122–130. All Rights Reserved.
Copyright © 1971 by D. Reidel Publishing Company, Dordrecht-Holland.

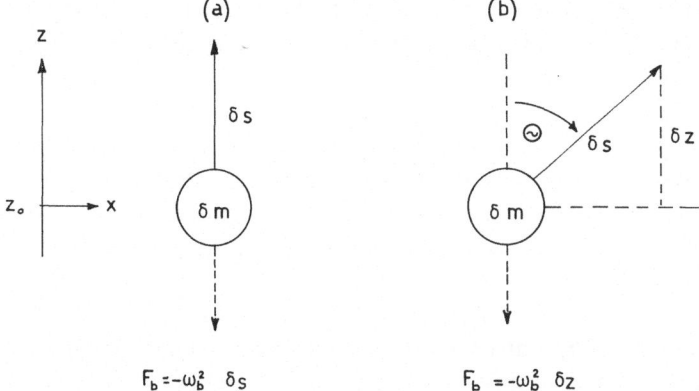

Fig. 1.   Schematic description of fluid elements, their displacements, and buoyancy forces per unit mass.

or

$$\frac{d^2 \, \delta s}{dt^2} = \frac{g}{\varrho(z_0)} \frac{d\varrho}{dz} \, \delta s. \tag{1}$$

In a stably stratified, incompressible fluid $d\varrho/dz < O$. Hence, Equation (1) describes a harmonic oscillation with a frequency ω_b given by

$$\omega_b^2 = - \frac{g}{\varrho(z_0)} \frac{d\varrho}{dz}. \tag{2}$$

ω_b is known as the Brunt-Väisäla frequency. The effect of compressibility is to change our expression for ω_b to the following

$$\omega_b^2 = \frac{g}{T} \left(\frac{dT}{dz} + \frac{g}{c_p} \right) \tag{3}$$

where T is the temperature of the ambient fluid.

Let us designate the buoyancy force per unit volume on a displaced fluid element as

$$F_b = - \omega_b^2 \, \delta s. \tag{4}$$

F_b is directed vertically. Now consider a fluid element that is somehow constrained to move at some angle θ with respect to the vertical (viz. Figure 1b). The force exerted on this fluid element will be the projection of the buoyancy force:

$$F = - \omega_b^2 \, \delta z \cos \theta = - \omega_b^2 \cos^2 \theta \, \delta s \tag{4a}$$

and the element will oscillate with a frequency, ω, given by

$$\omega^2 = \omega_b^2 \cos^2 \theta. \tag{5}$$

A real fluid is, of course, continuous and cannot be thought of in terms of isolatable elements. However, when we excite a real fluid at a frequency ω (where $\omega^2 < \omega_b^2$), the pressure forces do indeed constrain the fluid to move at an angle θ with respect to the vertical such that the projection of the buoyancy force corresponds to the restoring force appropriate to the frequency ω. The situation becomes more concrete when we consider waves excited by a corrugated bottom moving at a constant speed c (viz. Figure 2). The horizontal wavelength imposed by the corrugated bottom is L_H which can also be expressed in terms of horizontal wavenumber $k = 2\pi/L_H$. Moving the bottom at a speed c gives rise to a local oscillation with frequency

$$\omega = kc. \tag{6}$$

The angle θ is given by Equation (5). If the fluid flow shown in Figure 2 were to be plotted as a function of altitude at a fixed time and horizontal position then we would

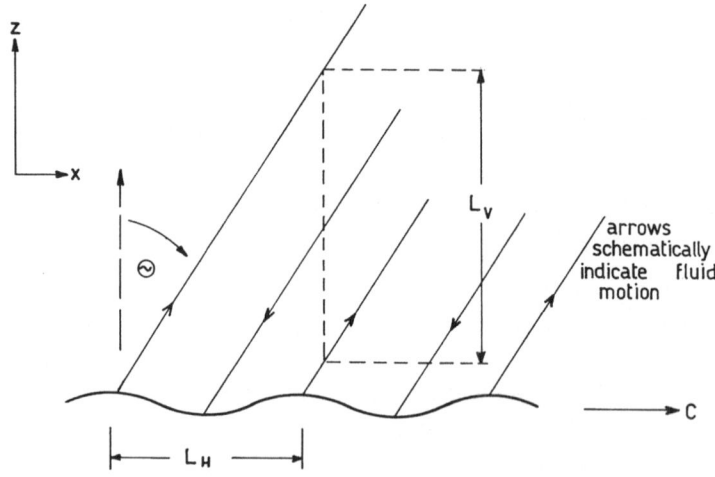

Fig. 2.   Schematic description of motions excited in a stably stratified fluid by the motion (at speed c) of a corrugated bottom.

see a sinusoidal structure with a vertical wavelength L_V (or vertical wavenumber $n = 2\pi/L_V$). Now

$$(L_H/L_V)^2 = (n/k)^2 = \tan^2 \theta. \tag{7}$$

Combining (7) and (5) gives, approximately, the correct dispersion relation for internal gravity waves:

$$\left(\frac{n}{k}\right)^2 = \frac{\sin^2 \theta}{\cos^2 \theta} = \frac{1 - \cos^2 \theta}{\cos^2 \theta} = \frac{1 - (\omega/\omega_b)^2}{(\omega/\omega_b)^2} \tag{8}$$

or

$$n^2 = \frac{1 - (\omega/\omega_b)^2}{(\omega/\omega_b)^2} k^2. \tag{8a}$$

Thus far we are considering oscillations of the form

$$e^{i(\omega t + kx + nz)} \tag{9}$$

where z =altitude, x =horizontal distance, and t =time.

Such oscillations are waves which propagate upward with (in the absence of dissipation or mean flow) constant energy density. Since density is decreasing with height, the constancy of energy density (consisting in terms like $\frac{1}{2}\varrho u^2$, where u =oscillatory horizontal velocity) requires that oscillatory fields increase in amplitude as $1/\varrho^{1/2}$. The relations obtained thus far are appropriate only for isothermal atmospheres, where ω_b^2 is a constant. Because of the requirement of energy constancy (9) must be replaced by

$$e^{z/2H} \, e^{i(\omega t + kx + nz)} \tag{10}$$

for an isothermal atmosphere where

$$H = \frac{RT}{g} \tag{11}$$

(R =gas constant for air, H is the atmospheric scale height). Because the z-dependence is no longer purely sinusoidal (8a) must be slightly modified

$$n^2 = \frac{1 - (\omega/\omega_b)^2}{(\omega/\omega_b)^2} k^2 - \frac{1}{4H^2}. \tag{12}$$

Incidentally, we may see from Figure 2 why *upward* energy flux for internal gravity waves implies *downward* phase progression. If the bottom is to do work on the fluid it must push the fluid, in which case the fluid motions must tilt to the right of the vertical if c is positive. In this case a given phase appears first aloft and progresses downward.

Essential to the existence of internal gravity waves is the ability of fluid elements to retain their buoyancy (adiabaticity) and to move freely *across* isobars. The latter is inhibited by various frictional processes and also by rotation which tends to cause a fluid to move *parallel* to isobars. In a plane rotating system the dispersion relation for internal gravity waves becomes

$$n^2 = \frac{\omega_b^2 \left(1 - (\omega/\omega_b)^2\right)}{\omega^2 - f^2} k^2 - \frac{1}{4H^2} \tag{13}$$

where $f = 2\Omega$, and Ω =the rotation rate of the fluid. From (13) we see that the vertical propagation of an internal gravity wave requires

(i) $\qquad f^2 < \omega^2 < \omega_b^2$

and

(ii) $\qquad k^2 > \frac{1}{4H^2} \frac{(\omega^2 - f^2)}{\omega_b^2 (1 - (\omega/\omega_b)^2)}.$

The requirement $\omega^2 > f^2$ is particularly important for atmospheric tides. On a rotating sphere we must consider $\Omega \sin \varphi$ (where φ =latitude), the vertical component of the rotation vector, in place of Ω. Thus f varies from zero at the equator to $2\pi/12$

hours at the poles. Hence, internal gravity waves with periods *shorter* than 12 hours can propagate vertically anywhere, but for longer periods they propagate vertically in regions increasingly confined to the vicinity of the equator. Thus the diurnal tidal modes consist in vertically propagating modes confined primarily to the region between $\pm 30°$ latitude and exponentially trapped modes polewards of these latitudes.

Finally, some comment is in order on the excitation of gravity waves (including tides). In our example we considered excitation by a moving, corrugated boundary. While pedagogically convenient, this was an obviously unrealistic example. Nevertheless, it has the element that is essential to any excitation – it causes time varying vertical displacements of material surfaces. More practical ways of doing this are by fluid motions over stationary orography*, by explosions and volcanic eruptions, or by the daily variations in the absorption of sunlight by the atmosphere due to the rotation of the earth. The list is hardly exhaustive. In a stably stratified atmosphere, where dissipation is not dominant, any excitation with frequency components between ω_b and f will excite internal gravity waves.

Certainly one of the more important developments in aeronomy in recent years has been the recognition by chemical aeronomers that collective transfer is important in determining atmospheric structure (dynamicists knew this all along). However, the same brave aeronomer who faces arrays of dozens of reactions, many of them completely unknown, has difficulty in comprehending transport in terms of anything more complicated than 'eddy diffusion' (an essentially fictitious process) and mean meridional circulations. We shall here concentrate on eddy diffusion. Initially, it was imagined that both the atmosphere and the ocean were filled with a ubiquitous turbulence which acted to enhance diffusion orders of magnitude beyond their molecular values. Things have since improved a bit for the oceans and the lower atmosphere; however aeronomy remains somewhat primitive. Samuel Johnson, about 200 years ago, remarked that 'patriotism is the last refuge of the scoundrel''. By merely changing 'patriotism' to 'turbulence', 'last' to 'first' and 'scoundrel' to 'aeronomer', we have an apt description of the present situation. The situation is gradually improving insofar as many aeronomers are coming to realize that 'eddy diffusion' is no more than an exceptionally crude parameterization of mixing by any fluid motions – turbulent and laminar. It is at this point that internal gravity waves (including tides) enter the picture. At the moment it appears that above 75 km, over most of the globe and during most of the year the motion of the atmosphere is dominated by internal gravity waves (viz. Haurwitz, 1964). Thus what turbulence there is in the upper atmosphere must be due to the instability of these internal waves, and in the absence of instabilities transport must be directly due to the waves themselves.** How are we to relate such transports to

---

* By means of a shift of coordinates, this is equivalent to our idealized excitation; examples in nature are waves in the lee of wind flowing over mountains. Although mountains do not consist in infinite sinusoidal corrugations, their effects can be simulated by Fourier synthesis.
** This statement may prove a bit too strong. As more observation stations with longer time series become available, it will become possible to ascertain better the importance of long period planetary waves and mean circulations. These could prove significant for transport, especially horizontal transport.

eddy coefficients? Eddy coefficients have the dimensions $[D]=[\text{Length}]^2 [\text{Time}]^{-1}$.

Mechanistically, an eddy coefficient for vertical transfer can be thought of as representing some *weighted* product of a characteristic vertical velocity and a characteristic vertical scale for the motion system under consideration. Depending on the particular motion system, the weighting factor may be as small as zero. For example, for a pure adiabatic, sinusoidally oscillating, coherent internal gravity wave any *passive tracer* carried up will also be carried down and the weighting factor will be zero. Of course, this idealized situation is never realized. Waves change in amplitude, irregular phase changes occur, 'passive' tracers are subject to some molecular diffusion, chemical alteration, etc. All these factors increase the weighting factor to a finite value. Figure 3 shows a simple example of how this might occur. The oscillation is

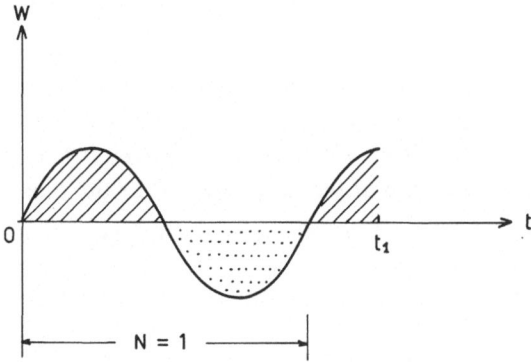

Fig. 3. Schematic (and idealized) illustration of an oscillation begun at $t=0$ and ended at $t=t_1$. The purpose of the illustration is to show how oscillations of finite duration may lead to net transports.

turned on at $t=0$ and shut off at $t=t_1$. If t_1 is incommensurate with the period of the oscillation, then a net transfer will have occurred. A plausible weighting factor in this case might be $0.5/N$, when N is the lifetime of the oscillation in cycles to the nearest integer, while the factor 0.5 been introduced to allow for a measure of randomness in the exact lifetime. A similar factor could be used to account for irregular changes in amplitude and phase, where N might then be the number of cycles over which such changes occur.*

The use of present knowledge of gravity waves in the upper atmosphere to estimate eddy coefficients is largely a matter of conjecture, but a rough attempt will be made. In the tropics we expect the diurnal tide to be of particular importance. The diurnal component of the vertical velocity increases with height as $1/\varrho^{1/2}$ (see earlier) to a maximum value of about 20 cm/sec at about 90 km. Larger values of w (the vertical velocity) lead to the instability of the tide (Lindzen, 1968). The vertical wavelength

---

* The effect of chemical action is somewhat different. If the dissociation of some compound XY increases rapidly with height then XY brought from below might dissociate before it is brought down again, thus effecting a net transfer. Such transport is not readily described in terms of eddy diffusion. We shall here concentrate on transports which can be roughly modelled by diffusive flux.

associated with this tide is about 25 km; a typical vertical scale is therefore 25 km/2π. We would therefore expect an eddy coefficient of the form

$$D \approx \frac{0.5}{N} \times 20 \text{ cm/s} \times 4 \times 10^5 \text{ cm} \times e^{(z-90 \text{ km})/2H}$$

$$\approx \frac{4 \times 10^6}{N} e^{(z-90 \text{ km})/2H} \text{ cm}^2/\text{s} \quad \text{for} \quad z < 90 \text{ km}, \tag{14}$$

where H=scale height ≈ 7 km.

We have little basis for choosing N; however, from an examination of variability of tides revealed in radio-meteor data we suggest $N \approx 10$–20. Thus (14) becomes (using $N=15$)

$$D \approx 2.7 \times 10^5 \, e^{(z-90 \text{ km})/2H} \text{ cm}^2/\text{s} \quad \text{for} \quad z \gtrsim 90 \text{ km}. \tag{15}$$

Above 90 km our conjectures become even more ad hoc. Over the equator, between about 90 km and 108 km, the diurnal tide will be unstable (Lindzen and Blake, 1970). We suggest that in this 'turbulent' region the average value of $|w|$ will remain at about 20 cm/s and that the vertical scale will remain at about 4×10^5 cm, but that N will drop to approximately unity because of the instability of the tide; i.e., we propose

$$D \approx 4 \times 10^6 \text{ cm}^2/\text{s} \quad \text{for} \quad 90 \text{ km} \gtrsim z \gtrsim 108 \text{ km}. \tag{16}$$

Our arguments, admittedly loose, suggest a marked increase in D near 90 km – at least near the equator.

At high latitudes and altitudes, tidal oscillations (and especially the diurnal tide) seem to be of less importance than they are in the tropics. From very limited rocket grenade data (Smith *et al.*, 1968) it appears that the upper atmosphere at high latitudes during summer is exceptionally quiescent. The data *do not offer much possibility of temperature oscillations with amplitudes greater than about 3 K.* If we *arbitrarily* associate such an oscillation with a period of about 6 h we may calculate that w will have an amplitude of about 10 cm/s. We can't be too far off in choosing a vertical scale of about 4×10^5 cm, in which case we obtain

$$D \gtrsim \frac{0.5}{N} \times 10 \text{ cm/s} \times 4 \times 10^5 \text{ cm}$$

$$\gtrsim \frac{2 \times 10^6}{N} \text{ cm}^2/\text{s} \tag{17}$$

for high latitudes in summer. For non-tidal gravity waves, excitation is likely to be irregular and N will be smaller than it is for tides. A plausible guess might be $N=5$ in which case

$$D \gtrsim 4 \times 10^5 \text{ cm}^2/\text{s}. \tag{18}$$

(18) represents an 'upper bound' and depends upon our assumed characteristic period, τ. If $\tau > 6$ h, the bound will be smaller and if $\tau < 6$ h it will be larger.

During winter the high latitude upper atmosphere appears to be very 'noisy' (Smith et al., 1968). Between 80 km and 95 km temperature fluctuations with amplitudes of about 30 K and *estimated* periods of about 6 h are observed. Below 80 km amplitudes appear to decrease, consistent with the $1/\varrho^{1/2}$ behavior of internal gravity wave amplitudes. Vertical wavelengths appear to be characteristically on the order of 20 km; vertical scales are $20 \text{ km}/2\pi \sim 3$ km. With such a vertical scale, a temperature oscillation of amplitude 30 K is marginally, convectively unstable. A temperature oscillation of 30 K with a period of 6 h is associated with a w of amplitude ~ 1 m/s. Thus we suggest (using arguments similar to those used for the tropics, while setting $N=5$ below 80 km) that

$$D \sim \frac{0.5}{5} \times 10^2 \text{ cm/s} \times 3 \times 10^5 \text{ cm } e^{(z-80 \text{ km})/2H}$$

$$\sim 3 \times 10^6 \text{ cm}^2/\text{s } e^{(z-80 \text{ km})/2H} \tag{19}$$

for $z \gtrsim 80$ km at high latitudes in the winter. Above 80 km we suggest

$$D \sim 1.5 \times 10^7 \text{ cm}^2/\text{s}. \tag{20}$$

I hope that this section has plausibly shown that internal waves can account for the eddy transports which aeronomers find necessary in their structure calculations. By considering the mechanism in some detail I have been able to speculatively suggest a latitude, altitude and seasonal structure for eddy coefficients which should prove interesting to test in future structure calculations.

As a final comment I would like to emphasize the fact that an eddy diffusion model is only appropriate to 'passive' quantities. By 'passive' I mean that something is not involved in the intrinsic physics of internal gravity waves. An example of a non-passive quantity is momentum. A vertically propagating internal gravity wave is associated with a mean vertical flux of horizontal momentum. As Eliassen and Palm (1960) have shown, even in the presence of mean shear this flux is independent of height (except at critical levels where the mean flow doppler shifts the wave frequency to zero). To represent such a flux as an eddy diffusion would be impossible.

Acknowledgements

This paper has been prepared with the support of the National Science Foundation under Grant No. GA-1622.

References

Booker, J. R. and Bretherton, F. P.: 1967, *J. Fluid. Mech.* **27**, 513–39.
Chapman, S. and Lindzen, R. S.: 1970, *Atmospheric Tides*, D. Reidel Publ. Co., Dordrecht, Holland, 200 pp.
Eliassen, A. and Palm, E.: 1961, *Geofys. Publik.* **22**, 1–23.
Haurwitz, B.: 1964, 'Tidal Phenomena in the Upper Atmosphere', *W.M.O. Rept.* No. 146, T.P. 69.
Hines, C. O.: 1960, *Can. J. Phys.* **38**, 1441–81.
Hines, C. O.: 1963, *Quart. J. Meteorol. Soc.* **89**, 1–42.

Lindzen, R. S.: 1968, *Proc. Roy. Soc.* **A303**, 299–316.

Lindzen, R. S.: 1970a, 'Internal Gravity Waves in Atmospheres with Realistic Dissipation and Temperature. Part 1', *Geophys. Fl. Dyn.* **1**, in press.

Lindzen, R. S.: 1970b, Part 3, submitted to *Geophys. Fl. Dyn.*

Lindzen, R. S. and Blake, Donna: 1970, Part 2, *Geophys. Fl. Dyn.* **1**, in press.

Pitteway, M. L. V. and Hines, C. O.: 1963, *Can. J. Phys.* **41**, 1935–48.

Smith, W. S., Katchen, L. B. and Theon, J. S.: 1968, *Meteorol. Monographs* **8**, 170–5.

THE METEOROLOGICAL STRUCTURE OF THE
MESOSPHERE INCLUDING SEASONAL AND
LATITUDINAL VARIATIONS

J. S. THEON and W. S. SMITH

Goddard Space Flight Center, Greenbelt, Md. 20771, U.S.A.

During 1960–68, the Goddard Space Flight Center has conducted 208 meteorological soundings of the mesosphere. The acoustic grenade technique provided a large portion of the data, consisting of vertical profiles of temperature, pressure, density and wind. These were supplemented by the pitot probe technique, which produces profiles of temperature, pressure, and density. The details of these experiments have been published by Nordberg and Smith (1964) and Horvath *et al.* (1962). The soundings included in the following analyses are tabulated individually and with complete error analyses by Smith *et al.* (1964, 1966, 1967, 1968, 1969, and 1970); they were carried out from the following sites: Ascension Islands (8°S); Natal, Brazil (6°S); Wallops Islands, Virginia, U.S.A. (38°N); Churchill, Manitoba, Canada (59°N); and Barrow, Alaska, U.S.A. (71°N). The individual soundings are relatively well distributed throughout the year with respect to season, and also with respect to the diurnal cycle. The number of observations is admittedly limited because obtaining *in situ* measurements in the mesosphere requires somewhat sophisticated, and therefore expensive, techniques: the analyses presented here represent all the available data for the sites discussed.

The standard atmosphere models, such as the CIRA (1965) and the U.S. Standard Atmosphere Supplements (1966), are based on the observations available prior to 1965. Additional data are now available, and the techniques for observing the meteorological structure of the mesosphere have been improved upon, revealing some differences and filling some voids in the model atmospheres. The purpose of this paper, then, is to produce a climatology of observed values of temperature, pressure, density, and wind at various latitudes and seasons which will update and supplement the standard atmospheres, and provide realistic inputs for computations involving neutral and ion composition, energy propagation, and transport processes.

The individual profiles from a given site were classed to produce monthly and seasonal means. Data taken during December, January, and February were combined in the winter mean and those taken during June, July, and August in the summer mean. Data from Ascension Island and Natal were combined into an annual mean since there is little annual variation in the temperature, pressure and density profiles in the tropics. The distribution of the soundings by month is given in Table I. Month to month smoothing was performed where data were scarce or nonexistent in the cross-sections.

As is the case with monthly and seasonal smoothing, the short term variations (i.e., periods less than a month) such as synoptic scale eddies, tides, and gravity waves are

Fiocco (ed.), Mesospheric Models and Related Experiments, 131–146. All Rights Reserved.
Copyright © 1971 by D. Reidel Publishing Company, Dordrecht-Holland.

TABLE I

Numbers of soundings included in the analyses shown in Figures 1–4

Site	J	F	M	A	M	J	J	A	S	O	N	D
Natal (incl. Ascension)	1	3	3	2	4	3	0	7	0	8	0	3
Wallops	4	14	7	6	6	2	8	13	4	6	7	6
Churchill	4	17	1	1	2	2	0	10	1	6	2	4
Barrow	2	10	0	4	4	2	0	8	1	6	2	2

removed, although these phenomena are believed to be very large in amplitude in certain individual soundings. The averaging procedure appears to be an effective method for removing short term variations, if the coherent geostrophic analyses that result from the mean values at all but the highest altitudes are any indicator.

The mean temperature profiles were plotted as a function of month for each of the four sites indicated in Table I. Isotherms were drawn for 10 K increments of temperature, and the results are shown in Figure 1a–d. Figure 1a, the mean temperature cross-section for Natal, is characterized by its uniformity below 75 km. The stratopause is slightly warmer during the summer and fall (January–May) than at other times of the year, but this change is only on the order of 10 K throughout the year. The lapse rates in the mesosphere do not vary significantly below 70 km. The upper mesosphere and lower thermosphere are subject to variations of the mean temperature of only 10–20 K throughout the year, and these probably result from the large variations in temperature caused by the thermal tides (Lindzen, 1967) rather than a genuine seasonal effect. Figure 1b, which shows the mean temperature cross-section for Wallops, indicates a slightly larger seasonal effect in the stratopause temperature with the maximum occurring in late spring – early summer. Note the variability in mesospheric lapse rates, say January ($1.4\,\mathrm{K\,km^{-1}}$) vs. July ($2.5\,\mathrm{K\,km^{-1}}$). Also there is a pronounced seasonal effect in the upper mesosphere. The mesopause in summer is some 40 K colder than it is in winter, and there are much larger temperature changes with time, larger than at tropical sites, as indicated by the larger horizontal temperature gradients especially in the 70–80 km region in winter.

Figure 1c shows the mean temperature cross-section for Churchill. Here the patterns that were suggested in the Wallops data are very pronounced. The stratopause temperature varies by almost 30 K with season and reaches its highest temperatures between May and August. The mesopause temperature has a mean value of 140 K in July, considerably colder than is found at lower latitudes. The seasonal temperature variation at the mesopause is two to three times larger than at the stratopause, varying from 140 K in summer to 220 K in winter. The very steep horizontal temperature gradients at the 80 km level in spring and late summer occur at a time when the individual profiles are smooth and almost without short term fluctuations, indicating a very swift but orderly transition from the disturbed thermal structure of winter.

Figure 1d gives the mean temperature cross-section for Barrow. The latitudinal

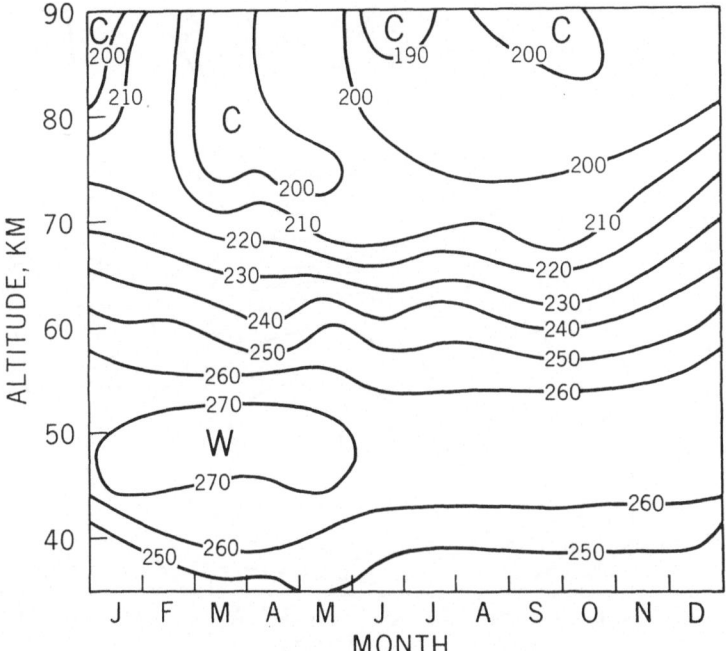

Fig. 1a. Temperature-time cross-section over Natal, Brazil (6°S) based on 34 soundings listed in Table I. Mean isotherms are in K.

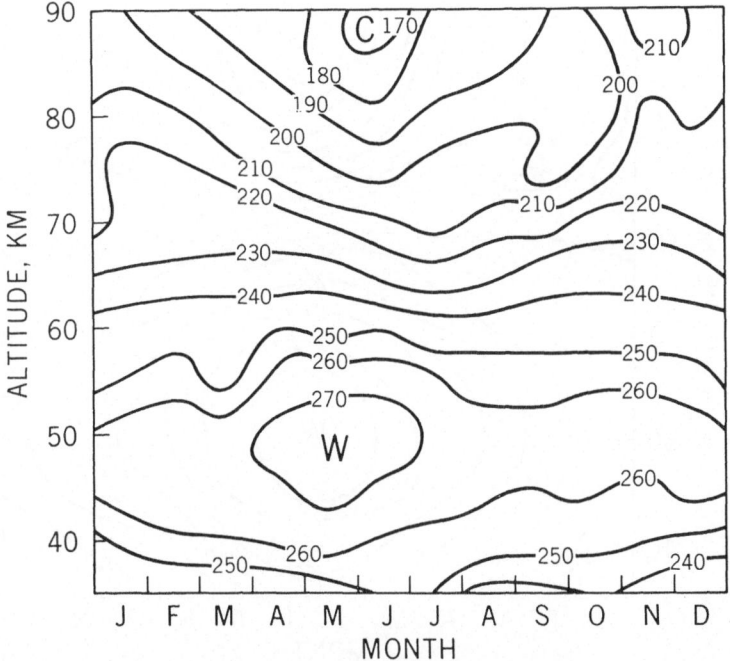

Fig. 1b. Temperature-time cross-section over Wallops Island, Va. (38°N) based on 83 soundings listed in Table I. Mean isotherms are in K.

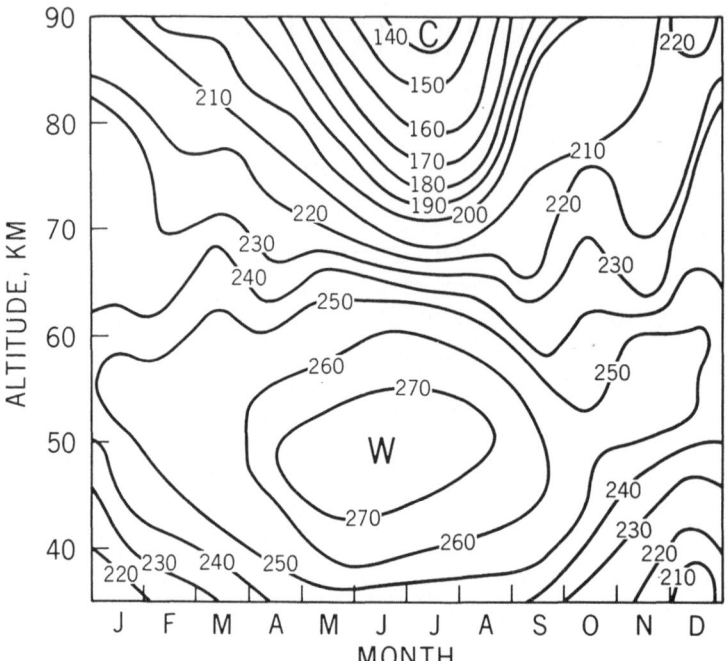

Fig. 1c.   Temperature-time cross-section over Churchill, Canada (59 °N) based on 50 soundings listed in Table I. Mean isotherms are in K.

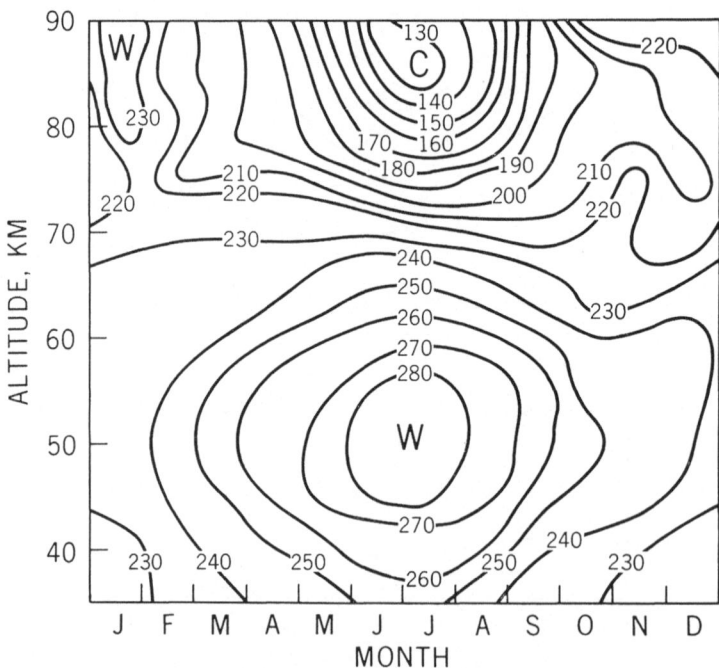

Fig. 1d.   Temperature-time cross-section over Barrow, Alaska (71 °N) based on 41 soundings listed in Table I. Mean isotherms are in K.

trend seen in Figures 1a–c is continued in the Barrow data, with a mean summer stratopause temperature of 280 K, and a winter value some 40 K lower. The coldest temperatures found in the Earth's atmosphere are seen at the summer mesopause with a June mean of about 130 K. In contrast, the winter mesosphere at Barrow is more nearly isothermal, especially in January. The mesopause temperature varies from a summer low of 130 K to 230 K in winter. This appears to be the largest seasonal variation of temperature that occurs in the atmosphere below the mesopause.

When the monthly mean temperatures are combined into seasonal means and a diagonal path traced across the North American continent from Barrow at the northwest through Churchill, and Wallops and southeasterly across the Atlantic to Natal in the southeast, a quasi-meridional cross-section results, as shown in Figure 2a. Here the seasonal mean profiles (i.e., winter and summer) have been plotted as a function of latitude for Barrow, Churchill and Wallops along with the annual mean profile for Natal. The resulting pattern is an expected one: a warm stratopause and a cold mesopause occur at high latitudes in summer; and a profile approaching the 230 K isotherm from 35 to 90 km at high latitudes in winter. These features are, in general terms, similar to the earlier models of Murgatroyd (1957) and Leovy (1964) except that the summer high latitude stratopause and mesopause are cooler here than in the models. Figure 2a also compares well with the cross-section given in the U.S. Standard Atmospheres Supplements, 1966, except that the winter mesosphere is cooler than in the standard model.

When the mean pressure profiles from all the sites are combined with a meridional cross-section for winter and summer, Figure 2b results. The values are plotted as percent deviation from the 1962 U.S. Standard Atmosphere, for convenience. Note that the zero percent deviation line is most nearly approximated by an equatorial pressure profile while there is a well developed low toward the winter pole and a high pressure region toward the summer pole. These differences, of course, drive the mean circulation in the mesosphere, as will be shown later. Note that the low pressure region in the wintertime at high latitudes underlies a higher pressure region, and that the high pressure region in the summertime at the same latitudes underlies a low pressure region. The areas with the strongest horizontal pressure gradients, near 60 km altitude at 45° latitude for example, are the regions where the most intense zonal winds occur.

Figure 2c is a meridional cross-section of the mean density. The isopleths are drawn for percent deviation from the 1962 U.S. Standard Atmosphere density profile. The variation of density with season and latitude is seen to resemble the pressure pattern, with low values over the winter pole and high values over the summer pole. The total seasonal change in density from winter to summer at Barrow is approximately 60 percent and should be taken into account in any computations involving density in the mesosphere.

The general circulation in the mesosphere has not been well documented since relatively few observations have been made compared to the lower regions of the atmosphere. The data presented here are the monthly and seasonal mean values and thus are intended to give only a gross picture of the circulation in the mesosphere. Figure 3

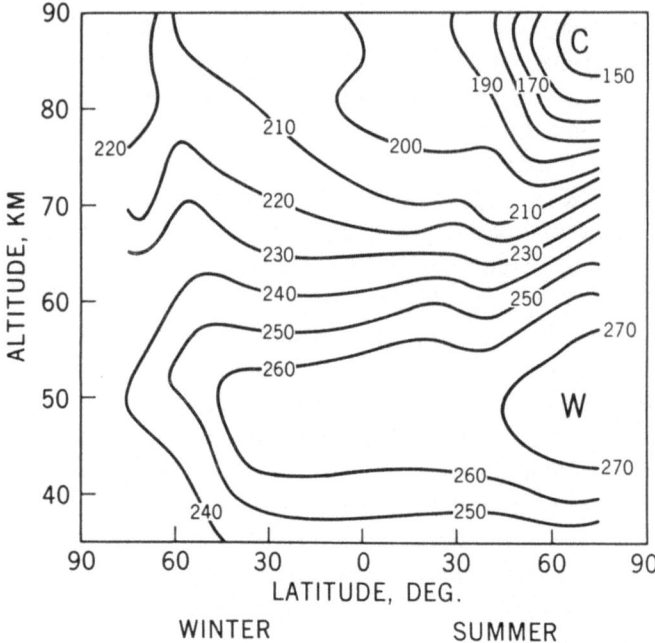

Fig. 2a.   Quasi-meridional cross-section of temperature. Mean isotherms are in K.

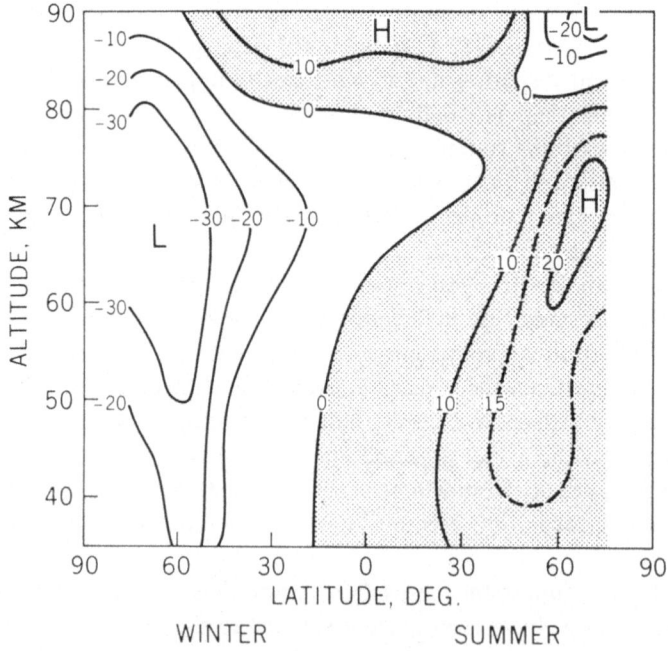

Fig. 2b.   Quasi-meridional cross-section of pressure. Isopleths are in percent deviation from 1962
U.S. Standard Atmosphere.

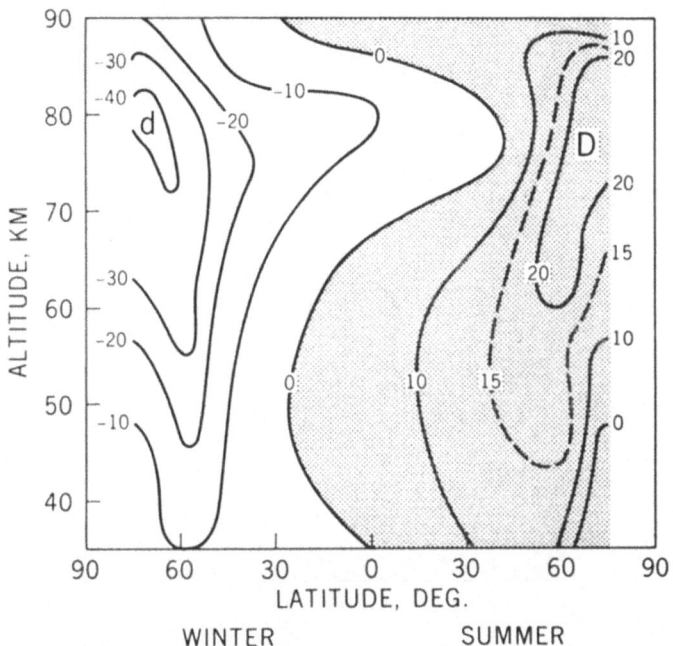

Fig. 2c. Quasi-meridional cross-section of density. Isopleths are in percent deviation from 1962 U.S. Standard Atmosphere.

presents time cross-sections of zonal and meridional components based upon monthly means for each of the four sites considered in this study. Figure 4 then combines the seasonal means for the three North American sites to produce map analyses.

In Figure 3a, the mean zonal winds at Natal are plotted as a function of month. The winds are primarily easterly below 60 km and above 70 km during the months October through February. Maximum easterly wind speeds of 60 mps occur at 40 km in January with 40 mps at 80 km in February. Westerly winds dominate the mesosphere circulation from February to October and maximum westerly wind speeds reach 60 mps in February at 65 km. There is a suggestion of a semiannual variation with the two easterly maxima at 80 km in February and October, and two reversals in the flow at 40–45 km.

In Figure 3b, which gives the time variation of the meridional winds in the mesosphere over Natal, no outstanding pattern exists. The meridional components are generally small and no seasonal effect is obvious. Also, the data are sparse, and though no conclusions can be drawn from this figure, it is included because it is typical.

Figure 3c, which gives the mean zonal winds for Wallops Island, shows a very well developed seasonal fluctuation. The flow is dominated at all levels by westerlies during the months September through April and easterlies during the summer months. The westerlies reach a maximum speed of 100 mps near 70 km in January and the most intense easterlies reach a speed of 60 mps near 70 km in June. The spring and fall

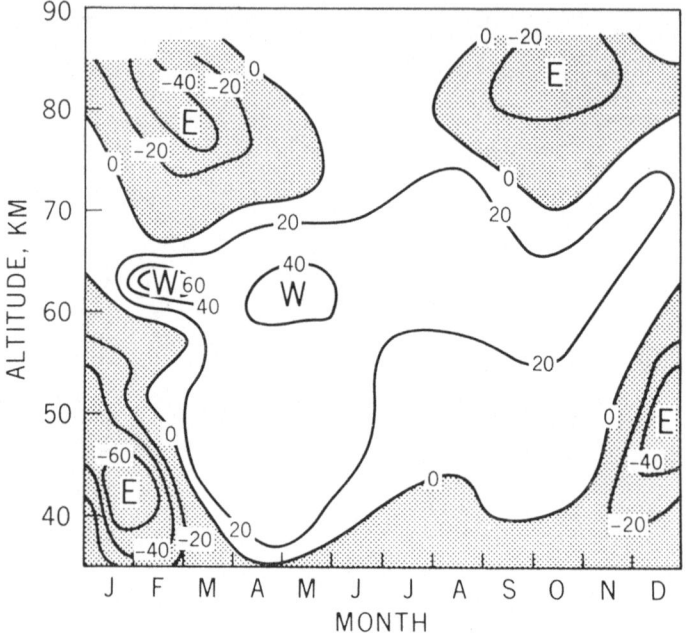

Fig. 3a.   Mean zonal winds above Natal (6°S). Isotachs are in mps.

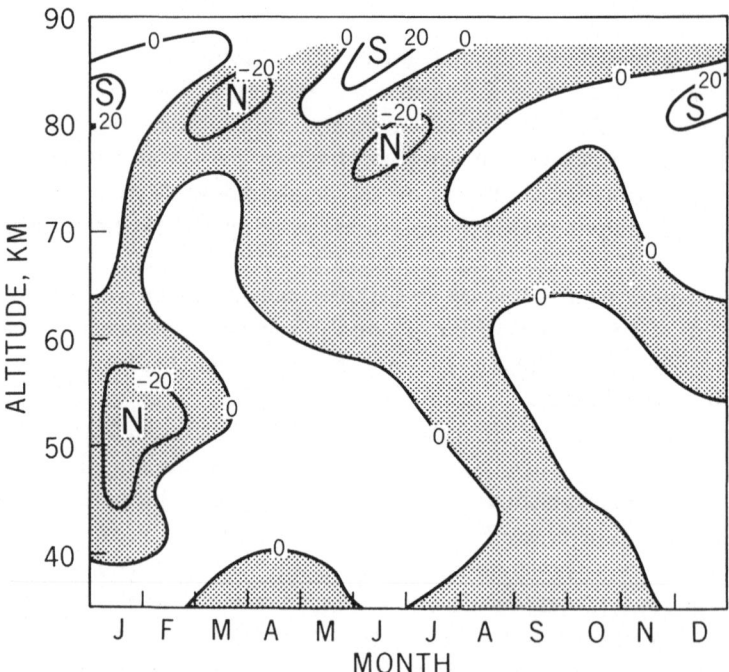

Fig. 3b.   Mean meridional winds above Natal (6°S). Isotachs are in mps.

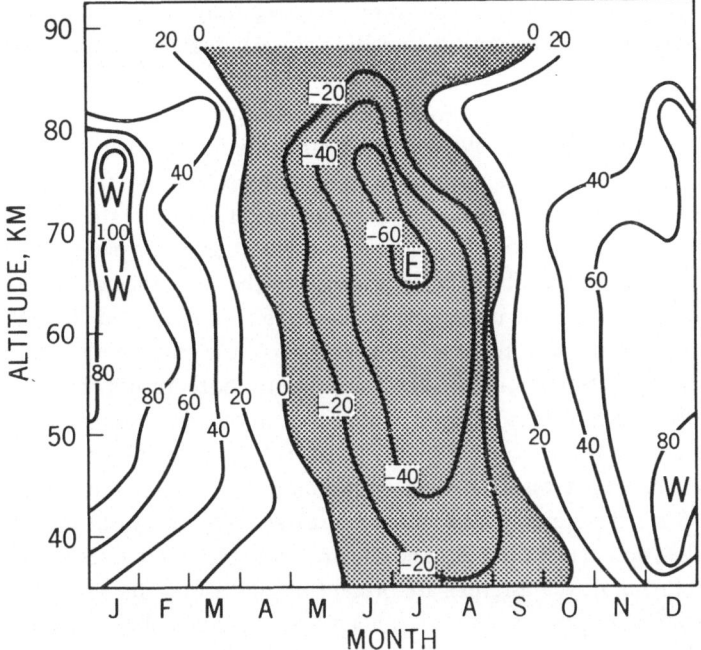

Fig. 3c. Mean zonal winds above Wallops (38°N). Isotachs are in mps.

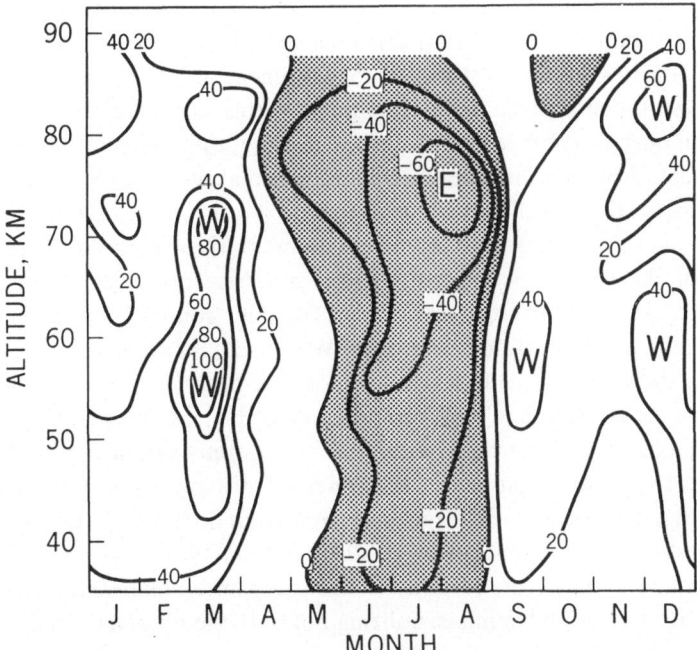

Fig. 3d. Mean zonal winds above Churchill (59°N). Isotachs are in mps.

transitions (i.e., the reversal of wind direction) begin at higher levels and appear to propagate downward. The spring transition is seen first at 85 km in March and it reaches the 35 km level by early June. The fall transition is not quite as orderly, appearing first at about 82 km in August and reaching the 35 km level by mid-October. Since the meridional winds above Wallops Islands display no identifiable pattern, they, as well as the Churchill meridional winds, will be omitted from this discussion.

Figure 3d gives the mean zonal winds above Churchill, where the seasonal pattern is well developed with westerlies prevailing in the winter mesosphere and easterlies in the summer mesosphere. The maximum speed in the westerlies of 100 mps occurs at 55 km in March while the maximum in the easterlies occurs at 75 km in July and August. The westerlies prevail for a larger fraction of the year at Churchill as compared with Wallops. The transitions above Churchill are more abrupt, but they occur first at higher altitudes and descend over a period of about one month.

The mean zonal winds above Barrow, given in Figure 3e, indicate that the seasonal relationship, in which westerly winds dominate the winter circulation and easterly winds dominate the summer flow, does not exist everywhere outside of the tropics. The westerlies do occur in winter, and they reach a maximum speed of about 60 mps near 45 km in January. Easterlies are present in mid winter, however, and the transition to easterlies begins as early as January. The easterly flow reaches a maximum intensity of 40 mps in mid summer in the 70–80 km region, and the easterlies dominate the flow from mid April until late August. The transitions first appear at 72 km in the spring and about 50 km in the fall. The fall transition is very abrupt, occurring at almost every level simultaneously. The easterly core centered near 60 km in December is also a significant departure from the other non-tropical sites. It is the geographic location of Barrow on the west side of the continent which distinguishes it from the other continental sites, since neither the polar vortex in winter nor the polar anticyclone in summer is symmetric with respect to longitude. Herein lies the danger in taking the data from one latitude and longitude and applying it to all longitudes of the same latitude. If the meridional winds over Barrow are analyzed, as is done in Fig. 3f, then their intense development becomes apparent. Northerly components prevail for most of the year, especially below 80 km. These northerly components reach 100 mps at 50 km in December, and except for light southerly winds in March and September, the northerly flow is unbroken. With mean meridional components of this magnitude, there are certainly large transports of heat, momentum, and trace constituents across latitude circles. There are, of course, compensating flows at other longitudes, but an asymmetric meridional circulation certainly contributes to the spatial and temporal variations in the observed distributions of these parameters.

That a seasonally dependent, large scale circulation does exist in the mesosphere has been well established (see Murgatroyd, 1957; Batten, 1961), but usually the data are presented as meridional cross-sections. In this study, mean maps were drawn at 60, 70, and 80 km for winter and summer utilizing not only the observed winds, but also the computed pressure at each site. Whenever possible, appropriate winds from the Meteorological Rocket Network (1968) were included to supplement the available

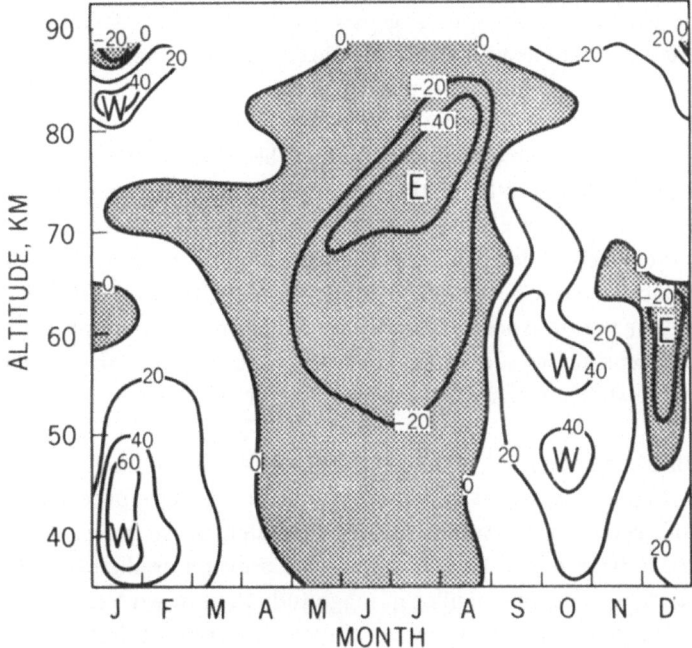

Fig. 3e. Mean zonal winds above Barrow (71 °N). Isotachs are in mps.

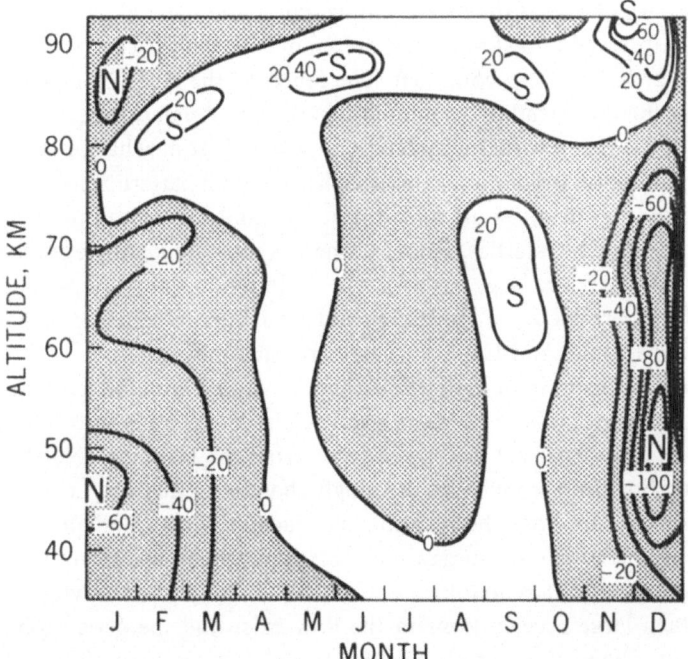

Fig. 3f. Mean meridional winds above Barrow (71 °N). Isotachs are in mps.

data. The winds were assumed to be geostrophic, so that the magnitude and direction of the pressure gradient as well as the pressure itself were combined with the wind for each site to produce the maps given in Figure 4.

The mean winter circulation at 60 km, shown in Figure 4a, indicates the dominance of the polar vortex in the mesospheric regime. The center of the vortex is located over north central Canada, far from the geographic pole indicated by the cross. This circulation produces strong westerly winds over most of the United States and predominantly northerly winds over Alaska. If the analysis is accurate, strong southerly components occur over Greenland. The polar asymmetry of the flow, then, provides a ready means for the transport of atmospheric properties across latitude circles. Note that in north central Canada where the low pressure of the vortex center occurs in winter, the pressure increases by almost a factor of two in summer, as shown in Figure 4b.

The mean summer circulation at 60 km (Figure 4b) is dominated by an anticyclone over the north, and lower pressure to the south. This pattern produces easterly winds over most of the continent. As was the case in winter, the strongest winds occur along the southern portion of the United States. The pressure gradients, and therefore the maximum wind speeds, are smaller in magnitude than in winter.

In Figure 4c, which gives the mean winter circulation at 70 km, the same vortex seen at 60 km persists. Most of the comments that applied at 60 km also apply to this level, so a detailed discussion is not necessary. A difference worth noting is that the maximum winds observed at 70 km are not as intense as those observed at 60 km.

Figure 4d, which shows the mean summer circulation at 70 km, is dominated by the anticyclone observed at 60 km in summer, and as a result, easterly winds prevail over the entire North American continent. The center of the anticyclone cannot be determined from the data available; it could be over the Asian continent.

Figure 4e, the mean winter circulation at 80 km, is somewhat unexpected since the individual synoptic maps do not exhibit such a coherent pattern. The averaging process evidently smooths out the short term variations, and the prevailing drift, in this case the vortex generated westerlies, remains. There is some indication even in this so-called mean flow that there are large variations in the individual soundings. The ridging over eastern Alaska producing a southerly wind over Barrow and the divergent flow between Wallops Island and Cape Kennedy are such indications.

The mean summer circulation at 80 km, shown in Figure 4f, presents a much different picture than was seen at lower levels. There is still a high pressure region to the north of the continent, but ridging occurs across the center of North America in a north-south direction, producing a seemingly chaotic circulation. Low pressure regions extend onshore from both the Atlantic and Pacific Oceans, and the flow is generally light (except Barrow) and disorganized. This picture may result from the fact that the geostrophic assumption is no longer applicable. If tides and/or gravity waves dominate the flow, then large accelerations in the flow exist and the geostrophic assumption breaks down. The map might also be unrealistic since the variability of the individual observations about the mean is very large and, considering the small number of ob-

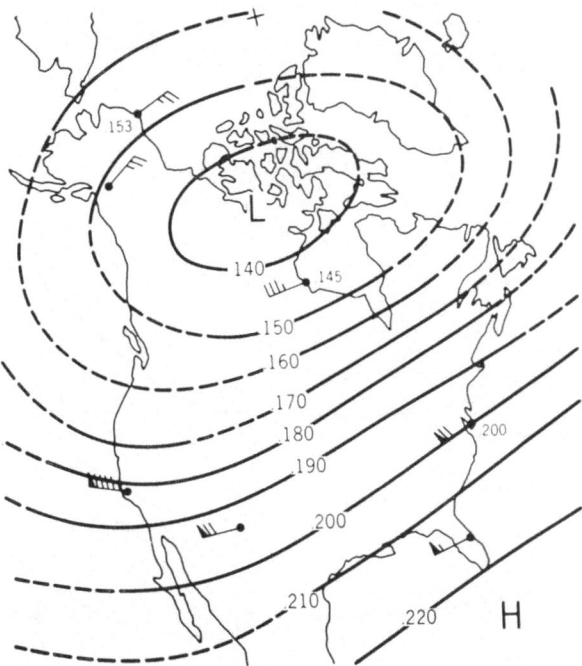

Fig. 4a.   Mean winter circulation at 60 km. Contours are in mb, wind speeds in mps.

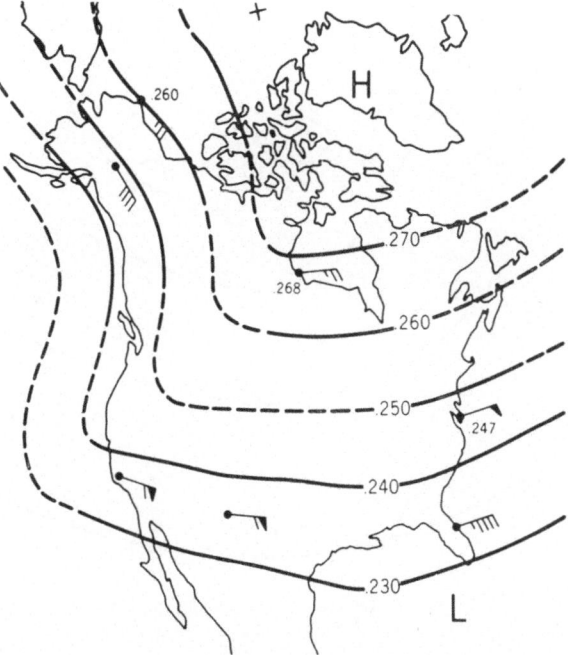

Fig. 4b.   Mean summer circulation at 60 km. Countours are in mb, wind speeds in mps.

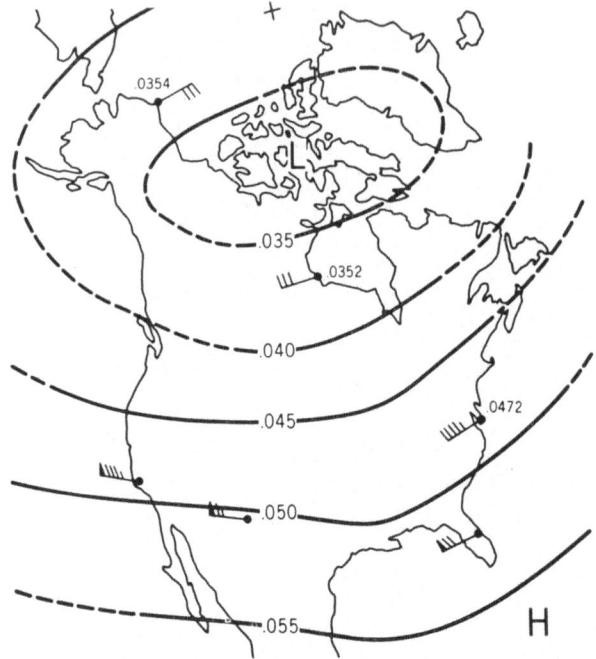

Fig. 4c.   Mean winter circulation at 70 km. Contours are in mb, wind speeds in mps.

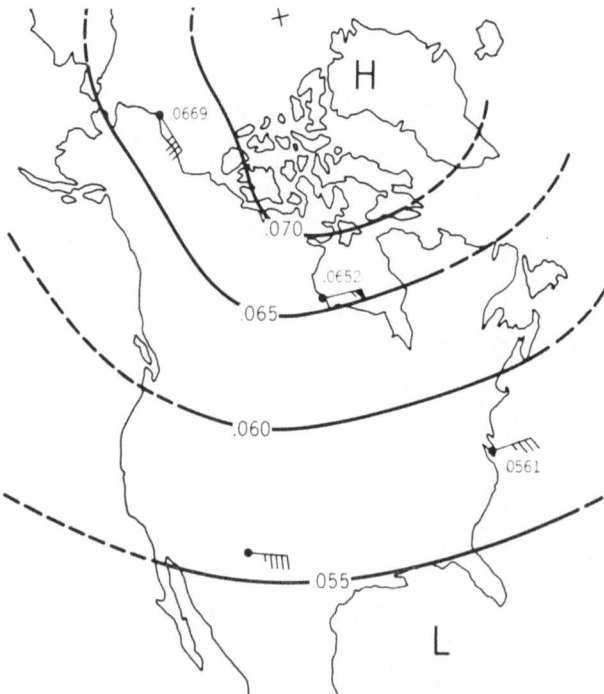

Fig. 4d.   Mean summer circulation at 70 km. Contours are in mb, wind speeds in mps.

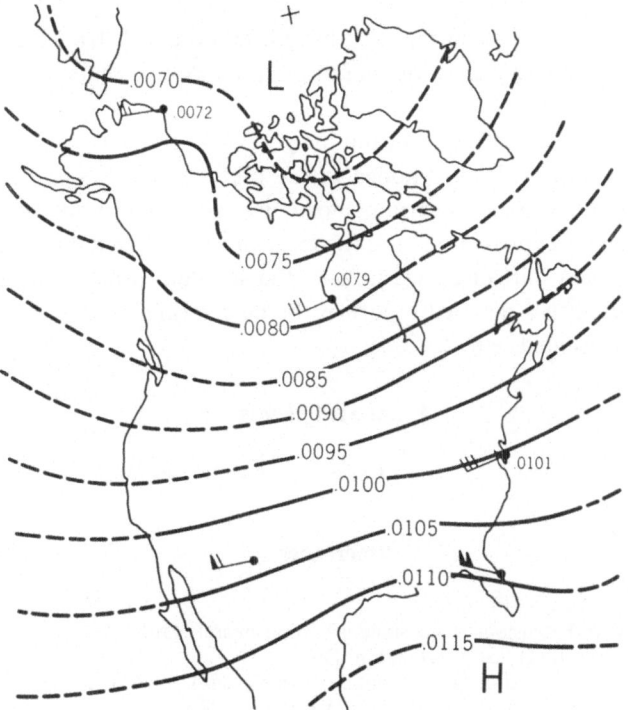

Fig. 4e.   Mean winter circulation at 80 km. Contours are in mb, wind speeds in maps.

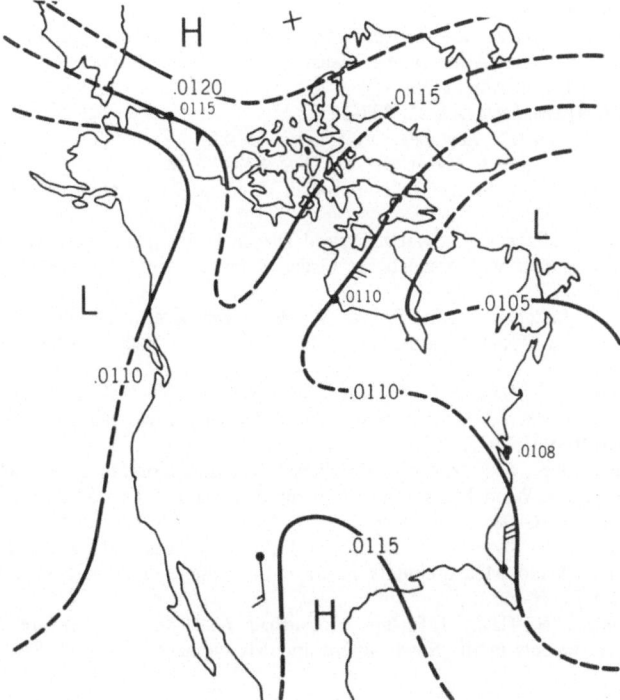

Fig. 4f.   Mean summer circulation at 80 km. Contours are in mb, wind speeds in maps.

servations involved, a true mean may not have been obtained. Nonetheless, the map points out the problems that exist in determining the mean circulation at 80 km and above.

In conclusion, the data observed during the years 1960–68 have been compiled into cross-sections and maps which it is hoped will be useful to workers in the mesosphere. Although these analyses contain certain shortcomings, such as scarcity of data, they are presented as a first approximation to the real atmosphere, and it is hoped that they may point out certain unrealistic features of the standard atmosphere models. The authors plan to publish a detailed report in the near future which will expand upon the information presented here.

Acknowledgments

The authors wish to express their thanks to J. F. Casey and B. R. Kirkwood for their roles in processing the data.

References

Batten, E. S.: 1961, *J. Meteor.* **18**, 283–91.
COESA (Committee on Extension to the Standard Atmosphere): 1962, *U.S. Standard Atmosphere, 1962*, U.S. Govt. Printing Office, Washington, D.C.
COESA (Committee on Extension to the Standard Atmosphere): 1966, *U.S. Standard Atmosphere Supplements, 1966*, U.S. Govt. Printing Office, Washington, D.C.
Committee on Space Research, *Cospar International Reference Atmosphere, 1965*, North-Holland Publishing Co., Amsterdam.
'Data Reports, Meteorological Rocket Network Firings,' World Data Center A, Vol. V, Nos. 1 to 12, January to December, 1968.
Horvath, J. J., Simmons, R. W., and Brace, L. H.: 1962, 'Theory and Implementation of the Pitot-Static Technique for Upper Atmospheric Measurements', Space Physics Res. Laboratory Report NS-1, Univ. of Michigan, Ann Arbor.
Leovy, C.: 1964, *J. Atmospheric Sci.* **21**, 327–341.
Lindzen, R. S.: 1967, *Quart. J. Roy. Meteor. Soc.* **93**, 18–42.
Murgatroyd, R. J.: 1957, *Quart. J. Roy. Meteor. Soc.* **83**, 417–58.
Nordberg, W. and Smith, W. S.: 1964, 'The Rocket Grenade Experiment', NASA Technical Note TND-2107.
Smith, W., Katchen, L., Sacher, P., Swartz, P., and Theon, J.: 1964, 'Temperature, Pressure, Density, and Wind Measurements with the Rocket Grenade Experiment, 1960–1963', NASA Technical Report TR R-211.
Smith, W., Theon, J., Katchen, L., and Swartz, P.: 1966, 'Temperature, Pressure, Density, and Wind Measurements in the Upper Stratosphere and Mesosphere, 1964', NASA Technical Report TR R-245.
Smith, W. S., Theon, J. S., Swartz, P. C., Katchen, L. B., and Horvath, J. J.: 1967, 'Temperature, Pressure, Density and Wind Measurements in the Upper Stratosphere and Mesosphere, 1965', NASA Technical Report TR-R-263.
Smith, W. S., Theon, J. S., Swartz, P. C., Katchen, L. B., and Horvath, J. J.: 1968, 'Temperature, Pressure, Density, and Wind Measurements in the Stratosphere and Mesosphere, 1966', NASA Technical Report TR R-288.
Smith, W. S., Theon, J. S., Swartz, P. C., Casey, J. F., and Horvath, J. J.: 1969, 'Temperature, Pressure, Density and Wind Measurements in the Stratosphere and Mesosphere, 1967', NASA Technical Report TR R-316.
Smith, W. S., Theon, J. S., Casey, J. F., and Horvath, J. J.: 1970, 'Temperature, Pressure, Density and Wind Measurements in the Stratosphere and Mesosphere, 1968', NASA Technical Report TR R-340.

NOCTURNAL VARIATIONS IN THE OH (8–3) BAND AIRGLOW

F. CONGEDUTI and G. FIOCCO

European Space Research Institute, Frascati, Italy

and

G. VISCONTI

Università de L'Aquila, Italy

Summary

Measurements have been carried out with a tilting interference filter photometer. The filter is slowly tilted to vary the wavelength of peak transmission from 7403 Å to 7300 Å and explore the OH (8.3) rotation vibration band. The filtered light is detected by a photon counting system. Compared to other scanning spectrophotometers this is a compact instrument with a high throughput. For the reduction of the data, synthetic spectra were constructed by considering the transfer function of the instrument and assuming that the lines within the same band were in thermal equilibrium. An isothermal atmosphere was assumed at first. The contribution of all lines of the (8.3) OH band was considered and the excitation temperature was varied in steps of 20 from 120 to 400 K.

The ratio between the P_1 and P_2 branch was fixed assuming that the doublet temperature coincided with the rotational temperature. A fit was then carried out between the synthetic spectra and the real spectra using the χ^2 criterion. The instrument was operated at two locations near L'Aquila (42° lat., 23° long.): at Campo Imperatore (altitude 2200 m) from July 1968 until August 1969, and at Preturo (altitude 700 m) for the period August–December 1969. Night variations of intensity and temperature were studied at intervals of one hour. For each night, the difference in temperature at successive hourly intervals showed, on the average, a decrease of about 20° over a period of about 8 hours. The hourly variation of the intensity of all the P lines relative to the minimum recorded in each night was also studied.

Our intensity data are in reasonable agreement with the diurnal intensity variation given by Hesstvedt (1970) and by Shimazaki and Laird (1970) and can be explained with the change in the concentration of H and O_3, species entering in the formation of OH* according to the reaction $H + O_3 \rightarrow OH^* + O_2$.

A discussion of the temperature variation should take into account the formation process and the mechanisms for populating the levels. A change in rotational temperature may reflect changes in the excitation process and in the height where the process takes place, as well as changes in ambient kinetic temperature.

To study the effect of the night variation of the emission profiles on the temperature obtained, we have constructed synthetic spectra by taking into account a vertical temperature profile (Theon and Smith, this symposium) for the fall and by weighting these data with the emission profiles predicted by various theoretical models for the OH emissions. A variation of about 10 K between sunset and midnight is obtained for the model of Shimazaki and Laird (1970), while the model of Hesstvedt gives a negligible change. In fact, in this model the change in intensity from sunset to midnight is accompanied by a change in the bottom height of the layer of only 1 km approximately.

A more detailed account of aspects of this research is given by Visconti *et al.* (1971); see also Fiocco *et al.* (1970).

References

Fiocco, G., Visconti, G., and Congeduti, F.: 1970, *Nature* **228**, 1079–80.
Hesstvedt, A.: 1970, *J. Geophys. Res.* **75**, 2337–39.
Shimazaki, T. and Laird, A. R.: 1970, *J. Geophys. Res.* **75**, 3221–35.
Visconti, G., Congeduti, F., and Fiocco, G.: 1971, in B. McCormac and A. Vallance Jones (eds.), 'The Radiating Atmosphere', D. Reidel, Dordrecht.

ULTRAVIOLET SOLAR RADIATION RELATED
TO MESOSPHERIC PROCESSES

M. ACKERMAN

Institut d'Aéronomie Spatiale de Belgique, Brussels, Belgium

1. Introduction

The atmospheric layer of strong negative temperature gradient called the mesosphere and located at altitudes between roughly 50 and 85 km is still part of the homosphere since the mean molecular mass is there maintained constant at all altitudes by the mixing of the major constituents. However, the chemical reactions are very efficient in this region and strongly affect the concentration of minor constituents. These processes are of course not strictly limited in altitude and continue to play a role in the neighbouring atmospheric regions, namely the stratosphere below and the thermosphere above, by transport processes.

These interferences lead to some difficulties in defining the solar ultraviolet radiation of interest to the mesosphere where the solar photons of sufficient energy to initiate chemical chain reactions, penetrate with different degrees of attenuation. The two main absorbing species involved in this attenuation are molecular oxygen and ozone of which the photodissociation produces atomic oxygen, providing the main oxidizing agent in the chemosphere. Typical total numbers of molecules above 50 and 85 km are respectively 4×10^{21} cm^{-2} and 2×10^{19} cm^{-2} for O_2 and 10^{17} cm^{-2} and 10^{13} cm^{-2} for O_3. Since maximum absorption cross sections are of the order of 10^{-17} cm^2, oxygen and to a lesser degree ozone will control the penetration of solar ultraviolet radiation in the mesosphere.

The data actually available on the solar ultraviolet flux from 1100 Å to 3500 Å will be discussed here. The short wavelength limit will allow inclusion of the Lyman α radiation which plays a role in the photodissociation of O_2, H_2O, CO_2, etc., and which constitutes the bridge between the chemosphere and the ionosphere owing to its important role in the formation of the D region by ionizing nitric oxide. The long wavelength limit has been fixed at the beginning of the ultraviolet absorption of ozone in the Huggins bands.

On the other hand, to allow the evaluation of the number of photons available at different altitudes, absorption cross section data on O_2 and O_3 will be discussed.

Finally, adopted values for the three parameters will be listed for fixed wavelength intervals suitable for photochemical computations.

2. Absorption Cross Section of Ozone

Between 3500 Å and 2000 Å, ozone absorbs in a wide band discovered by Hartley and peaking near 2500 Å. At this wavelength, unit optical depth for a typical ozone

Fiocco (ed.), Mesospheric Models and Related Experiments, 149–159. All Rights Reserved.
Copyright © 1971 by D. Reidel Publishing Company, Dordrecht-Holland.

distribution takes place at the lower boundary of the mesosphere for an overhead Sun and at 70 km for a solar zenith distance of 90°. This indicates the important mesospheric role of the Hartley band of ozone which has a bell shape presenting a diffuse structure corresponding to a complicated set of diffuse bands. On the long wavelength side, between 3100 Å and 3500 Å, the structure appears more clearly and was discovered by Huggins in the spectrum of Sirius near the horizon.

The photodissociation of ozone may lead to molecular and atomic oxygen in various excited states, as was already pointed out more than 30 years ago by Nicolet (1939). The absorption cross section of ozone has been well known since the work of Vigroux (1953), Watanabe *et al.* (1953) and Inn and Tanaka (1953). A slight disagreement between the values given by Vigroux (1953) and Inn and Tanaka (1953) at wavelengths smaller than 2700 Å prompted the measurements of Hearn (1961). These have been confirmed by Griggs (1968). New measurements of Vigroux (1969) have definitely shown the correctness of the data of Inn and Tanaka (1953). Various sets of data are represented in Figure 1. Below 2000 Å, only one set of data exists to our knowledge, that of Watanabe *et al.* (1953). Below 1800 Å ozone has so far not been shown to play a role, either in the photochemistry or in the transparency of the Earth atmosphere,

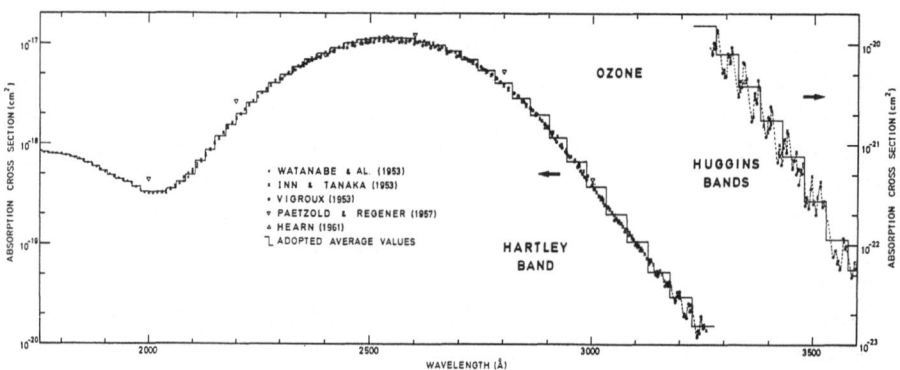

Fig. 1.   Absorption cross section of ozone vs wavelength in the Hartley and Huggins bands.

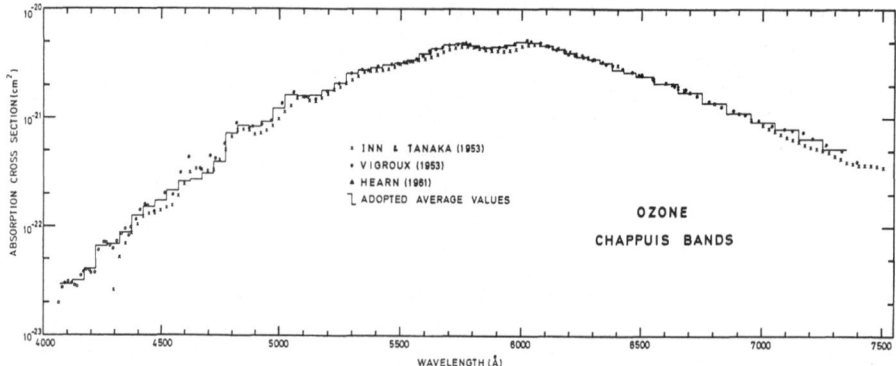

Fig. 2.   Absorption cross section of ozone vs wavelength in the Chappuis bands.

since oxygen is there the main absorbant and since the solar ultraviolet intensity is so small compared to that in the 2000 to 3500 Å region.

For the sake of completeness, absorption cross section values in the Chappuis bands are represented in Figure 2.

It appears that absorption cross sections of ozone are now very well known. However, the question about the quantum yield of the various photodissociation processes is not yet fully answered.

3. Absorption Cross Section of Molecular Oxygen

Many data exist on the absorption cross section of molecular oxygen. A critical analysis is necessary before adopting useful aeronomic values. This was done several times in the past, by Nicolet and Mange (1954) for instance. Many new measurements have been made in the last few years which justify a new analysis. This will be made for successive wavelength intervals.

1160 Å–1370 Å. The absorption by molecular oxygen changes rapidly with wavelength in this region, presenting optical windows that allow the penetration of the solar radiation in the mesosphere. Absorption cross sections have for instance been obtained by Watanabe *et al.* (1953, 1958). One optical window which is particularly well known for its fundamental importance in mesospheric problems allows the deep penetration of the solar Lyman α radiation. Values of absorption cross section of O_2 at 1215.67 Å have been obtained by various authors (Preston, 1940; Watanabe *et al.*, 1953; Ditchburn *et al.*, 1954; Lee, 1955; Watanabe *et al.*, 1958; Metzger and Cook, 1964; Ogawa, 1968; and Gailly, 1969) and range from 8.5×10^{-21} to 1.04×10^{-20} cm². This leads to unit optical atmospheric depth at about 75 km altitude for an overhead sun. The Lyman α line has been shown to be slightly on the short wavelength wing of the optical window and pressure effects on the absorption have not yet been satisfactorily interpreted.

1370 Å–1750 Å. This region is characterized by the Schumann-Runge continuum which presents high values of the absorption cross section measured by various authors (Landenburg and Van Voorhis, 1933; Schneider, 1940; Ditchburn and Heddle, 1953; Watanabe and Marmo, 1956; Huffman *et al.*, 1964; Metzger and Cook 1964; Hudson *et al.*, 1966) as shown in Figure 3, leading to a complete extinction of the solar radiation in the thermosphere.

1750 Å–2431 Å. Many absorption cross section values discussed recently by Ackerman *et al.* (1970) have been obtained since the beginning of this century for the Schumann-Runge bands of O_2 which extend from 1750 Å to 2000 Å. While they indicate the important role that the bands can play in the mesospheric photochemistry, they are not very valuable for precise aeronomical evaluations.

This situation has recently been considerably improved. New laboratory measure-

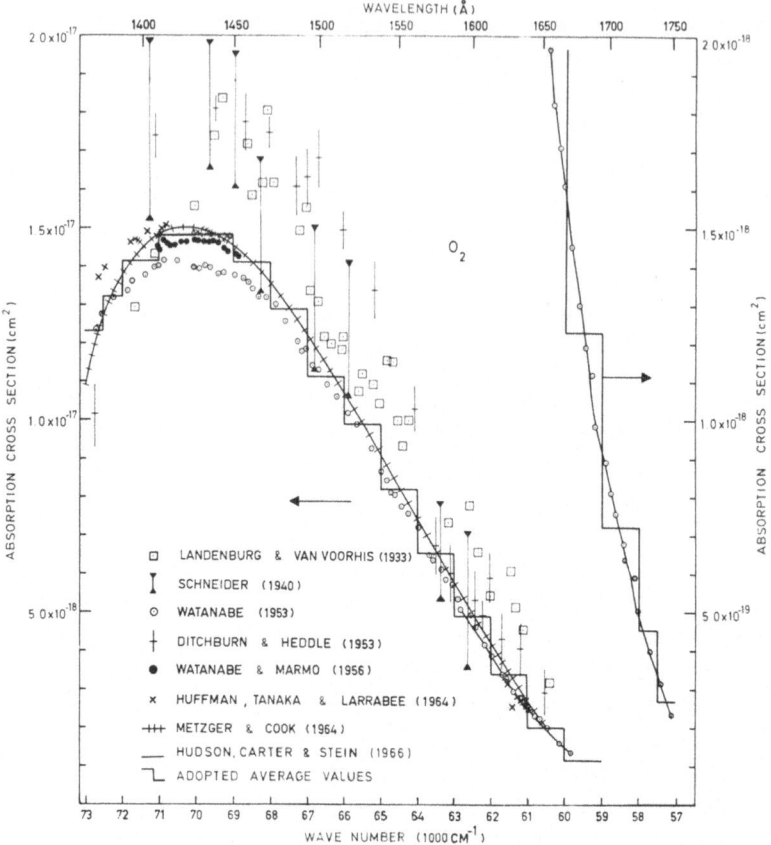

Fig. 3.   Absorption cross section of molecular oxygen vs wavelength and versus wavenumber in the
Schumann continuum.

ments of absorption properties by Ackerman *et al.* (1969), and of the structure of the
bands by Ackerman and Biaumé (1970), have led to a detailed evaluation of the absorption cross section taking temperature effects into account (Ackerman *et al.*, 1970).
Applications of these new data are presented by Kockarts (1971), this volume, p. 160.

Above 2000 Å the Herzberg continuum extends up to 2439 Å. Its absorption cross
section ranges from about 10^{-23} to 10^{-24} cm². Such low values imply that in this
wavelength region, molecular oxygen influences the atmospheric transparency only
in the stratosphere at heights around 25 km. In fact the strong absorption of ozone
limits the role of oxygen to the narrow optical window of 2000 Å which is essential in
the process of stratospheric ozone formation.

4. Solar Flux

From the first spectroscopic measurements of the solar ultraviolet radiation the conclusion was drawn that the considerable intensity decay taking place at 3000 Å towards shorter wavelengths was due to solar absorption itself. This phenomenon was attributed to the absorption by ozone after the presence of this atmospheric constituent was firmly established by Fowler and Lord Rayleigh in 1917, and after ground-based study of its absorption of the solar radiation at various zenith distances was performed by Fabry and Buisson in 1921.

At shorter wavelengths, rocket-borne experiments have brought information on the solar ultraviolet radiation since 1946. These data have already been reviewed several times (Hinteregger, 1965). In the wavelength range here considered the available values of solar radiation intensities were until recently those published in 1961 by Detwiler *et al.* (1961). In the range from 2000 Å to 1400 Å, these authors believed that the accuracy of their data was better than ±20% and were fairly sure that there were no errors greater than a factor of 1.5, whereas below 1300 Å errors as great as a factor of two or more were possible.

Using also photographic detection, Bonnet (1968) has obtained values of the solar flux from about 2000 Å to 3000 Å by integrating the flux measured at different locations of the solar disk using a rocket-borne instrument. His data indicated a much more important discontinuity, centered at 2085 Å, than had been previously measured. However, his values were practically in agreement with the previous ones outside of the discontinuity.

In 1968 also, Ackerman *et al.* (1968) reported observations made during the first flight of a balloon-borne photoelectric monochromator viewing the whole solar disk which indicated an even larger amplitude for the discontinuity. Their results showed also that the solar ultraviolet intensity below 2000 Å had been previously overestimated. This has since been confirmed by subsequent flights which are described elsewhere, one of them being discussed in detail by Frimout (1969).

Measurements in the solar spectrum between 1400 and 1875 Å with a photoelectric rocket-borne spectrometer have been published in 1969 by Parkinson and Reeves (1969) and correspond to a solar black body temperature of the order of 4500 K or less, namely much lower than the earlier values given by Detwiler *et al.* (1961).

This year finally, Widing *et al.* (1970), analysing rocket spectra photographed in 1966, have deduced new flux values which are situated as shown in Figure 4, between the two extremes. They are in agreement with the data of Ackerman *et al.* (1968) at 1950 Å.

5. Discussion and Conclusion

The solar ultraviolet of interest to the mesosphere may now be defined more easily. It appears to cover the whole wavelength interval that we have examined after excluding the range covered by the Schumann continuum. Below this feature, considering the absolute intensity, only the Lyman α line plays an important role. Since the sun is radiating more and more strongly towards longer wavelengths above 1200 Å, the solar photons have an increasingly important effect in the mesosphere. Below 2000 Å, the radiation has very much the character of a continuum, while above, it presents a

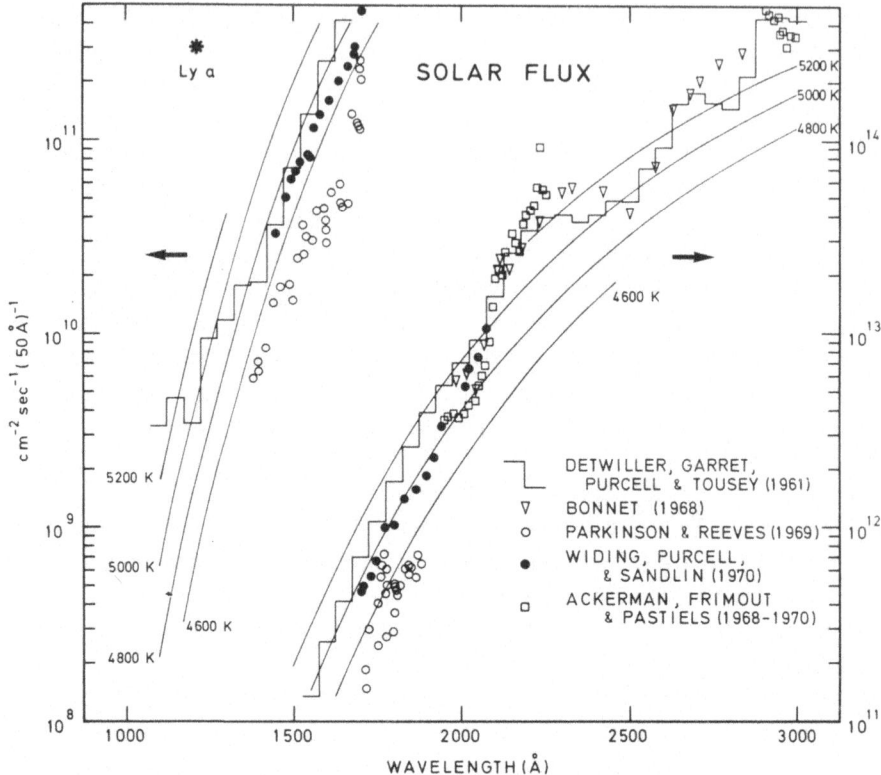

Fig. 4.   Flux of solar photons at one AU vs wavelength from 1000 Å to 3000 Å.

very complicated structure. In some applications intensity data at high resolution would be very useful.

Until two or three years ago the situation was rather comfortable from the aeronomical point of view since only one set of data existed. All authors referred to the work of Detwiler *et al.* (1961). New data have since been obtained showing that these authors were a little too optimistic about the accuracy of their measurements. The new values necessarily imply making a choice. This choice has to be precisely specified to allow comparison of the results obtained in photochemical evaluations, for instance. One set of adopted values is presented in Table I with values of absorption cross sections of O_2 and O_3. The adopted solar ultraviolet flux is based on the data given by Detwiler *et al.* (1961) from 1163 to 1428 Å, by Widing *et al.* (1970) from 1423 Å to 1942 Å, by Ackerman *et al.* (1968) and by Frimout (1969) from 1942 Å to 2299 Å, by Detwiler *et al.* (1961) and by Tousey (1963) from 2299 Å to 3625 Å, and by Johnson (1954) at longer wavelength. This is of course not a final choice and new data are now required.

The question of variability with time arises more strongly than ever. The dispersion of the experimental results shows at least that experimental techniques have to be

TABLE I

Flux of solar photons, q, at one AU, absorption cross section of O_2 and of O_3, $\sigma(O_2)$ and $\sigma(O_3)$, for wavelength intervals $\Delta\lambda$ and wavenumber intervals $\Delta\nu$ from Ly-α to 7300 Å

No.	$\Delta\lambda$(Å)	$\Delta\nu$(cm^{-1})	q(cm^{-2} s^{-1})	$\sigma(O_2)$ (cm^2)	$\sigma(O_3)$ (cm^2)
1	Ly α 1.215,67	82.259	3.00×10^{11}	1.00×10^{-20}	2.32×10^{-17}
2	1.170–1.163	85.500–86.000	1.03×10^8	2.00×10^{-20}	7.80×10^{-18}
3	1.176–1.170	85.000–85.500	2.66	1.25×10^{-18}	7.97
4	1.183–1.176	84.500–85.000	1.12	2.55×10^{-19}	8.66
5	1.190–1.183	84.000–84.500	1.24	3.00×10^{-20}	9.51
6	1.198–1.190	83.500–84.000	1.82	3.75×10^{-19}	1.25×10^{-17}
7	1.205–1.198	83.000–83.500	1.90	4.45×10^{-18}	1.84
8	1.212–1.205	82.500–83.000	7.40	8.35	2.19
9	1.220–1.212	82.000–82.500	2.28×10^9	6.00×10^{-19}	2.30
10	1.227–1.220	81.500–82.000	3.67	2.35	2.26
11	1.235–1.227	81.000–81.500	1.36	4.50	2.06
12	1.242–1.235	80.500–81.000	1.61	3.35	1.30
13	1.250–1.242	80.000–80.500	1.32	1.75×10^{-17}	8.91×10^{-18}
14	1.258–1.250	79.500–80.000	1.41	8.95×10^{-19}	7.24
15	1.266–1.258	79.000–79.500	3.11	4.30	6.09
16	1.274–1.266	78.500–79.000	1.06	1.10	5.66
17	1.282–1.274	78.000–78.500	1.37	2.05	5.87
18	1.290–1.282	77.500–78.000	1.02	4.43	6.47
19	1.299–1.290	77.000–77.500	1.14	5.55	8.14
20	1.307–1.299	76.500–77.000	7.29	4.20	1.24×10^{-17}
21	1.316–1.307	76.000–76.500	2.20	6.85	1.52
22	1.324–1.316	75.500–76.000	1.59	1.45×10^{-18}	1.47
23	1.333–1.324	75.000–75.500	2.21	2.25	1.51
24	1.342–1.333	74.500–75.000	1.24×10^{10}	2.30×10^{-18}	1.51×10^{-17}
25	1.351–1.342	74.000–74.500	1.99×10^9	4.55	1.65
26	1.360–1.351	73.500–74.000	3.09	7.23	1.54
27	1.370–1.360	73.000–73.500	2.57	9.50	1.35
28	1.379–1.370	72.500–73.000	2.74	1.23×10^{-17}	1.05
29	1.389–1.379	72.000–72.500	3.10	1.32	7.97×10^{-18}
30	1.408–1.389	71.000–72.000	7.60	1.36	7.17
31	1.428–1.408	70.000–71.000	1.01×10^{10}	1.40	6.28
32	1.449–1.428	69.000–70.000	1.30	1.48	5.66
33	1.470–1.449	68.000–69.000	1.82	1.41	5.23
34	1.492–1.470	67.000–68.000	2.33	1.29	4.47
35	1.515–1.492	66.000–67.000	2.66	1.15	3.69
36	1.538–1.515	65.000–66.000	2.90	9.91×10^{-18}	2.93
37	1.562–1.538	64.000–65.000	3.60	8.24	2.19
38	1.587–1.562	63.000–64.000	4.75	6.58	1.63
39	1.613–1.587	62.000–63.000	6.40	4.97	1.20
40	1.639–1.613	61.000–62.000	5.49	3.45	9.77×10^{-19}
41	1.667–1.639	60.000–61.000	1.19×10^{11}	2.08	8.66
42	1.695–1.667	59.000–60.000	1.76	1.23	8.14
43	1.724–1.695	58.000–59.000	2.32	7.22×10^{-19}	8.17
44	1.739–1.724	57.500–58.000	1.44	4.58	8.57
45	1.754–1.739	57.000–57.500	1.83	2.74	8.40
46	1.770–1.754	56.500–57.000	2.34	a	8.11
47	1.786–1.770	56.000–56.500	2.62	a	7.99
48	1.802–1.786	55.500–56.000	2.88×10^{11}	a	7.86×10^{-19}
49	1.818–1.802	55.000–55.500	3.14	a	7.63

a See Ackerman et al. (1970).

Table I (Continued)

No.	$\Delta\lambda$(Å)	$\Delta\nu$(cm^{-1})	q(cm^{-2} s^{-1})	$\sigma(O_2)$ (cm^2)	$\sigma(O_3)$ (cm^2)
50	1.835–1.818	54.500–55.000	3.81×10^{11}	a	7.29×10^{-19}
51	1.852–1.835	54.000–54.500	4.43	a	6.88
52	1.869–1.852	53.500–54.000	4.95	a	6.40
53	1.887–1.869	53.000–53.500	5.94	a	5.88
54	1.905–1.887	52.500–53.000	6.59	a	5.31
55	1.923–1.905	52.000–52.500	7.26	a	4.80
56	1.942–1.923	51.500–52.000	9.85	a	4.38
57	1.961–1.942	51.000–51.500	1.27×10^{12}	a	4.11
58	1.980–1.961	50.500–51.000	1.39	a	3.69
59	2.000–1.980	50.000–50.500	1.53	a	3.30
60	2.020–2.000	49.500–50.000	1.60	a	3.26
61	2.041–2.020	49.000–49.500	1.74	1.14×10^{-23}	3.26
62	2.062–2.041	48.500–49.000	2.31	1.05	3.51
63	2.083–2.062	48.000–48.500	4.20	1.00	4.11
64	2.105–2.083	47.500–48.000	7.30	9.55×10^{-24}	4.84
65	2.128–2.105	47.000–47.500	9.42	8.93	6.26
66	2.150–2.128	46.500–47.000	1.06×10^{13}	8.28	8.57
67	2.174–2.150	46.000–46.500	1.34	7.60	1.17×10^{-18}
68	2.198–2.174	45.500–46.000	1.32	6.92	1.52
69	2.222–2.198	45.000–45.500	1.73	6.28	1.97
70	2.247–2.222	44.500–45.000	1.80	5.65	2.55
71	2.273–2.247	44.000–44.500	1.82	5.03	3.24
72	2.299–2.273	43.500–44.000	2.26	4.40	4.00
73	2.326–2.299	43.000–43.500	2.40	3.76	4.83
74	2.353–2.326	42.500–43.000	2.25	3.09	5.79
75	2.381–2.353	42.000–42.500	2.21	2.44	6.86
76	2.410–2.381	41.500–42.000	2.32	1.75	7.97
77	2.439–2.410	41.000–41.500	2.50	6.74×10^{-25}	9.00
78	2.469–2.439	40.500–41.000	2.73		1.00×10^{-17}
79	2.500–2.469	40.000–40.500	2.88		1.07
80	2.532–2.500	39.500–40.000	3.02		1.11
81	2.564–2.532	39.000–39.500	3.97		1.12
82	2.597–2.564	38.500–39.000	7.13		1.11
83	2.632–2.597	38.000–38.500	4.37		1.03
84	2.667–2.632	37.500–38.000	1.12×10^{14}		9.43×10^{-18}
85	2.703–2.667	37.000–37.500	1.25		8.23
86	2.740–2.703	36.500–37.000	1.16		6.81
87	2.778–2.740	36.000–36.500	1.19		5.31
88	2.817–2.778	35.500–36.000	1.38		3.99
89	2.857–2.817	35.000–35.500	1.70		2.84
90	2.899–2.857	34.500–35.000	2.46		1.92
91	2.941–2.899	34.000–34.500	3.90		1.14
92	2.985–2.941	33.500–34.000	3.99		6.60×10^{-19}
93	3.030–2.985	33.000–33.500	3.86		3.69
94	3.077–3.030	32.500–33.000	5.08		1.97
95	3.100 ($\pm$25)	32.520–32.000	5.92		1.05
96	3.150	32.000–31.496	6.05		5.23×10^{-20}
97	3.200	31.496–31.008	6.94		2.91
98	3.250	31.008–30.534	8.12		1.50
99	3.300	30.534–30.075	9.71		7.78×10^{-21}
100	3.350	30.075–29.630	8.97		3.72

a See Ackerman *et al.* (1970).

Table I (Continued)

No.	$\Delta\lambda(\text{Å})$	$\Delta\nu(\text{cm}^{-1})$	$q(\text{cm}^{-2}\,\text{s}^{-1})$	$\sigma(O_2)\ (\text{cm}^2)$	$\sigma(O_3)\ (\text{cm}^2)$
101	3.400 ($\pm$ 25)	29.630–29.197	9.44×10^{14}		1.71×10^{-21}
102	3.450	29.197–28.777	1.01×10^{15}		7.46×10^{-22}
103	3.500	28.777–28.369	1.03		2.66
104	3.550	28.369–27.972	1.03		1.09
105	3.600	27.972–27.586	1.04		5.49×10^{-23}
106	3.650	27.586–27.211	1.18		–
107	3.700	27.211–26.846	1.23		–
108	3.750	26.846–26.490	1.24		–
109	3.800	26.490–26.144	1.17		–
110	3.850	26.144–25.806	1.11		–
111	3.900	25.806–25.478	1.09		–
112	3.950	25.478–25.157	1.19		–
113	4.000	25.157–24.845	1.54		–
114	4.050	24.845–24.540	1.90		–
115	4.100	24.540–24.242	1.99		2.91
116	4.150	24.242–23.952	1.99		3.14
117	4.200	23.952–23.669	2.02		3.99
118	4.250	23.669–23.392	2.01		6.54
119	4.300	23.392–23.121	1.94		6.83
120	4.350	23.121–22.851	1.98		8.66
121	4.400	22.851–22.599	2.25		1.25×10^{-22}
122	4.450	22.599–22.346	2.39		1.49
123	4.500	22.346–22.099	2.48		1.71
124	4.550	22.099–21.858	2.49		2.12
125	4.600	21.858–21.622	2.48		3.57
126	4.650	21.622–21.390	2.50		3.68
127	4.700	21.390–21.164	2.55		4.06
128	4.750	21.164–20.942	2.61		4.89
129	4.800	20.942–20.725	2.59		7.11
130	4.850	20.725–20.513	2.46		8.43
131	4.900	20.513–20.504	2.44		8.28
132	4.950	20.504–20.100	2.53		9.09
133	5.000	20.100–19.900	2.48		1.22×10^{-21}
134	5.050	19.900–19.704	2.49		1.62
135	5.100	19.704–19.512	2.50		1.58
136	5.150	19.512–19.324	2.43		1.60
137	5.200	19.324–19.139	2.43		1.78
138	5.250	19.139–18.957	2.52		2.07
139	5.300	18.957–18.779	2.58		2.55
140	5.350	18.779–18.605	2.64		2.74
141	5.400	18.605–18.433	2.67		2.88
142	5.450	18.433–18.265	2.70		3.07
143	5.500	18.265–18.100	2.68		3.17
144	5.550	18.100–17.937	2.66		3.36
145	5.600	17.937–17.778	2.66		3.88
146	5.650	17.778–17.621	2.67		4.31
147	5.700	17.621–17.467	2.67		4.67
148	5.750	17.467–17.316	2.69		4.75
149	5.800	17.316–17.667	2.71		4.55
150	5.850	17.667–17.021	2.71		4.35
151	5.900	17.021–16.878	2.71		4.42
152	5.950	16.878–16.736	2.72		4.61

Table I (Continued)

No.	$\Delta\lambda(\text{Å})$	$\Delta\nu(\text{cm}^{-1})$	$q(\text{cm}^{-2}\,\text{s}^{-1})$	$\sigma(\text{O}_2)\ (\text{cm}^2)$	$\sigma(\text{O}_3)\ (\text{cm}^2)$
153	6.000 ($\pm$ 25)	16.736–16.598	2.72×10^{15}		4.89×10^{-21}
154	6.050	16.598–16.461	2.71		4.84
155	6.100	16.461–16.326	2.70		4.54
156	6.150	16.326–16.194	2.70		4.24
157	6.200	16.194–16.064	2.70		3.90
158	6.250	16.064–15.936	2.69		3.60
159	6.300	15.936–15.810	2.68		3.43
160	6.350	15.810–15.686	2.67		3.17
161	6.400	15.686–15.564	2.66		2.74
162	6.450	15.564–15.444	2.65		2.61
163	6.500 ($\pm$ 50)	15.384–15.265	3.95		2.40
164	6.600	15.265–15.038	5.22		2.07
165	6.700	15.038–14.815	5.18		1.72
166	6.800	14.815–14.598	5.14		1.37
167	6.900	14.598–14.388	5.09		1.11
168	7.000	14.388–14.184	5.04		9.13×10^{-22}
169	7.100	14.184–13.986	4.99		7.93
170	7.200	13.986–13.793	4.94		6.40
171	7.300	13.793–13.605	4.90		5.14

considerably improved before an answer can be given. For Lyman α, a rather important variability may be taken as certain.

Finally, if ozone and oxygen have up to now been considered as the absorbers of the ultraviolet radiation related to the mesosphere, one has still to be cautious since other absorbers might be involved. Indications of such effects have been recently given in the 1300 Å region by Reid and Withbroe (1970) and in the 2000 Å region by Ackerman *et al.* (1968).

References

Ackerman, M. and Biaume, F.: 1970, *J. Mol. Spectr.* **35**, 73–82.
Ackerman, M., Biaume, F., and Kockarts, G.: 1970, 'Absorption Cross Sections of the Schumann-Runge Bands of Molecular Oxygen', *Planetary Space Sci.*, **18**, 1639–51.
Ackerman, M., Biaume, F., and Nicolet, M.: 1969, *Can. J. Chem.* **47**, 1834–40.
Ackerman, M., Frimout, D., and Pastiels, R.: 1968, *Ciel et Terre* **84**, 408–19.
Bonnet, R. M.: 1968, in *Space Res.* **7**, 458–72.
Detwiler, C. R., Garrett, D. L., Purcell, J. D., and Tousey, R.: 1961, *Ann. Geophys.* **17**, 9–18.
Ditchburn, R. W., Bradley, J. E. S., and Cannon, C. G., Munday, G.: 1954, in R. L. Boyd and M. J. Seaton (eds.), *Rocket Exploration of the Upper Atmosphere*, Pergamon Press, London, 327–34.
Ditchburn, R. W. and Heddle, D. W. O.: 1953, *Proc. Roy. Soc.* **A220**, 61–70.
Fabry, C. and Buisson, H.: 1921, *J. Phys. Radium* II S VI, 197–226.
Fowler, A. and Strutt, R. J.: *Proc. Roy. Soc.* **A93**, 577–86.
Frimout, D.: Thesis Rijksuniversiteit, Gent, Academiejaar 1969–1970.
Gailly, T. D.: 1969, *J. Opt. Soc. Amer.* **59**, 536–8.
Griggs, M.: 1968, *J. Chem. Phys.* **49**, 857–9.
Hearn, A. G.: 1961, *Proc. Phys. Soc.* **78**, 932–40.
Hinteregger, H. E.: 1965, *Space Sci. Rev.* **4**, 461–97.
Hudson, R. D., Carter, V. L., and Stein, J. A.: 1966, *J. Geophys. Res.* **71**, 2295–8.

Huffman, R. E., Tanaka, Y., and Larabee, J. C.: 1964, *Disc. Farad. Soc.* **37**, 159–65.
Inn, E. C. Y. and Tanaka, Y.: 1953, *J. Opt. Soc. Amer.* **43**, 870–3.
Johnson, F. S.: 1954, *J. Meteorol.* 431–9.
Kockarts, G.: 1971, this volume, pp. 160–76.
Landenburg, R. and Van Voorhis, C. C.: 1933, *Phys. Rev.* **43**, 315–21.
Lee, P.: 1955, *J. Opt. Soc. Amer.* **45**, 703–9.
Metzger, P. H. and Cook, R. G.: 1964, *J. Quant. Spectry. Radiative Transfer* **4**, 107–16.
Nicolet, M.: 1939, 'Le problème atomique dans l'atmosphère supérieure', *Mém. Inst. Météorol. Belg. XI*.
Nicolet, M. and Mange, P.: 1954, *J. Geophys. Res.* **59**, 15–45.
Ogawa, M.: 1968, *J. Geophys. Res.* **73**, 6759–63.
Parkinson, W. H. and Reeves, E. M.: 1969, *Solar Phys.* **10**, 342–7.
Preston, W. H.: 1940, *Phys. Rev.* **57**, 885–887.
Reid, R. H. G. and Withbroe, G. L.: 1970, *Harvard College Observatory TR-16*.
Schneider, E. G.: 1940, *J. Opt. Soc. Amer.* **30**, 128–32.
Tousey, R.: 1963, *Space Sci. Rev.* **2**, 3–69.
Vigroux, E.: 1953, *Contribution à l'étude expérimentale de l'ozone*, Masson et Cie, Paris.
Vigroux, E.: 1969, *Ann. Geophys.* **25**, 169–72.
Watanabe, K., Inn, E. C. Y., and Zelikoff, M.: 1953, *J. Chem. Phys.* **21**, 1026–30.
Watanabe, K. and Marmo, F. F.: 1956, *J. Chem. Phys.* **25**, 967–71.
Watanabe, K., Sakai, M., Mottl, J., and Nakayama, T.: 1958, '*Absorption cross section of* O_2, NO *and* NO_2 *with an improved photoelectric method*', Contribution No. 11, Hawaii Institute of Geophysics.
Watanabe, K., Zelikoff, M., and Inn, E. C. Y.: 1953, 'Absorption Coefficients of Several Atmospheric Gases', *Geophys. Res. Pap.* No. 21, A.F.C.R.L., Bedford, Mass.
Widing, K. G., Purcell, J. D., and Sandlin, G. D.: 1970, *Solar Phys.* **12**, 52–62.

PENETRATION OF SOLAR RADIATION IN
THE SCHUMANN-RUNGE BANDS OF MOLECULAR OXYGEN

G. KOCKARTS

Institut d'Aéronomie Spatiale de Belgique, Brussels, Belgium

Molecular oxygen is subject to photodissociation leading to a production of oxygen atoms which are involved in numerous aeronomic reactions below 100 km. Among several band systems, the Schumann-Runge bands $B\,^3\Sigma_u^- - X\,^3\Sigma_g^-$ are of fundamental importance in the chemosphere since they are situated in a wavelength region (1750 Å–2050 Å) where the solar radiation can penetrate deeply into the Earth's atmosphere.

In the thermosphere, atomic oxygen is produced mainly by photodissociation of O_2 in the Schumann-Runge continuum ($\lambda < 1750$ Å). In the stratosphere and mesosphere, molecular oxygen is photodissociated by radiation between 2424 Å and 1750 Å, i.e. in the spectral region of the Herzberg continuum and of the Schumann-Runge bands. Since predissociation occurs in the Schumann-Runge bands (Flory, 1936), an additional source of $O(^3P)$ atoms produced in the Herzberg continuum is available in the mesosphere. The effects of this process have been investigated by Hudson and Carter (1969), by Hudson *et al.* (1969) and by Brinkmann (1969). Recent absorption cross section measurements (Ackerman *et al.*, 1969) and the determination of the structure of the 0–0 to the 13–0 Schumann-Runge bands (Ackerman and Biaumé, 1970) made it possible to compute the O_2 absorption cross sections at very close intervals over the whole Schumann-Runge system (Ackerman *et al.*, 1970). The depth of the penetration of solar radiation into the chemosphere can be determined with these new data. It is of importance to know the fine structure of the absorption in the Schumann-Runge bands, especially when minor constituents are studied by absorption techniques. Jursa *et al.* (1959) attempted to detect nitric oxide in the altitude range 60 to 90 km and obtained in situ absorption spectra of the O_2 Schumann-Runge bands which clearly show the importance of the rotational structure. At 63 km altitude they even observed with certainty the 1–0 band and at 87 km the bands extend to $v' = 13$.

The numerical method and the experimental data used for the computation of the absorption cross section every 0.5 cm^{-1} from the 19–0 to the 0–0 band are described by Ackerman *et al.* (1970). Figure 1 gives a summary of the computations performed at 300 K. Experimental cross sections obtained by Ackerman *et al.* (1969) are indicated by crosses. It can be seen that the $v'' = 1$ bands have to be taken into account in the overall absorption. Also, the overlapping of the rotational lines within a specific band cannot be neglected in order to get theoretical cross sections which fit the experimental values obtained at precisely known wavelengths of silicon emission lines (Ackerman *et al.*, 1969). As the rotational line widths range between 0.5 cm^{-1} and 3.7 cm^{-1} (Ackerman and Biaumé, 1970), an experimental cross section can in fact be situated in the wing of a specific rotational line, and moreover the cross section changes

Fiocco (ed.), Mesospheric Models and Related Experiments, 160–176. All Rights Reserved.
Copyright © 1971 by D. Reidel Publishing Company, Dordrecht-Holland.

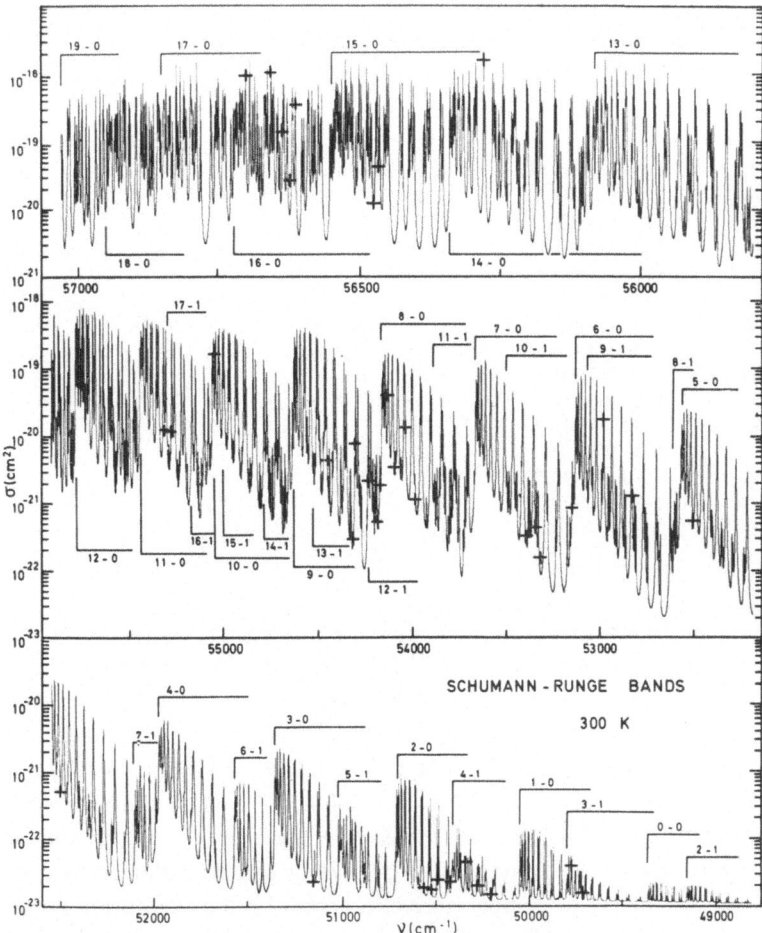

Fig. 1.   Absorption cross sections of molecular oxygen at 300 K. Experimental values of Ackerman *et al.* (1969) are indicated by crosses (+).

by more than an order of magnitude over a rotational line. It was therefore necessary to compute the cross section every 0.5 cm^{-1} between 57030.5 cm^{-1} (1753.45 Å) and 48 767.5 cm^{-1} (2050.55 Å) by taking into account the overlapping of all the lines having their maximum intensity within an interval of 150 cm^{-1} centred on each wave number. This means that the absorption contribution of the broadest rotational lines in the 4–0 band is taken into account up to 20 line widths from the center of the line. In such a way, more than 16 000 absorption cross sections were obtained in the range of the Schumann-Runge bands for different temperatures. Table I gives average values over 500 cm^{-1} (~20 Å) of the absorption cross sections for 160 K, 200 K and 300 K. The wave number intervals were chosen only for convenience of presentation in other sections of this paper. The contribution of the Herzberg continuum is included in the computations and Table I shows that this contribution becomes more and more

TABLE I

Average absorption cross sections σ for different temperatures

ν(cm^{-1})	λ(Å)	σ(cm^2) $T = 300$K	σ(cm^2) $T = 200$K	σ(cm^2) $T = 160$K
57000–56500	1754.4–1769.9	1.28×10^{-19}	1.50×10^{-19}	1.57×10^{-19}
56500–56000	1769.9–1785.7	1.18	1.19	1.18
56000–55500	1785.7–1801.8	7.37×10^{-20}	6.47×10^{-20}	6.06×10^{-20}
55500–55000	1801.8–1818.2	4.77	5.05	5.21
55000–54500	1818.2–1834.9	3.16	3.02	2.94
54500–54000	1834.9–1851.8	1.61	1.40	1.33
54000–53500	1851.8–1869.2	8.74×10^{-21}	7.57×10^{-21}	7.25×10^{-21}
53500–53000	1869.2–1886.8	4.19	3.48	3.40
53000–52500	1886.8–1904.8	1.90	1.44	1.37
52500–52000	1904.8–1923.1	9.48×10^{-22}	6.04×10^{-22}	4.84×10^{-22}
52000–51500	1923.1–1941.8	6.24	5.72	5.72
51500–51000	1941.8–1960.8	2.15	1.87	1.87
51000–50500	1960.8–1980.2	7.56×10^{-23}	5.40×10^{-23}	5.42×10^{-23}
50500–50000	1980.2–2000.0	3.06	1.83	1.77
50000–49500	2000.0–2020.2	1.94	1.54	1.49

important for $\nu < 51\,500$ cm^{-1} since in this wave number region the absorption cross section due to the continuum is of the order of 1.3×10^{-23} cm^2.

Within their limits of experimental error, Hudson and Carter (1969) could detect no significant change in the absorption cross sections when the temperature varies between 200 K and 300 K. Table I shows a ratio of the order of two for the average cross section in the interval $52\,500$ cm^{-1}–$52\,000$ cm^{-1} when the temperature decreases from 300 K to 160 K. A slight increase of the average cross section can even be seen towards shorter wavelengths. This behaviour can be explained by considering the factors responsible for a temperature effect. Firstly, the absorption cross section depends on the relative population of the first excited vibrational level $v'' = 1$ of the ground state. When the temperature decreases, the $v'' = 1$ bands shown in Figure 1 tend to disappear and, in the interval $52\,500$ cm^{-1}–$52\,000$ cm^{-1}, the 7–1 band is practically negligible at 160 K. Secondly, the absorption cross section is influenced by the relative intensities of the rotational lines of the P and R branches inside a specific band. When an *average* cross section is then obtained over a certain interval, this effect is practically smoothed out and the average value can even slightly increase due to a different rotational distribution. However, when the cross sections are considered with a wave number resolution of the order of 0.5 cm^{-1}, the temperature effect is quite large and, at certain specific wave numbers, changes of the order of two are obtained between 300 K and 200 K. Some of the experimental values of Ackerman *et al.* (1969) could be fitted only with a 300 K theoretical spectrum and not with a 200 K spectrum (Ackerman *et al.*, 1970). Temperature-dependent absorption cross sections will therefore be used in the subsequent sections of this paper.

The penetration of the solar radiation into the atmosphere is limited by the optical depth, which in turn depends on the nature and the total content of the absorbing species. In the lower thermosphere and mesosphere, molecular oxygen is the principal constituent responsible for the absorption in the Schumann-Runge spectral region. In the stratosphere, however, it is not possible to neglect the absorption by ozone since at 50 km the optical depth of ozone varies between 1×10^{-2} and 5×10^{-2} in the Schumann-Runge band region and increases rapidly in the Herzberg continuum. The presence of ozone affects strongly the rate of dissociation of O_2 below the stratopause (Nicolet, 1964) and the O_3 absorption has to be taken into account for the total optical depth in the Schumann-Runge bands. Table II gives the oxygen and ozone concentrations used in the following computations. This model has been deduced by Nicolet (1970) for the analysis of the ozone and hydrogen reactions in the chemosphere and it corresponds to daytime conditions.

TABLE II

Temperature, oxygen and ozone concentrations and their total contents

z (km)	T (K)	$n(O_2)$ (cm^{-3})	$n(O_3)$ (cm^{-3})	$\int_z^\infty n(O_2)\,dz$ (cm^{-2})	$\int_z^\infty n(O_3)\,dz$ (cm^{-2})
15	210.8	8.14×10^{17}	1.10×10^{12}	5.07×10^{23}	6.59×10^{18}
20	218.9	3.55	2.90	2.30	5.54
25	227.1	1.60	3.25	1.08	3.96
30	235.2	7.43×10^{16}	2.90	5.19×10^{22}	2.40
35	251.7	3.47	2.00	2.59	1.19
40	268.2	1.70	1.00	1.36	4.37×10^{17}
45	274.5	8.92×10^{15}	3.17×10^{11}	7.30×10^{21}	1.38
50	274.0	4.84	1.00	3.96	4.39×10^{16}
55	273.6	2.62	3.17×10^{10}	2.14	1.40
60	252.8	1.50	1.01	1.14	4.50×10^{15}
65	231.9	8.19×10^{14}	3.18×10^{9}	5.70×10^{20}	1.49
70	211.2	4.23	1.01	2.68	5.42×10^{14}
75	194.2	2.02	3.20×10^{8}	1.18	2.41
80	177.2	9.00×10^{13}	1.40	4.79×10^{19}	1.35
85	160.3	3.71	1.00	1.77	7.89×10^{13}
90	176.7	1.25	1.10	6.48×10^{18}	2.64
95	193.0	4.67×10^{12}	1.33×10^{7}	2.51	3.16×10^{12}
100	209.2	1.89	1.58×10^{6}	9.80×10^{17}	3.93×10^{11}
105	230.9	6.50×10^{11}	2.00×10^{5}	4.03	5.36×10^{10}
110	261.9	2.85	2.70×10^{4}	1.80	8.10×10^{9}
115	293.0	1.10	4.00×10^{3}	8.85×10^{16}	2.14

As molecular oxygen cross sections are now available every $0.5\ \mathrm{cm}^{-1}$ in the Schumann-Runge bands (Ackerman et al., 1970), it is possible to define an optical depth τ_i at height z for $0.5\ \mathrm{cm}^{-1}$ intervals by the relation

$$\tau_i = \int_z^\infty \sigma_i(O_2)\, n(O_2)\, dz + \sigma(O_3) \int_z^\infty n(O_3)\, dz \tag{1}$$

where $n(O_2)$ and $n(O_3)$ are the molecular oxygen and the ozone concentrations, respectively. The ozone absorption cross section $\sigma(O_3)$ is taken as a constant over each

500 cm^{-1} interval given in Table I. The numerical values for $\sigma(O_3)$ are those adopted by Ackerman (1971). Since the absorption cross section for molecular oxygen is temperature dependent, it is necessary to introduce $\sigma_i(O_2)$ in (1) under the integral sign. The optical depths defined by (1) correspond to an overhead sun and have been computed over the whole Schumann-Runge system with a wave number resolution of 0.5 cm^{-1} and a height resolution of 1 km. Figure 2 shows the optical depth obtained at 60 km altitude between 1818 Å and 1835 Å. The 10–0 and 9–0 band origins are respectively at 55050.90 cm^{-1} and 54622.17 cm^{-1} and the absorption structure due to the P and R branches of these bands is visible in Figure 2. In particular, the doublet structure in the 10–0 band results from the relative situation of the alternate triplets P and R: $7P$ at 54966.9 cm^{-1} and $9R$ at 54990.4 cm^{-1}; $9P$ at 54966.8 cm^{-1} and $11R$ at 54958.8 cm^{-1}; $11P$ at 54930.1 cm^{-1} and $13R$ at 54920.8 cm^{-1}. The 15–1 and 14–1 bands fall also in the wave number region of Figure 2, and their effect is strongly apparent between 54800 cm^{-1} and 54650 cm^{-1} where the optical depth is less than unity. The $v''=1$ bands practically disappear below 200 K, but at 60 km they contribute to the optical thickness since the principal contribution arises from a region extending to a few scale heights above 60 km where the temperature is high enough. It is only at mesopause levels ($T<200\,\mathrm{K}$) that the $v''=1$ bands are less efficient for the absorption, although it should be realized that the optical depth results from an effect which depends on an integration of the combined variation of $n(O_2)$ and $\sigma_i(O_2)$ over a range of heights. Figure 3 shows a similar behaviour at 40 km altitude between 1887 Å and 1905 Å. The structure results from the 6–0 and 5–0 bands for which the band origins are

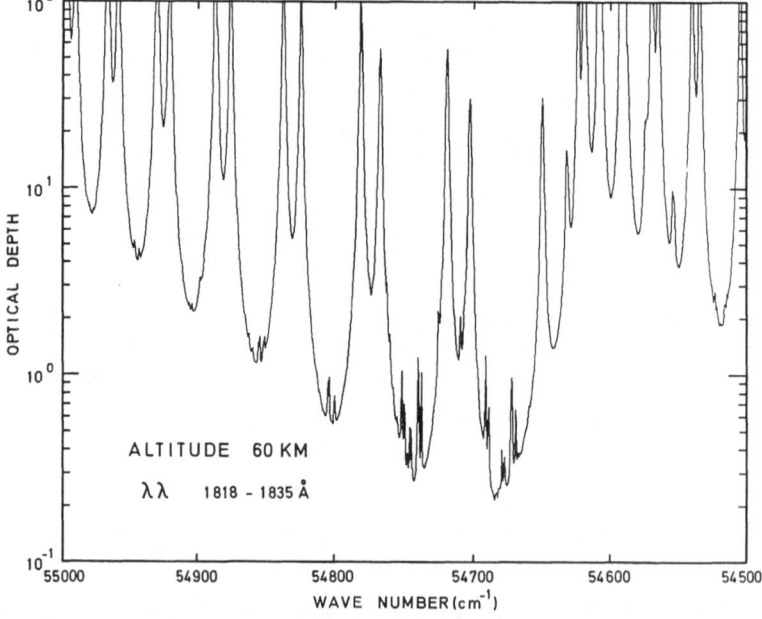

Fig. 2.   Optical depth between 1818 Å and 1835 Å at 60 km. Resolution 0.5 cm⁻¹.

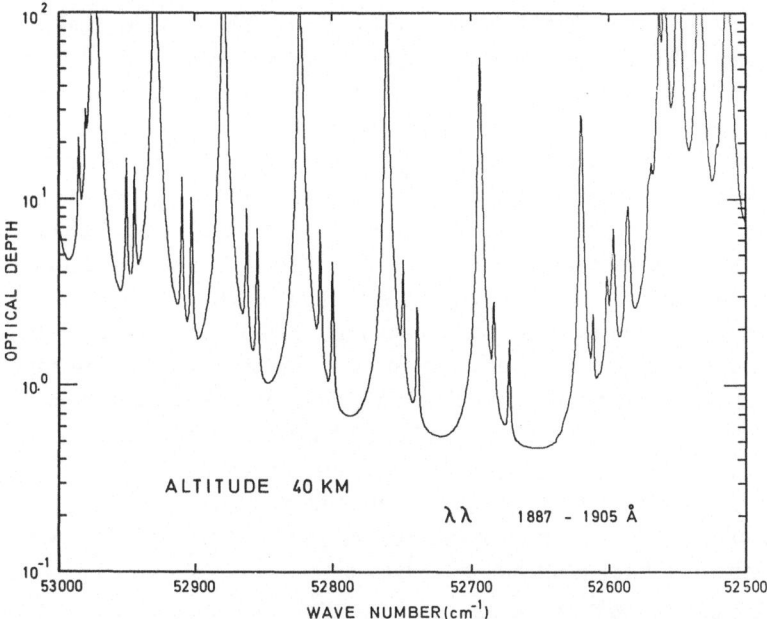

Fig. 3.   Optical depth between 1887 Å and 1950 Å at 40 km. Resolution 0.5 cm⁻¹.

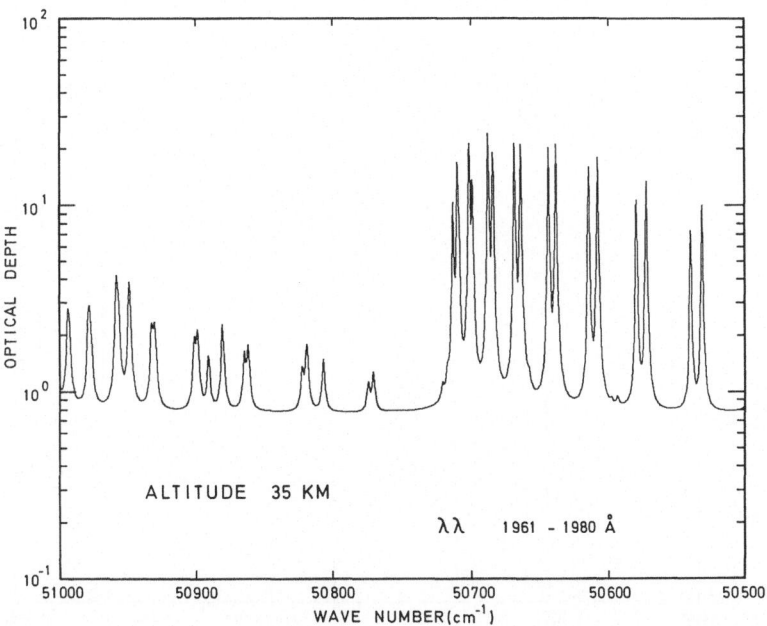

Fig. 4.   Optical depth between 1961 Å and 1980 Å at 35 km. Resolution 0.5 cm⁻¹.

at $53\,122.79\,\text{cm}^{-1}$ and $52\,561.39\,\text{cm}^{-1}$ respectively. The smallest peaks are due to the 9–1 and 8–1 bands. At 35 km altitude, Figure 4 gives the optical depth between 1961 Å and 1980 Å. In this figure, the effect of the Herzberg continuum and of the ozone absorption appears clearly since there is practically a constant optical depth background of 0.8 which results from τ (Herzberg)$=0.33$ and from $\tau(O_3)=0.44$. The peaks above $50\,750\,\text{cm}^{-1}$ are due to the 5–1 band and the larger peaks below $50\,750\,\text{cm}^{-1}$ arise from the 2–0 band which has its band origin at $50\,710.83\,\text{cm}^{-1}$. Within a few cm^{-1} the optical depth increases from 1 to 10 and such a feature should be detectable by balloon-borne instruments with high resolution.

It is clear from Figures 2, 3 and 4 that any high resolution absorption or fluorescence experiment should take into account the fine structure of the absorption in the Schumann-Runge bands. In particular, a high resolution investigation of the absorption or emission of a minor constituent can only be performed at wavelengths where the O_2 absorption is not too high. Such a situation has been described by Jursa *et al.* (1959) in their measurements of the $\delta(0, 0)$ band of nitric oxide.

If the solar flux at the top of the atmosphere is $\Phi_i(\infty)$ in a $0.5\,\text{cm}^{-1}$ wave number interval, the flux $\Phi_i(z)$ at altitude z is given by

$$\Phi_i(z) = \Phi_i(\infty) \exp(-\tau_i) \tag{2}$$

where τ_i is obtained by expression (1). At the present time, however, the solar flux is not known with a wave number resolution of $0.5\,\text{cm}^{-1}$ but the reduction factor $R_i(z)$ can be defined by

$$R_i(z) = \Phi_i(z)/\Phi_i(\infty). \tag{3}$$

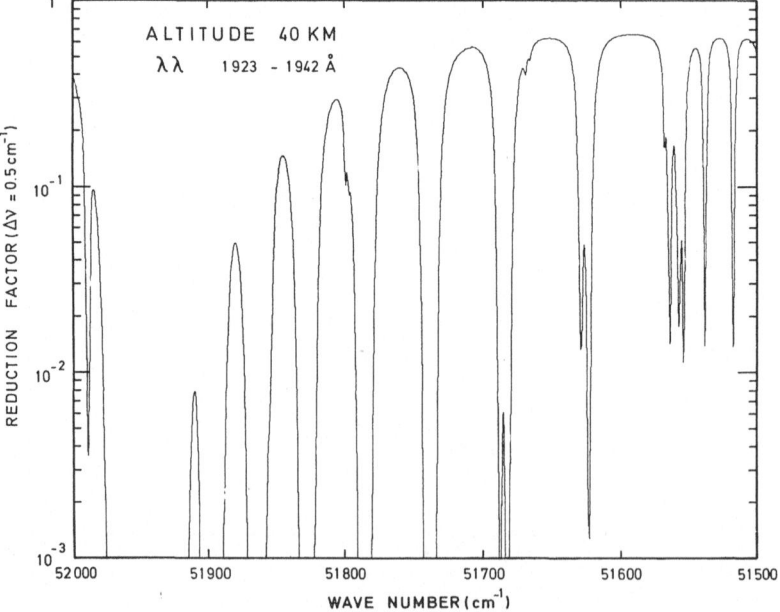

Fig. 5.   Reduction factor of the solar flux between 1923 Å and 1942 Å at 40 km.

Figure 5 shows, as an example, the reduction factor at 40 km altitude in the wave number interval 52000–51500 cm^{-1}. The solar flux distribution in that interval is actually modulated by the curve given in Figure 5. As a consequence of the band structure, the reduction factor $R_i(z)$ can change by more than a factor 1000 over a few cm^{-1} wave number interval.

A total reduction factor R is defined for every 500 cm^{-1} interval of Table I by the relation

$$R = \sum_{i=1}^{1000} R_i. \tag{4}$$

This reduction factor for an overhead sun is shown in Figure 6 for altitudes corresponding to the mesopause and stratopause, and down to a level of 30 km which can easily be reached by balloon-borne experiments. Although R and R_i are independent of the solar flux, the values presented in Figure 6 can however be used only when average solar fluxes are adopted over every 500 cm^{-1} wave number interval. For the 85 km values, there is a slight decrease of R between 55500 cm^{-1} and 55000 cm^{-1}; this effect is explained by the adopted line width of 1.7 cm^{-1} (Ackerman *et al.*, 1970)

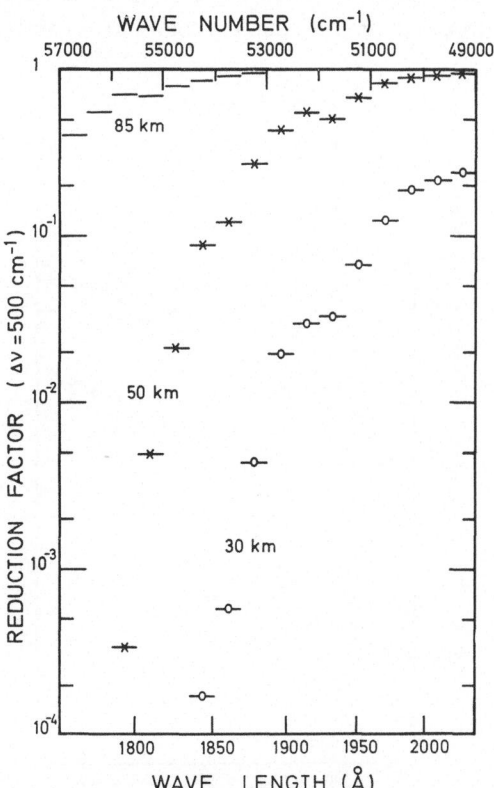

Fig. 6.   Reduction factor for 500 cm^{-1} intervals at the mesopause (85 km), at the stratopause (50 km) and at 30 km.

in the 11–0 band where there is a secondary maximum of predissociation. The effect is even more pronounced at 50 km between 52 000 cm^{-1} and 51 500 cm^{-1}, i.e. in the 4–0 band where the maximum of predissociation corresponds to a line width of 3.7 cm^{-1}. Any increase of the line width leads, in fact, to an increase of the mean absorption in the considered band. Predissociation implies therefore an increase of the total solar flux absorption.

The solar penetration in the chemosphere depends of course on the solar flux available at the top of the atmosphere. There is at the present time some discrepancy between measured fluxes in the Schumann-Runge wavelength region. This problem has been discussed by Ackerman (1971) and his suggested values will be adopted in the present computation. Figure 7 gives the solar flux at intervals of 500 cm^{-1} for several altitudes. The solar flux $\Phi(z)$ at height z for $\Delta v = 500$ cm^{-1} is given by

$$\Phi(z) = \sum_{i=1}^{1000} \Phi_i(\infty) \exp(-\tau_i) \tag{5}$$

where the optical depth τ_i is computed at intervals of 0.5 cm^{-1} according to expression (1). The fluxes $\Phi_i(\infty)$ at the top of the atmosphere have been obtained by dividing Ackerman's values by 1000 in order to get fluxes in photons cm^{-2} s^{-1} for $\Delta v = 0.5$ cm^{-1}. This procedure implies that the solar flux is constant over the 500 cm^{-1}

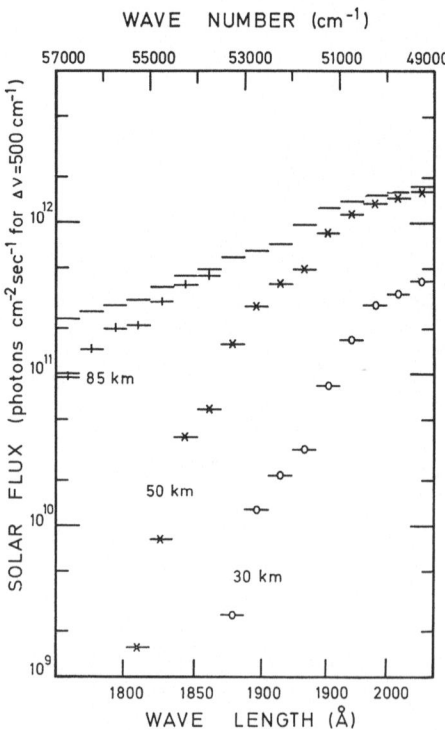

Fig. 7.   Solar flux available at 85 km, 50 km and 30 km altitude.

intervals. It is, however, known that a structure exists over every $500 \, \text{cm}^{-1}$ interval, and a detailed representation of the depth of solar penetration will only be possible when the solar fluxes are available with $0.5 \, \text{cm}^{-1}$ wave number resolution. It is therefore necessary to have a digitalized solar spectrum with good absolute values and a resolution which should preferably be higher than the spectrum discussed by Brinkmann *et al.* (1966). This spectrum has not been used in the present work, since the absolute values of the ultraviolet flux have to be changed (Ackerman, 1971). The results presented in Figure 7 can nevertheless be applied to an analysis of global effects which do not require high wave number resolution.

The absorption in the Schumann-Runge bands is not only important for the atomic oxygen production rate in the chemosphere but also for the photodissociation of minor constituents such as O_3, H_2O, CO_2 and N_2O. The photodissociation coefficient $J_i(z, X)$ in a $0.5 \, \text{cm}^{-1}$ wave number interval is given for a constituent X by

$$J_i(z, X) = K_i \Phi_i(z) \tag{6}$$

where $\Phi_i(z)$ is given by (2) and K_i is the photodissociation cross section for the constituent X. For every $500 \, \text{cm}^{-1}$ interval of Table I, the photodissociation coefficient is simply

$$J(z, X) = \sum_{i=1}^{1000} J_i(z, X). \tag{7}$$

The total photodissociation coefficient over the Schumann-Runge bands is then obtained by summing the $J(z, X)$ values given by (7).

The molecular photodissociation coefficient in the Schumann-Runge system depends of course on the existence of predissociation. According to Flory (1936) and to Hudson and Carter (1969), the upper vibrational levels with $v' > 2$ of the excited $B^3\Sigma_u^-$ electronic state are subject to predissociation. The measurements by Feast (1949) indicate no predissociation for the bands with $v' = 3$. Ackerman and Biaumé (1970) have, however, determined a total rotational line width of the order of $1 \, \text{cm}^{-1}$ in the $v' = 0$, 1 and 2 bands. Considering that the total Doppler broadening at $2000 \, \text{Å}$ is of the order of $0.1 \, \text{cm}^{-1}$ for a temperature of $300 \, \text{K}$, it can be suggested from the measurements of Ackerman and Biaumé (1970) that predissociation occurs even for $v' \geqslant 0$. Therefore, in the wavelength region where predissociation is taken into account, the photodissociation cross section K_i is composed of two terms: a contribution from the Schumann-Runge bands, and one from the Herzberg continuum which also leads to the production of $O(^3P)$ atoms. In order to show the effect of predissociation starting at $v' = 0$ or at $v' > 3$, Table III gives the O_2 photodissociation coefficients in the Schumann-Runge bands computed for the two cases. The last column is the total photodissociation coefficient due to Lyman α, to the Schumann-Runge continuum, to the Schumann-Runge bands and to the Herzberg continuum. The total value of $J(O_2)$ has been computed with predissociation for $v' > 3$. When complete predissociation occurs in the Schumann-Runge bands, Table III shows that $J(O_2)$, given in the last column, has to be multiplied by a factor 1.08 at 60 km. Above and below this altitude the difference between the values of $J(O_2)$ for total predissociation and for

G. KOCKARTS

TABLE III
Photodissociation coefficient (s^{-1}) for O$_2$

z(km)	J(S-R) Predissociation $v' \geqslant 0$	J(S-R) Predissociation $v' > 3$	J(O$_2$)
100	9.14×10^{-8}	9.14×10^{-8}	3.77×10^{-7}
95	6.35	6.32	1.56
90	3.65	3.62	5.45
85	1.76	1.73	2.14
80	9.07×10^{-9}	8.80×10^{-9}	1.18
75	4.66	4.40	6.50×10^{-9}
70	2.59	2.35	3.72
65	1.47	1.29	2.46
60	8.99×10^{-10}	7.47×10^{-10}	1.90
55	5.73	4.52	1.57
50	3.38	2.53	1.27
45	2.06	1.51	9.41×10^{-10}
40	1.19	8.88×10^{-11}	5.25
35	5.47×10^{-11}	4.45	2.14
30	2.05	1.79	7.24×10^{-11}

predissociation for $v' > 3$ decreases and becomes of the order of 1% at the mesopause. Despite the fact that there is some evidence for total predissociation, the slightly lower values J(O$_2$) are adopted, since the absolute values of the solar flux are not known with sufficient accuracy and also since the difference between the two possibilities does not lead to very important changes in the total photodissociation coefficient.

As an example, Figure 8 gives the fine structure of the photodissociation coefficient of O$_2$ between 1905 Å and 1923 Å. This wavelength interval has been chosen since there is no great variation in the structure of the solar radiation according to Brinkmann *et al.* (1966). The dashed line in Figure 8 gives the mean value for J(O$_2$) over the considered 500 cm^{-1} interval. Some smooth minima appear in Figure 8, particularly below 52 300 cm^{-1}. These features are due to the combined effect of the reduction factors R_i and the photodissociation cross sections K_i which are multiplied by each other in the photodissociation coefficients. At certain wavelengths, the decrease of the reduction factor R_i is compensated by the cross section K_i appearing in front of the exponential term. It will not be possible to present physically significant curves like that of Figure 8 until the solar spectrum is known with greater resolution than at present.

It is interesting to compare the present photodissociation coefficient of O$_2$ in the Schumann-Runge bands with the values deduced by Hudson *et al.* (1969) from their laboratory absorption measurements. It can be seen in Figure 9 that the agreement is quite satisfactory when the molecular oxygen total content varies between 10^{17} cm^{-2} and 10^{21} cm^{-2}. These values correspond roughly to the height region between 100 km and 60 km. In order to obtain a significant comparison, the computation presented in Figure 9 has been made using the solar fluxes of Brinkmann *et al.* (1966) averaged over the 500 cm^{-1} wave number intervals of Table I.

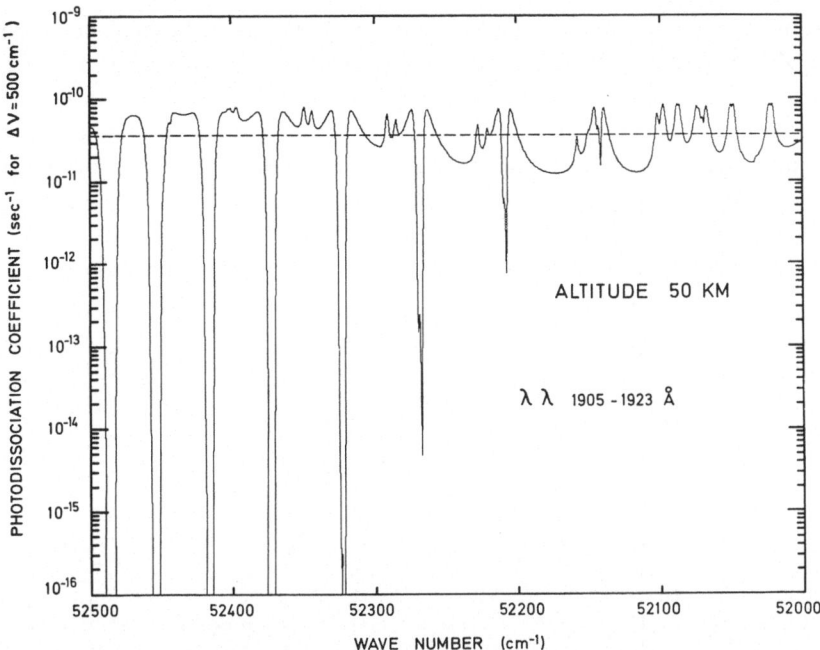

Fig. 8. Example of the structure of the molecular oxygen photodissociation coefficient at 50 km altitude between 1905 Å and 1923 Å.

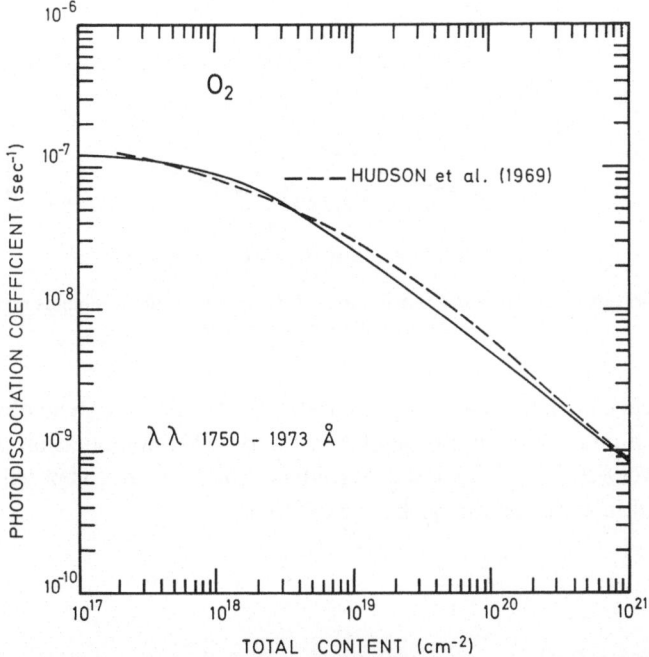

Fig. 9. Comparison between the calculated photodissociation coefficient of molecular oxygen for the spectral range 1750 Å–1973 Å and the values given by Hudson *et al.* (1969) between 2×10^{17} and 10^{21} molecules cm^{-2}.

The photodissociation rate of molecular oxygen depends on the whole solar spectrum of $\lambda < 2424$ Å. The importance of the Schumann-Runge bands is shown in Figure 10 where the total photodissociation rate is represented as well as the contribution due to the wavelength region between 1754 Å and 2020 Å. Between approximately 60 km and 90 km altitude, the predissociation in the Schumann-Runge bands is the major process responsible for O_2 dissociation for overhead sun conditions.

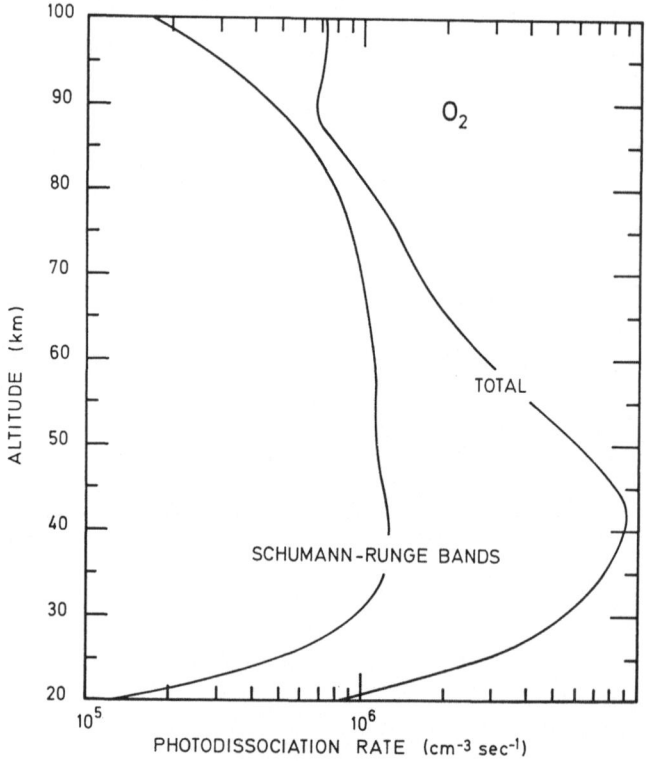

Fig. 10. O_2 photodissociation rate. Contribution of the Schumann-Runge bands region is important between 90 km and 60 km.

For practical calculations, it would be useful to have a set of mean absorption cross sections which could give results similar to the detailed computation described previously. In every 500 cm^{-1} wave number interval of Table I, it is possible to define a mean absorption cross section σ_m by the relation

$$\sigma_m = \frac{\sum \sigma_i \exp\left(-\tau_i\right)}{\sum \exp\left(-\tau_i\right)} \tag{7}$$

where the sums extend over the 1000 values included in the interval considered. The mean absorption cross sections obtained in this way are altitude dependent. Figure 11 shows σ_m versus wave number for different values of the total optical depth and it is

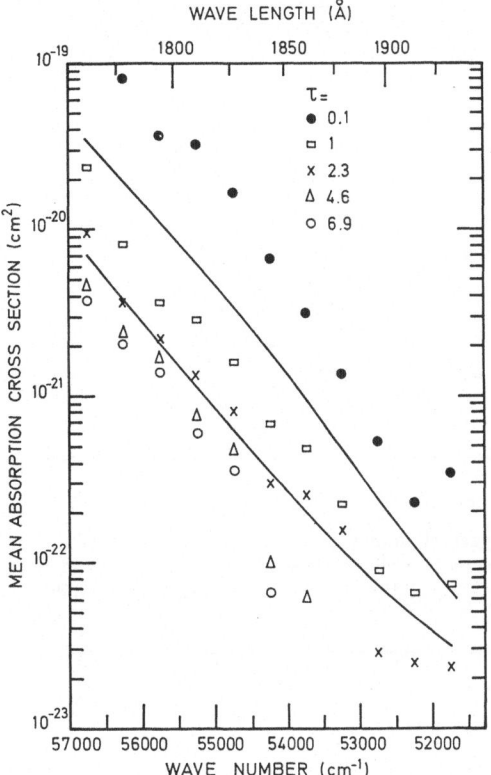

Fig. 11. Mean absorption cross sections for $\Delta v = 500$ cm^{-1} and for different values of the optical depth τ.

seen that σ_m decreases with increasing optical depth. The squares in Figure 11, corresponding to unit optical depth, clearly show three plateaux which are located in the bands where a maximum of predissociation occurs, i.e. at $v' = 4$, 7 and 11. This feature is also visible in Figure 6 showing the reduction factor.

From Figure 11 it is not possible to deduce a general law which is sufficiently precise for the exact representation of the variation of σ_m with wave number and with height. However, the two solid curves H and L of Figure 11 are an attempt to represent the mean absorption cross section for optical depths between 0 and 1 and for greater values of τ. In order to discuss the validity of this choice, Figure 12 gives the molecular oxygen photodissociation coefficient in the Schumann-Runge bands computed using the two sets of values. Curve H corresponds to the high values of Figure 11 and curve L corresponds to the low values. An analysis of Figure 12 indicates that the high values used by Nicolet (1970) give the best agreement with the exact computation represented by the full line. There is, however, a slight underestimation of the total photodissociation coefficient near 90 km when curve H is adopted. But the low values L cannot be used above 70 km for computing $J(O_2)$.

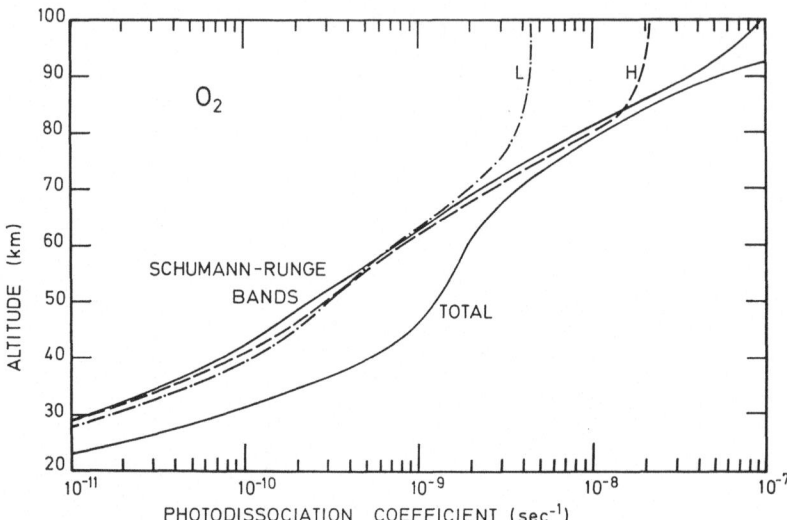

Fig. 12. Molecular oxygen photodissociation coefficient computed for low (L) and high (H) values of the mean absorption cross sections. The total $J(O_2)$ corresponds to the detailed computation (solid line) in the Schumann-Runge bands.

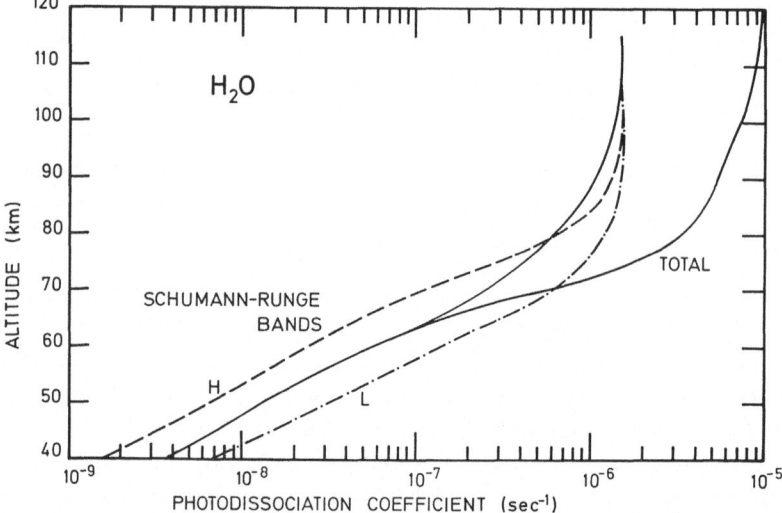

Fig. 13. Water vapour photodissociation coefficient computed for low (L) and high (H) values of the mean absorption cross sections. The total value corresponds to the detailed computation in the Schumann-Runge bands which give the major contribution below 70 km altitude.

It is now necessary to investigate the effect of the high and low values of σ_m on the photodissociation coefficients of minor constituents such as H_2O, CO_2, O_3 and N_2O, since the mean absorption cross section is not identical with the photodissociation cross sections K_i of Equation (5). Figure 13 shows the photodissociation coefficient for water vapour. Down to the lower mesosphere, Lyman α makes an important contribution to $J(H_2O)$ (Nicolet, 1970, 1971); the total $J(H_2O)$ is also shown in Figure 13 in order to indicate where the Schumann-Runge bands become the major component. The curve L and H correspond to the mean cross sections L and H of Figure 11. It is clear that neither the low values nor the high values can fit the detailed computation indicated by the solid curve. The low values (curve L) overestimate $J(H_2O)$ by a factor of 2 and the high values (curve H) underestimate $J(H_2O)$ by a factor of 2 below 70 km where the Schumann-Runge bands produce the major contribution to the total photodissociation coefficient. The comparison between Figures 12 and 13 indicates that it is not possible to adopt a unique set of mean absorption cross sections which simultaneously fit the effective dissociation coefficients of O_2 and of the minor constituents. It is therefore more suitable to use the reduction factors R described earlier for a computation of the different photodissociation coefficients for minor constituents. Moreover, the reduction factors can be adapted to any degree of resolution of the solar spectrum. It should however be pointed out that in the present work all the computations are made for an overhead sun and the reduction factors of Figure 6 cannot be applied to other zenith angles without recomputing all the fine structure reduction factors R_i.

High resolution absorption cross sections are now available for molecular oxygen in the Schumann-Runge band system. Since the cross sections are temperature dependent, the absorption of solar radiation is a function not only of the total content of the absorbing species but also of the atmospheric temperature.

The great variability of the optical thickness as a function of wavelength implies that any high resolution experiment designed should take into account the possibility of changes by a factor 100 in the optical depth over a very small wave number interval. A detailed study of the solar radiation absorption will be possible only when the solar spectrum is known with a better resolution. However it is possible at present to investigate the average solar penetration by means of the reduction factors which are independent of both the absolute value and the fine structure of the solar flux. The reduction factors show the line broadening effect related to the predissociation in the Schumann-Runge bands.

The comparison between the values of the exact photodissociation coefficients and those obtained with different mean cross sections shows that it is not possible to reconcile these values with a unique set of mean cross sections. This is due to the fact that the mean absorption cross section depends on the optical depth, i.e. on the altitude. A calibration must be made in each case in order to determine, in the spectral region of the Schumann-Runge bands, the mean absorption cross section which must be adopted. Finally, another investigation will be necessary in order to study the effect of various solar zenith angles.

Acknowledgement

I would like to express my gratitude to Prof. M. Nicolet for his valuable advice during the preparation of this work.

References

Ackerman, M.: 1971, this volume, pp. 149–59.

Ackerman, M. and Biaume, F.: 1970, *J. Mol. Spectr.* **35**, 73–82.

Ackerman, M., Biaume, F., and Nicolet, M.: 1969, *Can. J. Chem.* **47**, 1834–40.

Ackerman, M., Biaume, F., and Kockarts, G.: 1970, *Planetary Space Sci.* **18**, 1639–51.

Brinkmann, R. T.: 1969, *J. Geophys. Res.* **74**, 5355–68.

Brinkmann, R. T., Green, A. S., and Barth, C. A.: 1966, 'A Digitalized Solar Ultraviolet Spectrum', Technical Report No. 32–951, Jet Propulsion Laboratory, Pasadena, California.

Feast, M. W.: 1949, *Proc. Phys. Soc.* **A62**, 114–21.

Flory, P. J.: 1937, *J. Chem. Phys.* **4**, 23–7.

Hudson, R. D. and Carter, V. L.: *Can. J. Chem.* **47**, 1840–4.

Hudson, R. D., Carter, V. L., and Breig, E. L.: 1969, *J. Geophys. Res.* **74**, 4079–86.

Jursa, A. S., Tanaka, Y., and Le Blanc, F.: 1959, *Planetary Space Sci.* **1**, 161–72.

Nicolet, M.: 1964, *Disc. Faraday Soc.* **37**, 7–20.

Nicolet, M.: 1970, *Ann. Geophys.* **26**, 531–46.

Nicolet, M.: 1971, this volume, pp. 1–51.

RELAXATION OF THE 2.7μ AND 4.3μ BANDS
OF CARBON DIOXIDE

A. P. WILLIAMS

Clarendon Laboratory, Oxford, England

A significant amount of solar radiation is absorbed by the 2.7μ and 4.3μ vibration-rotation bands of CO_2 although the solar flux is comparatively small at these wavelengths. If the excited molecules were to relax only by collisions, all energy would go directly into kinetic energy at the level of initial absorption.

In the lower regions of the atmosphere collisional excitation and de-excitation are the dominant processes; the atmosphere is in a state of local thermodynamic equilibrium, with the populations of excited states governed by a Boltzmann distribution. The time between collisions is inversely proportional to pressure, so that as we go higher in the atmosphere, radiative deactivation becomes more important. The emitted radiation may be lost from the atmosphere or absorbed in other regions.

Thus, in order to determine the net heating rate, the radiation emitted – and that absorbed from other regions of the atmosphere – must be considered.

The starting point of all radiation calculations is Schwarzschild's equation of radiative transfer (Goody, 1964):

$$\cos(\theta)\frac{dI_\nu(z)}{dz} = -k_\nu[I_\nu(z) - J_\nu(z)]. \tag{1}$$

This equation shows the change in monochromatic radiative intensity (I_ν) when passing through a layer of atmosphere of thickness dz, owing to absorption and emission. θ is the angle between the direction of I_ν and the vertical co-ordinate axis (z). k_ν is the monochromatic volume absorption coefficient and J_ν the source function. It will be shown that J_ν depends simply on the population density of the upper state involved in the radiative processes.

We may solve Equation (1) for I_ν, and hence determine the flux of energy (F) at any height (z) in the atmosphere by integrating over solid angle and frequency:

$$F = \int_{\Delta\nu}\int_0^{\pi/2} I_\nu(z)\cos(\theta)\,2\pi\sin(\theta)\,d\theta\,d\nu, \tag{2}$$

where $\Delta\nu$ is the width of the spectral band. Such solutions for the upward and downward flux at height (z) are shown below.

$$\frac{F^\uparrow(z)}{\pi\Delta\nu} = J(z) - \int_0^z T_F(z - z')\frac{dJ(z')}{dz'}\,dz' \tag{3a}$$

Fiocco (ed.), Mesospheric Models and Related Experiments, 177–187. All Rights Reserved.

$$\frac{F^{\downarrow}(z)}{\pi\Delta v} = J(z) + \int_z^Z T_F(z - z') \frac{dJ(z')}{dz'} dz' + \frac{1}{\pi} S_f \cos(\psi) T_{sf}(z - Z). \quad (3b)$$

The last term in Equation (3b) is the flux due to the direct solar beam and the other terms account for the subsequent transfer of radiation.

In deriving these equations, I have assumed that the source function is isotropic and varies little over the extent of the band. $J(z)$ is then the source function at the band centre, $J_{v_0}(z)$. Hence the integrations over θ and v occur only in the term, $T_F(z-z')$. This is the flux transmission for the complete band, expressed by

$$T_F(z - z') = \frac{1}{\Delta v} \int_{\Delta v} \left[2 \int_0^{\pi/2} \exp\left[- k_v(z-z') \sec(\theta) \right] \cos(\theta) \sin(\theta) \, d\theta \right] dv.$$
$$(4a)$$

The central term, $\exp[-k_v(z-z')\sec(\theta)]$, is simply the slant path monochromatic transmission between heights z and z'. I have removed the integration over θ, using the Elsasser (1942) diffusivity factor of 1.66, giving

$$T_F(z - z') = \frac{1}{\Delta v} \int_{\Delta v} \exp\left[- k_v(z - z') \times 1.66 \right] dv. \quad (4b)$$

The solar flux transmission

$$T_{sf}(z - Z) = \frac{1}{\Delta v} \int_{\Delta v} \exp\left[- k_v(z - Z) \sec(\psi) \right] dv \quad (4c)$$

is required for a particular solar angle (ψ), and contains an integration over frequency only.

Since, at all heights of interest, the spectral lines in a band do not overlap, the integration over frequency in Equations (4b) and (4c) is replaced by a summation over individual spectral lines. The transmission is expressed in terms of integrated absorption, and this is calculated, for each spectral line, taking into account combined Lorentz and Doppler broadening.

There remains the height integration in Equations (3a) and (3b). I have used the method suggested by Curtis (1956) of replacing the integration, by a sum over levels. By taking the differences of upward and downward fluxes at neighbouring levels, the heating rate at level z may be written

$$H_z = \sum_{k=0}^{z} C_{z,k} J_k + S_z \quad (5)$$

where each element, $C_{z,k}$, contains the transmission properties of the atmosphere and S_z is the heating rate due to initial absorption of solar radiation.

If Equation (1) is derived from first principles we find (Shved, 1965) that

$$J = \frac{2hv_0^3 c^2}{g\left(n(l)/n(u)\right) - 1}$$

where $n(u)$ and $n(l)$ are the population densities of the upper and lower states respectively and g, the ratio of their statistical weights. If the atmosphere is in local
thermodynamic equilibrium, $g\left(n(l)/n(u)\right) = \exp\left(hcv_0/kT\right)$ and the source function is
equal to the Planck function. In any case, the population of the upper state is always
smaller than that of the lower state and a valid approximation for the source function
at height level k is

$$J_k = \frac{2hv_0^3 c^2}{gn_k(l)} \times n_k(u). \tag{6}$$

Thus, in order to include the effect of transfer of radiation on the heating rate
(Equation (5)), we need to know the population density of the excited state. This
requires detailed investigation of the processes which influence it.

Houghton (1969) describes a system of collisional processes whereby energy relaxes
from the 2.7μ and 4.3μ excited states of CO_2. The scheme involves vibrationally
excited states of N_2, O_2, H_2O and other excited states of CO_2. Of these, the H_2O and
CO_2 states may relax, in addition, by radiative processes.

The energy levels involved are shown in Figure 1. The quantities H represent the
net heating rate due to a particular band; for example, H_3 includes not only the initial
solar absorption by the 4.3μ fundamental band of CO_2 but also the result of the subsequent transfer of energy by re-emission and reabsorption processes. Similarly,
H_4 is the net heating rate due to the 2.7μ band. The CO_2 $(v_3 + v_1, v_3 + 2v_2)$ state is
more likely to radiate the v_3 part of its energy, than to re-radiate at 2.7μ. This gives
rise to radiation in the 4.3μ hot band, and since the population of the CO_2 $(v_1, 2v_2)$
state is relatively small, little of this radiation will be reabsorbed. Thus we expect the
net 'heating' in this band, H_5, to be negative, representing energy lost from the system.

Although the 4.3μ fundamental and hot bands occur in the same part of the
spectrum, the individual spectral lines do not overlap. Thus radiation in the hot band
cannot be absorbed by CO_2 in the ground state.

Both the 2.7μ and 4.3μ energy levels suffer resonance exchange of energy with the
first vibrational level of N_2.

$$CO_2\,(v_3) + N_2 \overset{k_{17}}{\rightleftharpoons} CO_2 + N_2^* + K.E.\,(18\text{ cm}^{-1}) \tag{7}$$

$$CO_2\,(v_3 + v_1, v_3 + 2v_2) + N_2 \overset{K_{17}}{\to} CO_2\,(v_1, 2v_2) + N_2^* + K.E. \tag{8}$$

Collisions with O_2, giving the first vibrationally excited state, are also of importance.

$$CO_2\,(v_3) + O_2 \overset{k_{18}}{\to} CO_2\,(v_2) + O_2^* + K.E.\,(128\text{ cm}^{-1}) \tag{9}$$

$$CO_2\,(v_3 + v_1, v_3 + 2v_2) + O_2 \overset{K_{18}}{\to} CO_2\,(v_1, 2v_2) + O_2^* + K.E. \tag{10}$$

ENERGY LEVELS

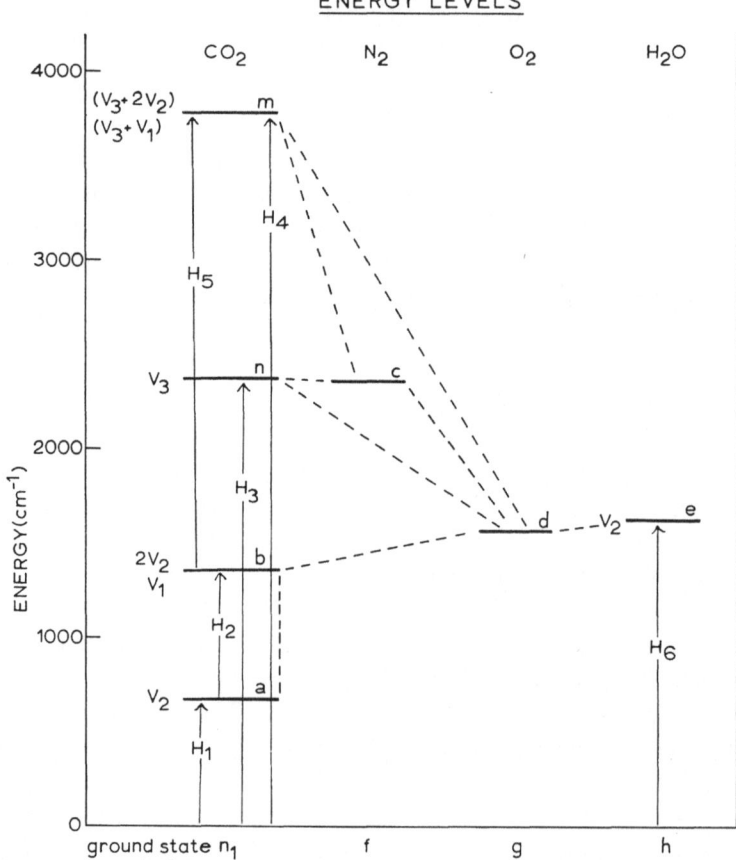

Fig. 1. The molecules (and their vibrational energy levels) involved in the relaxation of the 2.7μ ($v_3 + v_1$, $v_3 + 2v_2$) and 4.3μ (v_3) states of CO_2. The dashed lines show the main paths by which energy is exchanged by collisional processes.

The amount of kinetic energy released by processes (8) and (10) is not specified because there are really two levels of different energy associated with the combination of the v_1 and $2v_2$ vibrations. Fermi resonance leads to a perturbation and a separation of about $100\ \mathrm{cm}^{-1}$.

We might expect that the final state of the CO_2 molecule due to process (10) is more likely to be $(v_1 + v_2, 3v_2)$ instead of $(v_1, 2v_2)$. However, since the object of this work is to demonstrate the results of more accurate radiative transfer calculations, I have used the same collisional processes as Houghton.

Apart from the reverse reaction of (7), the only process of importance in relaxing N_2^* is collisions with O_2, producing the first vibrationally excited state.

$$N_2^* + O_2 \xrightarrow{k_{20}} N_2 + O_2^* + \mathrm{K.E.}\,(777\ \mathrm{cm}^{-1}).\tag{11}$$

Excited O_2, produced by reactions (9), (10) and (11) may exchange energy in the following ways:

$$O_2^* + \text{Air} \xrightarrow{k_{26A}} O_2 + \text{Air} + \text{K.E.} \,(1554 \text{ cm}^{-1}) \tag{12}$$

$$O_2^* + H_2O \underset{}{\overset{k_{27}}{\rightleftharpoons}} O_2 + H_2O \,(\nu_2) - \text{K.E.} \,(41 \text{ cm}^{-1}) \tag{13}$$

$$O_2^* + CO_2 \xrightarrow{k_{28}} O_2 + CO_2 \,(\nu_1, 2\nu_2) + \text{K.E.} \,(209 \text{ cm}^{-1}). \tag{14}$$

The excited H_2O may relax by radiating at 6.3μ; this energy loss from the system is represented in Figure 1 by H_6, which will be negative. Energy may also be converted into kinetic energy by collisions with air molecules.

$$H_2O \,(\nu_2) + \text{Air} \xrightarrow{k_{29A}} H_2O + \text{Air} + \text{K.E.} \,(1595 \text{ cm}^{-1}). \tag{15}$$

CO_2 in the $(\nu_1, 2\nu_2)$ state is formed by radiative relaxation of the $(\nu_3+\nu_1, \nu_3+2\nu_2)$ state, and by collisional processes (8), (10), and (14). According to Houghton the reaction

$$CO_2 \,(\nu_1, 2\nu_2) + CO_2 \xrightarrow{k} 2CO_2 \,(\nu_2) \tag{16}$$

occurs very rapidly, and he assumes that all energy of the $CO_2 \,(\nu_1, 2\nu_2)$ molecule transfers to $CO_2 \,(\nu_2)$ molecules. To simulate this I have used a reaction rate of 10^{10} s^{-1} for process (16); that is, every collision is effective. With a reaction rate of 10^7 s^{-1}, more energy will be radiated out in the 15μ hot band,

$$CO_2 \,(\nu_1, 2\nu_2) \rightarrow CO_2 \,(\nu_2) + hc\nu \,(15\,\mu). \tag{17}$$

If the reaction rate of process (16) is only 10^7 s^{-1}, then collisions with air molecules may be significant;

$$CO_2 \,(\nu_1, 2\nu_2) + \text{Air} \rightarrow CO_2 + \text{Air} + \text{K.E.}$$

This reaction has not been included in the present calculations.

Finally, the $CO_2 \,(\nu_2)$ molecules formed by the reactions (9) and (16), and the radiative process (17), may relax by colliding with air molecules,

$$CO_2 \,(\nu_2) + \text{Air} \, k_1 \, CO_2 + \text{Air} + \text{K.E.} \,(667 \text{ cm}^{-1}), \tag{18}$$

or by radiating in the 15μ fundamental band.

All the microscopic processes of energy exchange and transfer considered here are faster than any macroscopic changes of atmospheric temperature. Hence, the reaction system will be in equilibrium. Thus for each level in Figure 1, we may equate its excitation and de-excitation processes.

Consider first the $CO_2 \,(\nu_3+\nu_1, \nu_3+2\nu_2)$ level. The only excitation processes are H_4 and H_5 (shown in Figure 1). The number of de-excitation processes per second per unit volume are from Equations (8) and (10) $K_{17}mf + K_{18}mg$. The notation used is shown in Figure 1. If H represents the net number of quanta of energy absorbed per second per unit volume from the radiation field, the kinetic equation may be written,

$$H_4 + H_5 = K_{17} m f + K_{18} m g.$$

Similarly, for the other levels:

$CO_2 (v_3)$,	$H_3 + k'_{17} n_1 c = k_{17} n f + k_{18} n g$
N_2^*,	$k_{17} n f + K_{17} m f = k'_{17} n_1 c + k_{20} c g$
O_2^*,	$k_{18} n g + k_{20} c g + k'_{27} g e + K_{18} m g =$
	$= k_{26A} d n (A) + k_{27} d h + k_{28} d n_1$
$H_2O (v_2)$,	$H_6 + k_{27} d h = k'_{27} g e + k_{29A} e n (A)$
$CO_2 (v_1, 2v_2)$,	$H_2 + k_{28} d n_1 + K_{17} m f + K_{18} m g = H_5 + k b n_1$
$CO (v_2)$,	$H_1 + k_{18} n g + 2 k b n_1 = H_2 + k_1 a n (A).$

The rate constants for the reverse reactions in Equations (7) and (13) are denoted by k'_{17} and k'_{27} respectively. The values of the reaction rates, simply the rate constant multiplied by the number density of air molecules at standard temperature and pressure, are given in Table I.

TABLE I

Values of the reaction rates, corresponding to the rate constants of collisional processes

Rate constants	Reaction rate (s^{-1})
k_{17} and K_{17}	1.0_{10^7}
k'_{17}	0.88_{10^7}
k_{18} and K_{18}	1.0_{10^5}
k_{20}	50
k_{26A}	8.5
k_{27}	1.0_{10^8}
k'_{27}	1.33_{10^8}
k_{28}	0.7_{10^5}
k_{29A}	1.0_{10^5}
k_1	2.0_{10^5}
k	$1.0_{10^{10}} - 1.0_{10^7}$

We may solve the above set of kinetic equations for the unknown population densities, in terms of the quantities H. The solution may be expressed by

$$n_k^i = \sum_{j=1}^{6} \tau_k^{i, j} H_k^j \quad (i = 1, ..., 5) \tag{19}$$

where i refers to one of the 5 levels of the system which may radiate; j refers to one of the 6 heating quantities (H); and k refers to a particular height level in the atmosphere. $\tau_k^{i, j}$ contains known quantities, i.e. rate constants and ground state population densities, and is height dependent. Assuming that the volume mixing ratios of the ground state population densities are constant with height, τ contains only the term $[(p_0/p)\,(T/T_0)]_k$, describing the variation of air density with height. The temperature dependence of the rate constants is not included.

We are now in a position to solve for the net heating rate due to each band.

From Equations (6) and (19) we may write down an expression for the source function for energy level i at height k.

$$J_k^i = \left[\frac{2hv_0^3c^2}{gn\,(l)}\right]_k^i \sum_{j=1}^6 \tau_k^{i,\,j} H_k^j$$

and, taking the term in square brackets inside the summation,

$$J_k^i = \sum_{j=1}^6 E_k^{i,\,j} H_k^j\,.$$

This may be written in matrix notation

$$\mathbf{J} = \mathbf{E} \times \mathbf{H}, \tag{20}$$

with each element of the column vectors $\mathbf{J}$ and $\mathbf{H}$ referring to a particular band and a particular height.

Equation (5) applies to each spectral band of the system, so we may write it in the form

$$H_z^i = \sum_{k=0}^z C_{z,\,k}^i J_k^i + S_z^i\,.$$

In matrix notation, this is

$$\mathbf{H} = \mathbf{C} \times \mathbf{J} + \mathbf{S}.$$

Substituting for $\mathbf{J}$ from Equation (20) gives

$$\mathbf{H} = \mathbf{C} \times \mathbf{E} \times \mathbf{H} + \mathbf{S}$$
$$\mathbf{H} - \mathbf{C} \times \mathbf{E} \times \mathbf{H} = \mathbf{S}$$
$$(\mathbf{I} - \mathbf{C} \times \mathbf{E}) \times \mathbf{H} = \mathbf{S}$$
$$\mathbf{H} = (\mathbf{I} - \mathbf{C} \times \mathbf{E})^{-1} \times \mathbf{S}.$$

Thus we see that by means of a matrix inversion method we obtain the heating rate due to any band at all heights. Using Equation (19), we can also obtain the population density of each excited state at any height.

The net heating rate for each band at heights between 60 and 100 km is shown in Figure 2. Heating rates are calculated, assuming that the Sun is overhead for 12 h. Results for absorption of solar radiation by the 2.7μ and 4.3μ fundamental bands are shown separately. In the case of the 4.3μ absorption, the values of H_4 and H_5 must be zero. The results for H_6, the heating rate due to radiative processes in the 6.3μ H_2O band, are not shown. Using a value for the mass mixing ratio of H_2O of 10^{-6}, H_6

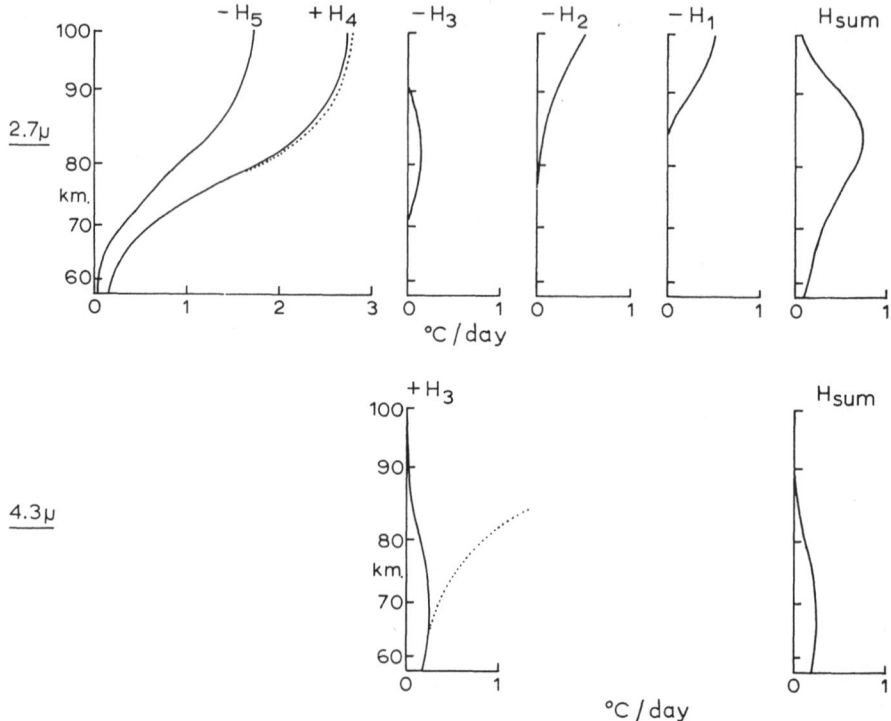

Fig. 2. The net heating rate, when significant, for each band using the notation indicated in Figure 1. Results for absorption in the 2.7μ and 4.3μ bands are shown separately. The dotted curves show the initial absorption in each case.

was found to represent a cooling with a maximum value of only $0.01\,°C/day$. The dotted lines show the initial absorption of solar radiation.

In the case of the 2.7μ absorption, H_4 is close to the dotted curve, showing that little energy is lost from the system due to re-emission at 2.7μ. Considerable energy is lost from the CO_2 $(v_3 + v_1, v_3 + 2v_2)$ level by radiation in the 4.3μ hot band and the net cooling which results $(-H_5)$ is shown. The difference between the curves for H_4 and $-H_5$ represents the amount of energy getting into the N_2^* and CO_2 $(v_1, 2v_2)$ states. Of this, the amount lost by radiative processes in the 4.3μ fundamental band $(-H_3)$ and in the two 15μ bands $(-H_2$ and $-H_1)$ is shown. The actual amount of energy which goes into kinetic energy at any height is given by the sum of all these curves and is shown by H_{sum}.

For the 4.3μ absorption, a considerable amount of energy is lost by re-radiation, even as low as 70 km. This is shown by the fact that H_3 is much less than the initial absorption. Results for H_2 and H_1 are not shown, since cooling for the 15μ CO_2 bands was very small.

If the two curves for H_{sum} are added to give the total heating rate due to the 2.7μ and 4.3μ bands, curve a in Figure 3 is obtained. Curve b is the final result if the reaction

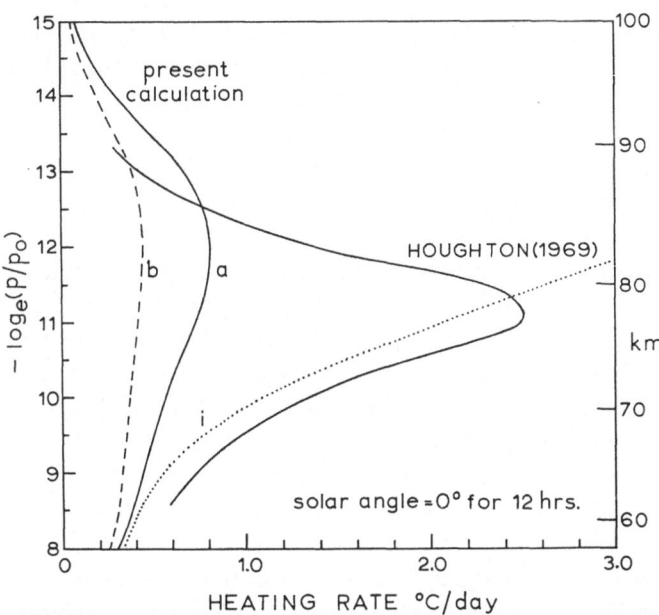

Fig. 3. The total net heating rate due to solar absorption by the 2.7μ and 4.3μ bands, compared with the results of Houghton (1969). Curves *a* and *b* are with the reaction rate of process (16) equal to 10^{10} s^{-1} and 10^7 s^{-1} respectively. The dotted line shows the initial solar absorption.

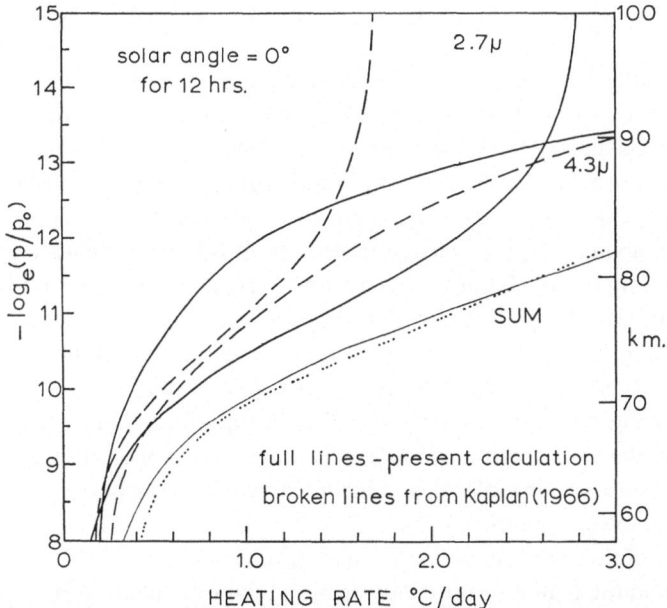

Fig. 4. The initial solar absorption; that is, the heating rate which would result if all energy went into kinetic energy at the level of initial absorption. Results are shown for the 2.7μ and 4.3μ bands separately, and their sum. The broken lines are the result of Kaplan (1966).

rate of Equation (16) is 10^7 s^{-1}. As can be seen, the heating rate found using more accurate radiative transfer calculations is significantly smaller than Houghton's result.

For the initial absorption of solar radiation, the curves from Kaplan (1966) are shown in Figure 4, together with my results. The discrepancy between the curves for the 4.3μ band could be explained if Kaplan included the effect of solar absorption by the weaker hot bands. The difference in our results for the 2.7μ absorption is only partly explained by my use of larger values of the band strength and solar flux. These discrepancies account for the differences between Houghton's curve and my calculation (a in Figure 3) only in the upper and lower limits of the height range.

The relative effect of initial absorption of solar radiation by the 4.3μ hot bands is largest in the neighbourhood of 80 km. The rate constants of the collisional relaxation processes of the upper states produced are not well known. Further, any deviation of the population densities of the lower absorbing states from a Boltzmann distribution, modifies the absorption coefficients appropriate to the hot bands. In any case, absorption by hot bands is not expected to increase the heating rate, even at 80 km, significantly.

The main reason for the large difference in the 65 to 85 km region lies in the methods used to describe the re-absorption of radiation. Houghton used a single representative spectral line in order to calculate the median free path of a photon between emission and absorption, and assumes that half the flux at any level is absorbed in that distance. In reality transfer occurs in spectral lines of widely differing strengths; radiation in weak lines and in the wings of strong lines will penetrate greater depths of atmosphere. Thus more radiation is lost from the heights of initial absorption and the heating rate is reduced.

A further difference between our calculations arises in the treatment of the v_3 part of the energy of the CO_2 ($v_3 + v_1$, $v_3 + 2v_2$) state, produced by insolation at 2.7μ. Houghton treats the v_3 part of the energy in the same way as that of the CO_2 (v_3) state. The collisional processes in Equations (8) and (10) are the same, but he allows radiation from the CO_2 ($v_3 + v_1$, $v_3 + 2v_2$) state to be absorbed by CO_2 molecules in the ground state, giving CO_2 (v_3) excited molecules. If this were included in my calculations, the cooling above 80 km represented by $-H_5$ in Figure 2 would be reduced by an amount equal to that getting into CO_2 (v_3). However, at these heights little CO_2 (v_3) energy goes into kinetic energy in any case, so a simulation of Houghton's scheme is not expected to produce a significantly larger heating rate.

Curve b in Figure 2 shows an even smaller heating rate at all levels. These results are correct in the upper region of the height range, but collisional deactivation of the CO_2 (v_1, $2v_2$) state by air molecules, if included, would cause curve b to coincide with curve a in the lower regions.

The results of this work show that the energy of excitation of CO_2 molecules is more easily lost from the region of initial absorption than estimated by Houghton. The net heating rate due to the 2.7μ and 4.3μ absorption was found to have a maximum value of $0.8\,°C$/day at about 83 km. The maximum value for the 4.3μ fundamental band alone is $0.25\,°C$/day at 68 km.

Since energy is readily lost by radiation in the 4.3μ band, the amount going via O_2^* to $H_2O(v_2)$ is small. Therefore, the effect of changing the H_2O mixing ratio will not be very marked. With a mass mixing ratio as large as 10^{-4}, the maximum cooling from $H_2O(v_2)$ in the 6.3μ band is only $0.07\,°C/day$ at about 75 km.

The accuracy of these results is limited by our inadequate knowledge of the reaction rates involved, and indeed the system of reaction equations may be incomplete. The temperature dependence of the reaction rates has not been included. I have already expressed a doubt about the final state of the CO_2 molecule in Equation (10). Also, we have seen that with a more realistic reaction rate of $10^7\ s^{-1}$ for process (16), the energy of the $CO_2(v_1, 2v_2)$ state will be lost by radiative relaxation at high levels (curve b in Figure 2), but at lower levels, may still go into kinetic energy via collisions with air molecules.

Bearing in mind the approximations made, we find that the net heating rate due to absorption of solar radiation by the 2.7μ and 4.3μ CO_2 bands has a maximum value in the region of 80 km of between 0.5 and $1.0\,°C/day$.

The heating rates in this work are calculated assuming the Sun overhead for 12 h. When the variation of solar angle with time of day, latitude, and season is introduced, it is found that the results quoted above are comparable with those for latitude 50° in mid-summer.

References

Curtis, A. R.: 1956, *Proc. Roy. Soc.* **A236**, 156–9.

Elsasser, W. M.: 1942, 'Heat Transfer by Infrared Radiation in the Atmosphere', *Harvard Meteorological Studies No. 6*, Harvard University Press.

Goody, R. M.: 1964, *Atmospheric Radiation*, Oxford University Press.

Houghton, J. T.: 1969, *Quart. J. Roy. Meteorol. Soc.* **95**, 1–20.

Kaplan, L. D.: 1966, 'The absorption of solar radiation by CO₂', in *Les problèmes météorologiques de la stratosphère et de la mésosphère*, Publications de C.N.E.S., Presses Universitaires de France, Paris, pp. 307–12.

Shved, G. M.: 1965, *Bulletin of the Leningrad University, Series Phys. and Chem.*, Issue 1, 4, 67–79. (English translation: The RAND Corporation, P-3597, May 1967.)

LABORATORY MEASUREMENTS OF *D*-REGION
ION-MOLECULE REACTIONS

E. E. FERGUSON

Aeronomy Laboratory, ESSA Research Laboratories, Boulder, Colorado 80302, U.S.A.

I have chosen to concentrate this discussion of laboratory measurements of ion-neutral reactions on *D*-region problems because that is where the most rapid progress is occurring. There have, to be sure, been advances in *E*- and *F*-region data since that subject was reviewed (Ferguson, 1969a) at the Madrid IAGA meeting in September, 1969. Some information about the energy dependences of rate constants has been obtained (Johnsen *et al.*, 1970; Kaneko *et al.*, 1970), some rate constant measurements have been improved (Fehsenfeld *et al.*, 1970), a theory of the $O^+ + N_2 \rightarrow NO^+ + N$ reaction has appeared (O'Malley, 1970), etc. These new data are all quantitative refinements however and have not significantly altered our broad view of the *E*- and *F*-region ion chemistry. The *D*-region ion chemistry was also reviewed (Ferguson, 1970) at the Madrid IAGA meeting but progress is occurring at such a rate that our views on *D*-region ion chemistry are constantly changing in major ways.

In addition to the rapidly increasing quantity of laboratory rate data, a vital role in increased understanding is being provided by direct ionospheric observations, particularly those of Narcisi (1970) on the *D*-region negative ions which have recently been observed for the first time. All in all, our understanding of *D*-region ion chemistry has increased manifold in just the past few years and this evolution in understanding has not yet abated.

As laboratory results are of more interest than laboratory techniques at this symposium, I shall discuss the latter only briefly. Figure 1 shows the flowing afterglow

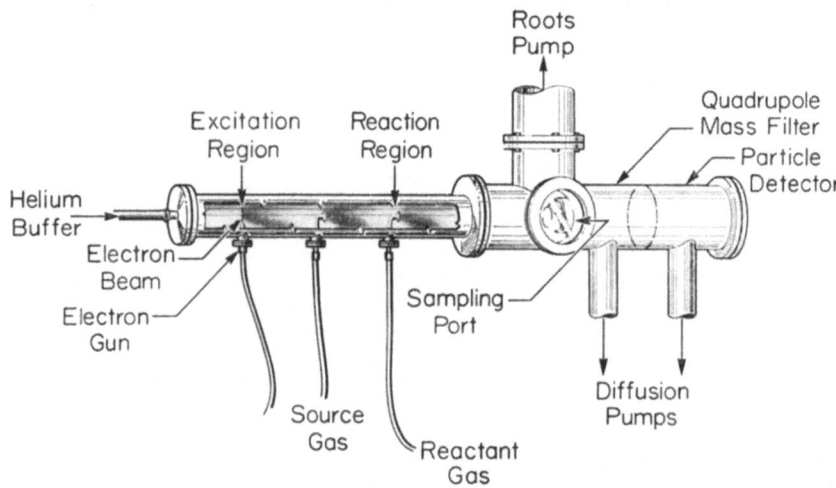

Fig. 1. ESSA Flowing Afterglow System.

Fiocco (ed.), Mesospheric Models and Related Experiments, 188–197. All Rights Reserved.
Copyright © 1971 by D. Reidel Publishing Company, Dordrecht-Holland.

technique developed for ion-molecule reaction rate measurements in our laboratory in Boulder (Ferguson *et al.*, 1969a). This technique is substantially more versatile than any other method available and has accordingly produced a large fraction of the available ionospheric ion-molecule reaction rate constants. Very few rate constants were reliably known prior to the introduction of the flowing afterglow technique in 1964.

Ions are produced in the front end of a meter-long tube of ~ 8 cm diameter and swept down the tube by a fast flowing buffer gas (e.g. helium). At a position downstream a neutral reactant is added. At the end of the tube the ion composition is monitored by a quadrupole mass spectrometer. The decrease of reactant ion as a function of added neutral reactant yields quantitative reaction rate constants. In 'chemically' favorable circumstances (e.g. using reactant gases whose flow is easily measured) we believe that binary reaction rate constants can now be measured to about $\pm 10\%$. This has not been true in earlier years and of course many of the most important ionospheric reactions are not 'chemically' favorable.

The versatility of the method has allowed measurements on such ions as O_4^+, $O_2^+ \cdot H_2O$, Fe^+, SiO^+, O_2^-, O_3^-, CO_3^-, NO_2^-, and many ions which for the most part have not been otherwise studied. Ions have been reacted with unstable neutrals such as O, N, H, OH, and O_3, and H_2O. Farragher *et al.* (1969) at Pittsburgh have extended the flowing afterglow technique to the charge transfer of O_2^+, N_2^+, and NO^+ with Na.

The flowing afterglow technique has been extended to three-body reaction studies (Bohme *et al.*, 1969) and the temperature has been varied from 80–600 K (Dunkin *et al.*, 1968; Ferguson *et al.*, 1969b). Three-body reactions are of great importance in the *D*-region and their rate constants have marked temperature dependences so that measurements in the ~ 200 K *D*-region temperature range are important. The afterglow technique suffers from lack of versatility of third bodies which can be used; the third body in general must be different from the ion and neutral reactant, e.g. the reactions $O_2^\pm + O_2 + O_2 \rightarrow O_4^\pm + O_2$ cannot be measured in this way.

In the last few years other laboratory techniques have been developed for (or extended to) *D*-region ion-neutral reactions. High pressure mass spectrometer ion source methods as practiced by Kebarle and his students (Good *et al.*, 1970) have led to reaction rate constants, equilibrium constants and bond energies for the important *D*-region water ion clusters. The stationary afterglow technique has been successfully applied to *D*-region problems by Lineberger and Puckett (1969). Drift tubes are providing three-body ion-neutral association rate constants without the afterglow restrictions on third body.

The problem of *D*-region positive ion chemistry was given life by the 1963 ion composition measurement of Narcisi and Bailey (1965). The startling finding of Narcisi and Bailey was that below ~ 80 km the major positive ion had a mass 37^+, which was interpreted as being $H_5O_2^+$ (or $H_3O^+ \cdot H_2O$). This result was repeatedly verified over the following years under a variety of geophysical conditions (night and day, during an aurora, meteor shower, and solar eclipse). It remained unexplained until the Third Illinois Aeronomy Conference (Ferguson, 1969) in September, 1968, when the author reported that the major primary *D*-region positive ions produced,

NO^+ and O_2^+, will proceed through reaction sequences that yield hydrated hydronium ions, $H_3O^+ (H_2O)_n$.

Investigations (Fehsenfeld and Ferguson, 1969; Ferguson and Fehsenfeld, 1969) in the ESSA flowing afterglow system established that in the case of O_2^+, $n=1$ and for NO^+, $n=2$.

The O_2^+ reaction scheme is given in Table I. This reaction scheme has been independently verified by Good et al. (1970). Reaction (4b) is probably endothermic and arises from vibrationally excited $O_2^+ \cdot H_2O$ so formed in reaction (3). Current disputes in the literature on the proton affinity of water, and lack of knowledge of the $O_2^+ \cdot H_2O$ bond energy, make this conclusion somewhat uncertain. Reaction (3) is a more important source of $O_2^+ \cdot H_2O$ in the D-region than reaction (2).

Using the reaction scheme (1)–(6), plus dissociative recombination of the ions and O_2^+ loss by

8. $\qquad O_2^+ + NO \rightarrow NO^+ + O_2, \qquad k_8 = 6.3 \times 10^{-10} \ cm^3/s,$

$H_5O_2^+$ ((37$^+$) ion concentration profiles were calculated (Ferguson and Fehsenfeld, 1969) to compare with the Narcisi and Bailey observations. The O_2^+ source is photoionization of $O_2(^1\Delta_g)$, first proposed by Hunten and McElroy (1968).

Two new factors have recently been discovered which alter this problem significantly. Firstly, Fehsenfeld has discovered that the reaction

9. $\qquad O_4^+ + O \rightarrow O_2^+ + O_3$

is fast and this interrupts the $O_2^+ \rightarrow H_5O_2^+$ sequence in an obvious way. A rate constant for (9) has not as yet been determined but it appears likely that $k_9 \sim 10^{-10} \ cm^3/s$. The effect of including reaction (9) is to produce the steep topside $H_5O_2^+$ profile that

TABLE I
D-region $O^+{}_2$ reaction scheme

Reaction	Rate Constant[a]
1. $\qquad O_2^+ + O_2 + O_2 \rightarrow O_4^+ + O_2$	2.8×10^{-30} [a][b]
2. $\qquad O_2^+ + H_2O + N_2 \rightarrow O_2^+ \cdot H_2O + N_2$	2.8×10^{-28} [b]
3. $\qquad O_4^+ + H_2O \rightarrow O_2^+ \cdot H_2O + O_2$	2.2×10^{-9} [b], 1.3×10^{-9} [c]
4a. $\qquad O_2^+ \cdot H_2O + H_2O \rightarrow H_3O^+ \cdot OH + O_2$	1.9×10^{-9} [b], 0.9×10^{-9} [c]
4b. $\qquad \rightarrow H_3O^+ + OH + O_2$	0.3×10^{-9} [b], 0.3×10^{-9} [c]
5. $\qquad H_3O^+ \cdot OH + H_2O \rightarrow H_3O^+ \cdot H_2O + OH$	3.2×10^{-9} [b], $\geqslant 1 \times 10^{-9}$ [c]
6. $\quad H_3O^+ \cdot H_2O + H_2O + N_2 \rightarrow H_3O^+ (H_2O)_2 + N_2$	2.3×10^{-27} [d][c]
7. $\quad H_3O^+ (H_2O)_2 + H_2O + N_2 \rightarrow H_3O^+ (H_2O)_3 + N_2$	2.4×10^{-27} [d], 2.0×10^{-27} [b][c]

[a] All reactions measured near 300 K, units are cm^6/s for three-body reactions, cm^3/s for binary reactions.
[b] k_1 is estimated to be $\sim 10^{-29}$ cm^6/s at D-region temperatures.
[c] k_6 and k_7 are about the same for O_2 and N_2 third body, ref. [c].
[a] Durden, Kebarle, and Good: 1969, J. Chem. Phys. 50, 805.
[b] Fehsenfeld, Mosesman, and Ferguson: 1971, J. Chem. Phys. 55.
[c] Good, Durden, and Kebarle: 1970, J. Chem. Phys. 52, 222.
[d] Good, Durden, and Kebarle: 1970, J. Chem. Phys. 52, 212.

has repeatedly been observed. Figure 2 shows this effect on some calculated profiles using the Hunten and McElroy O_2^+ production rates and 1 ppm H_2O. The fit is really quite good with $k_9 \approx 10^{10}$ cm^3/s, and if this rate constant is found to be about correct the problem of the steep topside dropoff would seem to be solved. The agreement with the rest of the profile is perhaps all right as well; it is known that the Narcisi and Bailey ion densities were too large, because of normalization to probe measurements, and we also believe the lower altitude densities are grossly exaggerated by the method of data reduction and by breakup of heavier water cluster ions in the rocket spectrometer sampling process.

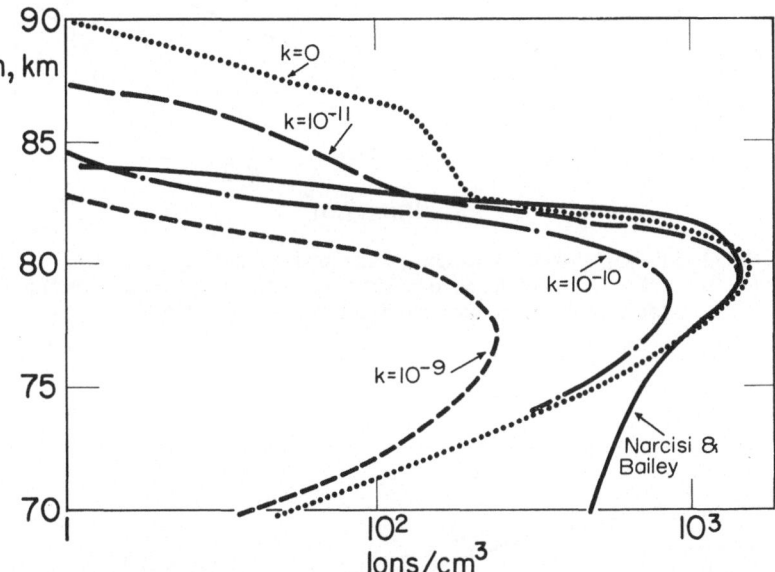

Fig. 2. Calculated $H_5O_2^+$ (37$^+$) profiles assuming rate constants, 0, 10^{-11}, 10^{-10} and 10^{-9} cm^3/s for the reaction $O_4^+ + O \rightarrow O_2^+ + O_3$. Reported 37$^+$ profile of Narcisi and Bailey (1965) shown for comparison. One ppm water mixing ratio (by number), Hunten and McElroy (1968) O_2^+ production.

The value of k_9 also influences the O_2^+ concentration in the D-region. In principle, k_9 could be determined by fitting O_2^+ concentration profiles, but present uncertainties in O_2^+ production rates, D-region O_2^+ concentrations and other factors make this unsatisfactory.

The topside $H_5O_2^+$ profile can be used to put some constraints upon the water concentration in the 80–90 km range. Figure 3 shows some calculations with 10^{-5}, 10^{-6}, 10^{-7}, and 10^{-8} water mixing ratios. The 10^{-5} value (or 10 ppm) is too high to give the steep falloff, the 10^{-7} and 10^{-8} values put the $H_5O_2^+$ ion peak too low. The 10^{-6} value (1 ppm) puts the peak at a reasonable height. The 10^{-6} profile does not differ significantly from a water profile of Anderson's (1970) as shown in Figure 4. Anderson had a 5 ppm water mixing ratio at 50 km and used an eddy diffusion coefficient of 4×10^6 cm^2/s for this particular calculation. This leads to 3 ppm water at

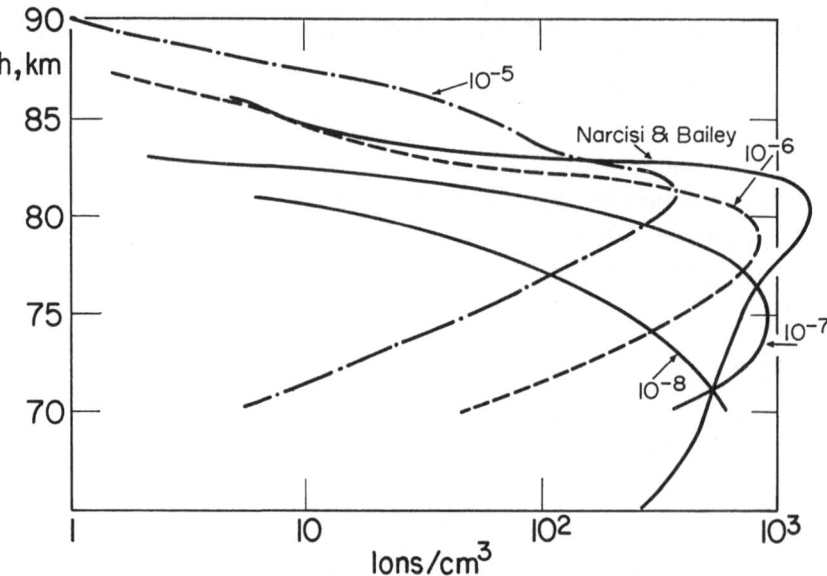

Fig. 3. Calculated $H_5O_2^+$ profiles for assumed water mixing ratios 10^{-5}, 10^{-6}, 10^{-7}, and 10^{-8}. Rate constant for $O_4^+ + O \rightarrow O_2^+ + O_3$ taken to be 10^{-10} cm^3/s. Narcisi and Bailey (1965) profile shown for comparison. Hunten and McElroy (1968) O_2^+ production.

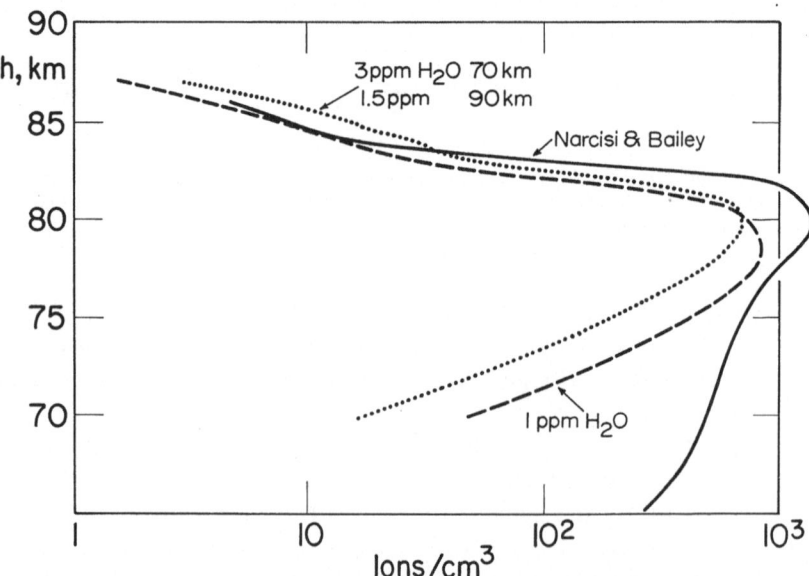

Fig. 4. Calculated $H_5O_2^+$ profile for 1 ppm H_2O mixing ratio and for one of Anderson's (1970) photochemical transport models based on 5 ppm water mixing ratio at 50 km and an eddy diffusion coefficient of 4×10^6 cm^2/s, leading to 3 ppm H_2O at 70 km and 1.5 ppm at 90 km. Rate constant for $O_4^+ + O \rightarrow O_2^+ + O_3$ taken to be 10^{-10} cm^3/s. Hunten and McElroy (1968) O_2^+ production.

70 km, decreasing to 1.5 ppm at 90 km. The need to have about 1 ppm water at 80 km to fit the $H_5O_2^+$ profile implies a lower limit on the eddy diffusion coefficient of $\sim 10^6$ cm^2/s in the Anderson analysis, which incorporates water photodissociation down to 50 km.

The second new discovery in this problem is the recent finding by Huffman *et al.* (1970) that CO_2 and O_2 absorption lower the O_2^+ production rates from the values given by Hunten and McElroy (1968) quite substantially. Figure 5 shows a calculation with O_2^+ production rates supplied by these authors. The lower O_2^+ production now leads to a severe problem in producing a significant $H_5O_2^+$ concentration at the lower

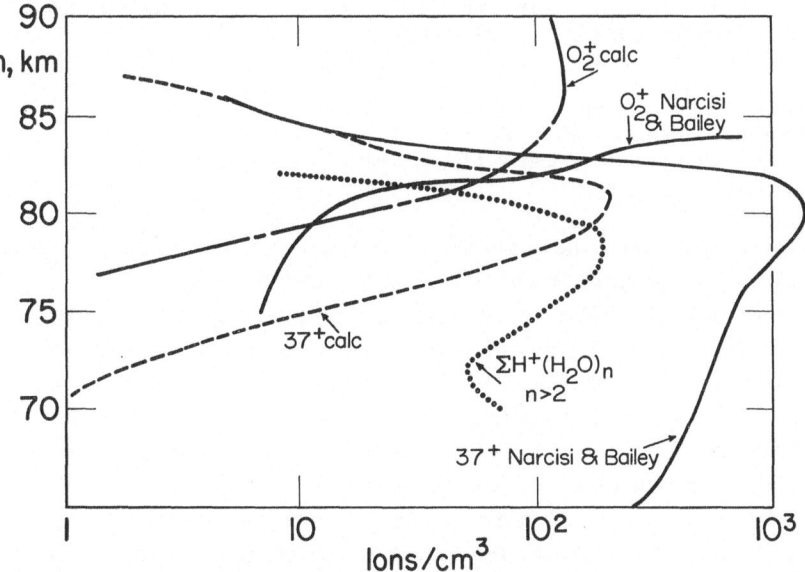

Fig. 5. Calculated O_2^+ and $H_5O_2^+$ (37+) profiles with Huffman *et al.* (1970) O_2^+ production. Narcisi and Bailey (1965) profiles shown for comparison. Dotted curve is an estimate of total concentration of water cluster ions heavier than 37+ produced from O_2^+. Same water profile and rate constants as for Figure 4.

altitudes ($h \lesssim 70$ km). In the light of this development, we will reexamine in the following section the role of NO^+ in leading to water cluster ions. It is believed that NO^+ leads to $H_7O_3^+$ production and not to $H_5O_2^+$ production, and an essential question to be answered before much further progress can be made is what is the true ambient ion composition below ~ 75 km. Specifically, to what extent is the $H_5O_2^+$ concentration that has been reported as low as 65 km representative of the ionospheric composition, and to what extent has it been due to breakup of ambient $H_7O_3^+$ ions in the rocket sampling process.

A major problem of the D-region ion chemistry now is the question of NO^+ production and loss. Accepting currently reported NO^+ production rates would require a faster NO^+ loss than we have been able to come up with. Reid (1970) has recently

TABLE II
D-region NO^+ reaction rates

Reaction	Rate Constant
10. $NO^+ + H_2O + N_2 \rightarrow NO^+ \cdot H_2O + N_2$	1.6×10^{-28} [a][a] 295 K
11. $NO^+ \cdot H_2O + H_2O + N_2 \rightarrow NO^+ (H_2O)_2 + N_2$	1.0×10^{-27} [a] 295 K
12. $NO^+ (H_2O)_2 + H_2O + N_2 \rightarrow NO^+ (H_2O)_3 + N_2$	2.0×10^{-27} [a] 295 K
13. $NO^+ (H_2O)_3 + H_2O \rightarrow H_3O^+ (H_2O)_2 + HNO_2$	0.8×10^{-10} [a] 295 K
14. $NO^+ + O_2 + He \rightarrow NO^+ \cdot O_2 + He$	$< 6 \times 10^{-34}$ [b] 200 K
15. $NO^+ + N_2 + He \rightarrow NO^+ \cdot N_2 + He$	$< 0.5 \times 10^{-32}$ [b] 200 K
16. $NO^+ + CO_2 + He \rightarrow NO^+ \cdot CO_2 + He$	$\sim 1 \times 10^{-29}$ [b] 200 K
17. $NO^+ + CO_2 + N_2 \rightarrow NO^+ \cdot CO_2 + N_2$	$\sim 3 \times 10^{-29}$ [b] 200 K
18. $NO^+ + CO_2 + Ar \rightarrow NO^+ \cdot CO_2 + Ar$	$\sim 3 \times 10^{-29}$ [b] 200 K
19. $NO^+ \cdot CO_2 + H_2O \rightarrow NO^+ \cdot H_2O + CO_2$	Fast [b] 200 K

[a] Lineberger and Puckett (1969) find $k_{10} = 1.5 \times 10^{-28}$ for NO third body.

[a] Fehsenfeld, Mosesman, and Ferguson: 1971, *J. Chem. Phys.* **55**.

[b] Dunkin, Fehsenfeld, and Ferguson: 1971, *J. Chem. Phys.* **54**, 3817.

discussed this problem in the literature and will be discussing it further at this meeting, so I will simply report the recent relevant laboratory measurements.

Table II gives the water reaction scheme proposed by Fehsenfeld and myself (reactions 10–14, 1969) along with some recent quantitative rate constant measurements. Lineberger and Puckett (1969) have also deduced this reaction scheme from stationary afterglow studies and that group is continuing to make quantitative studies of this system. The important point to note so far as hydrated water cluster ion, $H_3O^+ (H_2O)_n$, production is concerned is that the smallest ion produced is $H_7O_3^+$, whereas the major reported water cluster ion reported in the D-region has been $H_5O_2^+$. As mentioned above, the extent to which the breakup $H_7O_3^+ + 1\ eV \rightarrow$ $\rightarrow H_5O_2^+ + H_2O$ occurs in rocket sampling is uncertain. This problem is being actively investigated both at Air Force Cambridge Research Laboratories and at NASA's Goddard Space Flight Center.

In an attempt to determine the major D-region NO^+ loss process, the reactions (14) through (18) are under study in our laboratory. Preliminary indications are that the $NO^+ \cdot O_2$ and $NO^+ \cdot N_2$ clusters are too weakly bound to be significant for NO^+ loss. The main loss process thus far discovered is the association with CO_2,

17. $\quad NO^+ + CO_2 + N_2 \rightarrow NO^+ \cdot CO_2 + N_2$.

The relatively large CO_2 abundance ($\sim 3 \times 10^{-4}$) makes (17) faster than dissociative recombination

20. $\quad NO^+ + e \rightarrow N + O$

below 80 km. We are confident that the reaction

19. $\quad NO^+ \cdot CO_2 + H_2O \rightarrow NO^+ \cdot H_2O + CO_2$

is very fast, $k_{19} \sim 10^{-9}$ cm^3/sec, from qualitative observations and from a generalization from our observations on many such reactions (Adams *et al.*, 1970). Reaction (17) followed by (19) is also probably faster than direct NO$^+$ hydration, since $[CO_2] \approx 10^2$ [H$_2$O], while $k_{17} > 10^{-2} k_{10}$.

The history of the D-region negative ion chemistry has been much different from that of the positive ion chemistry. In the negative ion case, laboratory measurements led to predictions concerning the negative ions which should be dominant prior to direct ionospheric observation. A series of flowing afterglow investigations (Fehsenfeld *et al.*, 1967; Fehsenfeld and Ferguson, 1968; Fehsenfeld *et al.*, 1969) led to the conclusion that NO$_3^-$ (and its hydrates) should be the dominant negative ion, and recent observations by Narcisi (1970) support this for an undisturbed D-region. The reaction scheme involved is outlined in Figure 6.

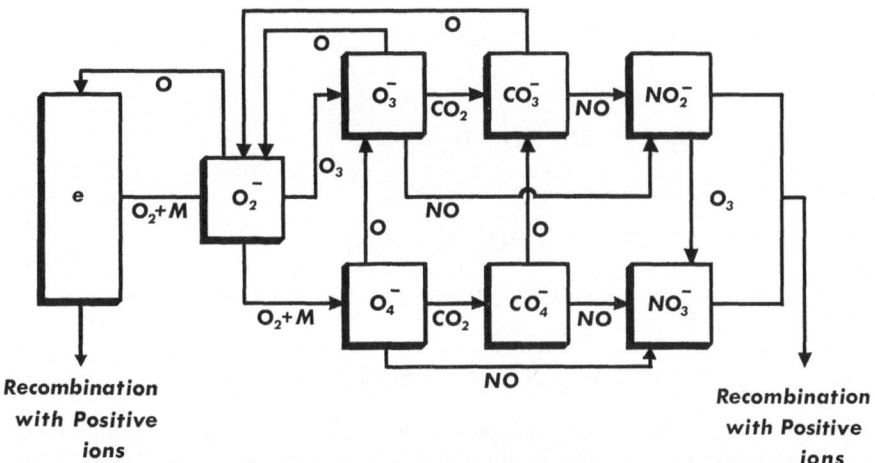

Fig. 6.   Schematic flow diagram of D-region negative ion chemistry, not including water hydration.

Electron attachment and detachment processes were recently reviewed by Phelps (1969) and negative ion-neutral reactions by the author (1969c). The starting point for negative ion chemistry is the three-body attachment of electrons to O$_2$,

21. $$e + 2O_2 \rightarrow O_2^- + O_2.$$

There is then a competition between three possible reactions of O$_2^-$,

22. $$O_2^- + O \rightarrow O_3 + e$$

23. $$O_2^- + O_3 \rightarrow O_3^- + O_2$$

24. $$O_2^- + O_2 + M \rightarrow O_4^- + M$$

The O$_3^-$ and O$_4^-$ ions undergo a rather complex series of reactions, which ultimately lead to NO$_3^-$ as the stable negative ion. There are various routes which may lead to electron detachment along the way, however. The reaction scheme is complex and depends sensitively on the O, O_3, and NO concentrations, which are not precisely known in the D-region, to predict ionospheric negative ion concentrations. The observational data are only qualitative to date as well and it must be said that D-region negative ion chemistry is still in a qualitative stage. The most detailed negative ion composition calculations are those of Reid (1970).

Reaction rate constants are not in very good shape yet either; for example (24) has not been measured at the appropriate D-region temperature. Three-body reactions have a much stronger temperature dependence than two-body reactions in general, the three-body rate constants increasing with decreasing temperature. Pack and Phelps (1970) have recently reported k_{24} to be 4×10^{-31} cm^6/s at 300 K with $M = O_2$.

It is by now clear that any long-lived positive or negative ion in the D-region will become hydrated but the mechanism for that is not yet determined. It is possible, for example, that the sequence of two reactions

25. $NO_3^- + O_2 + M \rightarrow NO_3^- \cdot O_2 + M$

and

26. $NO_3^- \cdot O_2 + H_2O \rightarrow NO_3^- \cdot H_2O + O_2$

might be more effective than direct hydration,

27. $NO_3^- + H_2O + M \rightarrow NO_3^- \cdot H_2O + M$.

In addition to O_2, either N_2 or CO_2 might play this intermediary role.

It is of interest that the reaction

28. $O_2^- \cdot H_2O + CO_2 \rightarrow O_2^- \cdot CO_2 + H_2O$

is rapid, $k_{28} = 4.3 \times 10^{-10}$ cm^3/s (Adams $et\ al.$, 1970). Because of the high CO_2 concentration in the D-region one would not expect to ever observe $O_2^- \cdot H_2O$. Reactions (22), (23), and (24) also preclude the presence of O_2^- in large abundance, of course. Pack and Phelps (1970) have also measured

29. $O_2^- + H_2O + O_2 \rightarrow O_2^- \cdot H_2O + O_2$, $k_{29} = 3 \times 10^{-28}$ cm^6/s

and

30. $O_4^- + H_2O \rightarrow O_2^- \cdot H_2O + O_2$, $k_{30} = 1 \pm 0.3 \times 10^{-9}$ cm^3/s

at 300 K. These should not be too significant in the D-region because of the relatively small H_2O concentration compared to other reactive species.

An important point not yet resolved is whether NO_3^- is truly a 'terminal' ion in the sense that it does not undergo further chemical reaction to yield a different negative ion. It is not known for example whether the reaction

31. $NO_3^- + O \rightarrow O_2^- + NO_2$

is endothermic or exothermic, and we have only established (Fehsenfeld $et\ al.$, 1969) that $k_{31} < 10^{-11}$ cm^3/s while a much smaller value of k_{31} could be significant. Only when quantitative ionospheric negative ion concentrations become available will detailed tests of the laboratory derived reaction scheme become possible.

References

Adams, N. G., Bohme, D. K., Dunkin, D. B., Fehsenfeld, F. C., and Ferguson, E. E.: 1970, *J. Chem. Phys.* **52**, 3133.

Anderson, J. G.: 1970, Thesis, Univ. of Colorado, Dept. of Astrogeophysics; see also 1970, *Trans. Am. Geophys. Union*, **51**, 366.

Bohme, D. K., Dunkin, D. B., Fehsenfeld, F. C., and Ferguson, E. E.: 1969, *J. Chem. Phys.* **51**, 863.

Dunkin, D. B., Fehsenfeld, Schmeltekopf, A. L., and Ferguson, E. E.: 1968, *J. Chem. Phys.* **49**, 1365.

Farragher, A. L., Peden, J. A., and Fite, W. L.: 1969, *J. Chem. Phys.* **50**, 287.

Fehsenfeld, F. C., Dunkin, D. B., and Ferguson, E. E.: 1970, *Planetary Space Sci.*, **18**, 1267.

Fehsenfeld, F. C. and Ferguson, E. E.: 1968, *Planetary Space Sci.* **16**, 70.

Fehsenfeld, F. C. and Ferguson, E. E.: 1969, *J. Geophys. Res.* **74**, 2217.

Fehsenfeld, F. C., Ferguson, E. E., and Bohme, D. K.: 1969, *Planetary Space Sci.* **17**, 1759.

Fehsenfeld, F. C., Schmeltekopf, A. L., Schiff, H. I., and Ferguson, E. E.: 1967, *Planetary Space Sci.* **15**, 373.

Ferguson, E. E.: 1969a, *Ann. Geophys.* **25**, 819.

Ferguson, E. E.: 1969b, Aeronomy Report No. 32, University of Illinois, April 1, 1969, C. F. Sechrist, Jr. (ed.), p. 319.

Ferguson, E. E.: 1969c, *Can. J. Chem.* **47**, 1815.

Ferguson, E. E.: 1970, *Ann. Geophys.* **26**, 589.

Ferguson, E. E. and Fehsenfeld, F. C.: 1969, *J. Geophys. Res.* **74**, 5743.

Ferguson, E. E., Fehsenfeld, F. C., and Schmeltekopf, A. L.: 1969a, in D. R. Bates and I. Estermann (eds.), *Adv. Atomic Molecular Phys.* **5**, 1.

Ferguson, E. E., Bohme, D. K., Fehsenfeld, F. C., and Dunkin, D. B.: 1969b, *J. Chem. Phys.* **50**, 5039.

Good, A., Durden, D. A., and Kebarle, P.: 1970, *J. Chem. Phys.* **52**, 222.

Huffman, R. E., Paulsen, D. E., Larrabee, J. C., and Cairns, R. B.: 1970, *J. Geophys. Res.* **76**, 1028.

Hunten, D. M. and McElroy, M. B.: 1968, *J. Geophys. Res.* **73**, 2421.

Johnsen, R., Brown, H. L., and Biondi, M. A.: 1970, *J. Chem. Phys.* **52**, 5080.

Kaneko, Y., Kobayashi, N., and Kanomata, I.: 1970, *Mass Spectroscopy* **18**, 920.

Lineberger, W. C. and Puckett, L. J.: 1969, *Phys. Rev.* **187**, 286.

Narcisi, R. S.: 1970, *Bull. Amer. Phys. Soc.* **15**, 518.

Narcisi, R. S. and Bailey, A. D.: 1965, *J. Geophys. Res.* **70**, 3687.

O'Malley, T. F.: 1970, *J. Chem. Phys.* **52**, 3269.

Pack, J. L. and Phelps, A. V.: 1970, Symposium on Physics and Chemistry of the Upper Atmosphere, Philadelphia, Penn., June 24–26. *Bull. Amer. Phys. Soc.*, 1971, **16**, 214.

Phelps, A. V.: 1969, *Can. J. Chem.* **47**, 1783.

Reid, G. C.: 1970, *J. Geophys. Res.* **75**, 2551.

THE ROLES OF WATER VAPOR AND NITRIC OXIDE
IN DETERMINING ELECTRON DENSITIES
IN THE D-REGION

GEORGE C. REID

Theoretical Studies Group, ESSA Research Laboratories, Boulder, Colorado 80302, U.S.A.

Many of the basic problems connected with the formation and loss of ionization in the D-region of the ionosphere remain unanswered as yet, despite many years of intensive study. Until quite recently, there was little information on even such a basic quantity as the electron density profile of the normal undisturbed mid-latitude D-region, but marked improvements have been made in the techniques, both rocket-borne and ground-based, for measuring these profiles, and a fairly large number of apparently reliable profiles have been published for a variety of locations, seasons, local times, and solar zenith angles. Unfortunately, the key to understanding these profiles lies in the ion chemistry of the region, and accurate and reliable profiles of ion densities cannot yet be produced, though there have been major advances in the area of determining the relative ion composition through rocket-borne mass-spectrometer techniques. These measurements have shown that there is a large amount of complexity and detail in the ion composition that is not present in the much simpler E and F regions, and that many of the mysterious features of the behavior of the D-region electron-density profiles are probably closely related to the complexity of ion chemistry, which therefore must be understood in as great detail as possible.

Four electron-density profiles obtained at the same solar zenith angle (60°) and geographical location (Wallops Island) for different seasons by Mechtly and Smith (1968) are shown in Figure 1. While differing among themselves in detail, these profiles all show a very steep gradient in electron density (indicated by heavy shading on the profiles) that generally occupies an altitude interval of about 2 km and lies somewhere between 80 and 90 km. In what follows, this feature will be referred to as the 'ledge' in electron density. Recently it has been shown (Reid, 1970) that electron densities above the ledge are generally consistent with a region in which the dominant positive ions are simple molecular species with dissociative recombination coefficients of a few times 10^{-7} cm^3 s^{-1}, but below the ledge the electron densities of the order of a few hundred per cubic centimeter can only be understood in terms of dominant positive ions having recombination coefficients in excess of 10^{-5} cm^3 s^{-1}, unless some very powerful unknown attachment reaction is operating to remove electrons. The negative-ion chemistry of the region, as we presently understand it, leads to the conclusion that attachment processes are not likely to be important during daytime at altitudes above about 70–75 km, well below the location of the ledge. It was suggested that the ledge in electron density coincided with the sharp change in nature of the dominant positive ion species from NO$^+$ above the ledge to the water-cluster ion

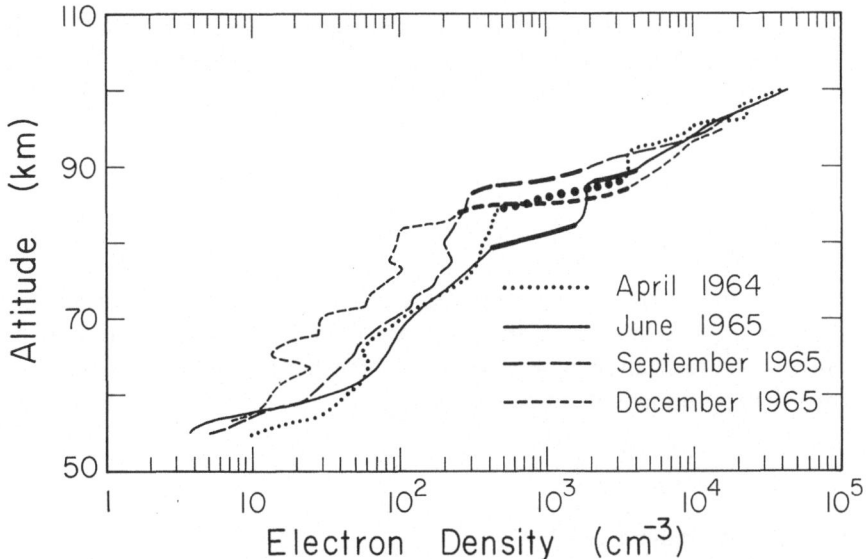

Fig. 1. *D*-region electron density profiles obtained over Wallops Island, Virginia, for a constant solar zenith angle of 60° (after Mechtly and Smith, 1968).

$H_3O^+ \cdot H_2O$ below the ledge as seen by ion mass-spectrometer measurements (Narcisi and Bailey, 1965), and that the water-cluster ions do indeed have extremely large recombination coefficients of the order of 10^{-5} cm^3 s^{-1}.

In this paper I shall extend the work by first of all discussing these high recombination coefficients in a qualitative way, and then by using some recent laboratory measurements of recombination coefficients of water-cluster positive ions to examine conditions in the neighborhood of 80 km with emphasis on the problem of NO^+ in relation to the water-cluster ions.

The large recombination coefficients that are apparently demanded by the water-cluster ions suggest the possibility that these ions recombine with electrons in a way that is different from the recombination of simple molecular ions. The important point is probably the existence of relatively weak bonds, of order 1 eV, in the cluster ions, as opposed to the strong bonds, of order 10 eV, in molecular ions. In a complex cluster ion, where several such weak bonds exist, there are many degrees of freedom, as was first pointed out by Massey and Burhop (1952), and the change in internal energy brought about by the arrival of an electron can be shared among these degrees of freedom. Put in another way, there are a large number of potential-energy curve crossings available to change the ion-electron system into a neutral system. Presumably as the ion grows more complex, a state is ultimately reached in which the recombination process is similar to that of simple Coulomb capture of an electron by a large positively charged surface. In that latter case, it is easy to obtain an estimate for the 'recombination' coefficient, which can then be regarded as an absolute upper limit on cluster ion recombination coefficients.

In order to be attracted to a positively charged body, an electron must suffer a change in energy due to the Coulomb attraction that is comparable in magnitude with its thermal energy; otherwise it will travel past the positive charge without being aware of its existence. Thus its minimum distance from the charge must be less than r, where

$$e^2/r \sim kT \tag{1}$$

i.e.,

$$r \sim e^2/kT. \tag{2}$$

If we assume that every passage within this distance results in recombination, then the effective cross-section, σ, of the positive charge is πr^2, and the recombination coefficient, α, is then given by σv, where v is the thermal velocity of the electron. Inserting the appropriate quantities, we find that

$$\alpha = \pi e^4 \left(\frac{3}{mk^3 T^3} \right)^{1/2} \tag{3}$$

Numerically, this reduces to

$$\alpha = 5.6 T^{-3/2} \tag{4}$$

and for a D-region temperature of $200\,K$ we find an upper limit on the recombination coefficient of $2 \times 10^{-3}\,cm^3\,s^{-1}$. Thus the recombination coefficients needed to explain the electron densities found below the ledge are still about two orders of magnitude less than the theoretical upper limit. The simple theory also predicts a $T^{-3/2}$ temperature variation for the recombination coefficient of very complex ions, a point that we shall use later.

The treatment so far has made use of the observed fact that below the ledge the dominant positive-ion species is normally found to be a water-cluster ion (actually $H_3O^+ \cdot H_2O$ (mass 37)), though this particular ion may at least partially result from the breaking up of more complex water-cluster ions in the shock wave ahead of the rocket or in the aperture field of the mass spectrometer (Narcisi, 1970). The rates of the various reactions leading to the formation of water-cluster ions in the D-region have been determined by Fehsenfeld and Ferguson (1969) and by Good et al. (1970). The water-cluster ions derive mainly from the primary ion O_2^+, which suffers a rapid clustering reaction with neutral O_2 forming the cluster ion O_4^+, which in turn reacts with the water vapor present in the D-region leading to the formation of H_3O^+ and its hydrates. Ferguson and Fehsenfeld (1969) have also shown that the relative concentrations of O_2^+ and the cluster ions found by Narcisi and Bailey (1965) can be satisfactorily explained in this way.

This still leaves a major problem unanswered, however, in that recent measurements of the neutral NO density profile (Barth, 1966; Pearce, 1969; Meira, 1970) have indicated that NO^+ is produced directly in the D-region in greater quantities than O_2^+, in addition to being formed indirectly by charge exchange with O_2^+. While NO^+ can give rise to the water-cluster ions (Fehsenfeld and Ferguson, 1969), the reaction chain is not nearly as efficient or rapid as that originating with O_2^+, and it appears qualitatively as though the dominant positive ions in the vicinity of 80 km are likely to be NO^+ and its first two hydrates $NO^+ \cdot H_2O$ (mass 48) and $NO^+ \cdot (H_2O)_2$ (mass 66). The recombination coefficients of the NO^+ hydrates are unknown, but it is probably safe to assume that they are comparable with those of the H_3O^+ hydrates, so that their presence in large quantities might not pose too much of a problem in explaining the electron densities, but dominance of the NO^+ derivatives would certainly be contrary to the mass-spectrometer results.

In order to examine this question quantitatively as far as possible, a number of computations have been carried out using the positive-ion reaction scheme shown in Table I. Most of the reaction rates appearing there have been measured at room temperature (300 K), and the variation with temperature is unknown in most cases; in a few cases (involving ions of relatively minor importance) the rates quoted are merely estimates based on the rates measured for similar reactions involving similar ion species. The complete scheme involves 16 species of positive ions, but many of these are merely transient species that are destroyed almost as soon as they are formed, resulting in a negligible contribution to the ambient ion composition. Notable among these is N_2^+, which is formed as a primary ion by ionization of neutral N_2 by solar

TABLE I
Positive-ion reactions in the D-region

Reaction	Rate[a] $(cm^3 \ s^{-1}$ or $cm^6 \ s^{-1})$	Source
$N_2^+ + O \rightarrow NO^+ + N$	2.5×10^{-10}	Ferguson *et al.* (1965)
$N_2^+ + O_2 \rightarrow O_2^+ + N_2$	1.0×10^{-10}	Ferguson (1967)
$O_2^+ + NO \rightarrow NO^+ + O_2$	8.0×10^{-10}	Ferguson *et al.* (1965)
$N_2^+ + NO \rightarrow NO^+ + N_2$	5.0×10^{-10}	Ferguson *et al.* (1965)
$NO^+ + H_2O + M \rightarrow NO^+ \cdot H_2O + M$	5.0×10^{-28}	Ferguson (1970)
$NO^+ \cdot H_2O + H_2O + M \rightarrow NO^+ \cdot (H_2O)_2 + M$	1.0×10^{-27}	Ferguson (1970)
$NO^+ \cdot (H_2O)_2 + H_2O + M \rightarrow NO^+ \cdot (H_2O)_3 + M$	3.0×10^{-27}	Ferguson (1970)
$NO^+ \cdot (H_2O)_3 + H_2O \rightarrow H_3O^+ \cdot (H_2O)_2 + HNO_2$	6.0×10^{-10}	Ferguson (1970)
$O_2^+ + O_2 + M \rightarrow O_2^+ \cdot O_2 + M$	1.0×10^{-29}	Durden *et al.* (1969)
$O_2^+ \cdot O_2 + H_2O \rightarrow O_2^+ \cdot H_2O + O_2$	1.6×10^{-9}	Ferguson (1970)
$O_2^+ \cdot H_2O + H_2O \rightarrow H_3O^+ + OH + O_2$	1.2×10^{-10}	Ferguson (1970)
$\rightarrow H_3O^+ \cdot OH + O_2$	1.2×10^{-9}	Ferguson (1970)
$H_3O^+ \cdot OH + H_2O \rightarrow H_3O^+ \cdot H_2O + OH$	1.2×10^{-9}	Ferguson (1970)
$H_3O^+ \cdot OH + O_2 + M \rightarrow H_3O^+ \cdot OH \cdot O_2 + M$	1.0×10^{-29}	(estimate)
$H_3O^+ \cdot OH \cdot O_2 + H_2O \rightarrow H_3O^+ \cdot OH \cdot H_2O + O_2$	1.0×10^{-9}	(estimate)
$H_3O^+ \cdot OH \cdot H_2O + H_2O \rightarrow H_3O^+ \cdot (H_2O)_2 + OH$	1.0×10^{-9}	(estimate)
$H_3O^+ + H_2O + M \rightarrow H_3O^+ \cdot H_2O + M$	5.0×10^{-27}	Ferguson (1970)
$H_3O^+ \cdot H_2O + H_2O + M \rightarrow H_3O^+ \cdot (H_2O)_2 + M$	3.0×10^{-27}	Ferguson (1970)
$H_3O^+ \cdot (H_2O)_2 + H_2O + M \rightarrow H_3O^+ \cdot (H_2O)_3 + M$	4.0×10^{-27}	Ferguson (1970)
$NO^+ + CO_2 + M \rightarrow NO^+ \cdot CO_2 + M$	4.0×10^{-29}	Ferguson (1970)
$NO^+ \cdot CO_2 + H_2O \rightarrow NO^+ \cdot H_2O + CO_2$	1.0×10^{-9}	Ferguson (1970)
$O_2^+ + H_2O + M \rightarrow O_2^+ \cdot H_2O + M$	6.0×10^{-28}	Ferguson (1970)

[a] Several of the rates quoted here are somewhat larger than those appearing in the References. The increase represents an attempt to extrapolate to mesospheric conditions ($T \sim 200 K$) from laboratory conditions ($T \sim 300 K$).

EUV and X-radiation and by cosmic rays. The rapidity of the charge exchange reaction with O_2, however, results in the lifetime of an N_2^+ ion at 80 km being only about 10^{-4} s, so that N_2^+ is a completely negligible constituent, and for most purposes its rate of production can be considered as equivalent to a small addition to the O_2^+ production rate.

In order to compute the ambient number densities of all these ion species, we must known the altitude profiles of the various neutral constituents entering the reactions, and the recombination coefficients of the individual ions. These are both areas of considerable uncertainty, and for that reason the results of the computation can only be regarded as a rough quantitative investigation, aimed at looking for major features, rather than a numerically accurate calculation of the ion composition. For the sake of brevity we shall confine ourselves in this paper to an altitude of 80 km, which lies below the normal location of the ledge; similar considerations will, however, apply to all altitudes between the ledge and the top of the negative-ion region, which we presently believe to be in the vicinity of 70 km.

The third body entering into all the three-body reactions in Table I was assumed to be either N_2 or O_2, so that the third-body number density, $[M]$, is simply the total particle number density. This, and the density of O_2, were taken from the CIRA 1965 model atmosphere. The density of atomic oxygen, which is relatively unimportant in these calculations, was taken from a recent time-dependent calculation including both photo-chemicel and transport effects by Shimazaki and Laird (1970), and the density of neutral nitric oxide from a recent measurement by Meira (1970). The water-vapor number density was based on an assumed mixing ratio of 3 parts per million, consistent with current estimates, though the effect of varying this mixing ratio will be discussed later. Carbon dioxide was assumed to have a mixing ratio of 3×10^{-4}, similar to that found in the lower atmosphere.

The recombination coefficients adopted for the various species are listed in Table II, and require a brief explanation. Reliable laboratory measurements of the recombination coefficients at mesospheric temperatures have been reported only for the molecular species N_2^+ (Kasner and Biondi, 1965), O_2^+ (Kasner and Biondi, 1963), and NO^+ (Gunton and Shaw, 1965), and for the cluster species O_4^+ (Kasner and Biondi, 1968). These values are quoted in Table II. Recently preliminary measurements of the species H_3O^+, $H_3O^+ \cdot H_2O$, and $H_3O^+ \cdot (H_2O)_2$ have been carried out by the University of Pittsburgh group (Biondi, 1970) at a temperature of 400K; the values quoted in Table II for these species represent an estimate of the appropriate values at mesospheric temperatures, assuming the variation with temperature to be proportional to $T^{-3/2}$ as was suggested by the simple analysis in the previous section based on Coulomb attraction between an electron and a massive complex positive charge. The recombination coefficients adopted for the remaining ions are simply estimates based on the observational fact that cluster ions show values in excess of 10^{-6} cm^3 s^{-1}, and that the recombination coefficient increases as the clusters become more complex. In all the calculations carried out, the eight ion species O_2^+, NO^+, $NO^+ \cdot H_2O$, $NO^+ \cdot (H_2O)_2$, H_3O^+, $H_3O^+ \cdot H_2O$, $H_3O^+ \cdot (H_2O)_2$, and $H_3O^+ \cdot (H_2O)_3$ account for more than

TABLE II

Species	Recombination coefficient $(cm^3\ s^{-1})$	Source
N_2^+	3.0×10^{-7}	Kasner and Biondi (1965)
O_2^+	3.5×10^{-7}	Kasner and Biondi (1968)
NO^+	7.5×10^{-7}	Gunton and Shaw (1965)
$O_2^+ \cdot O_2$	2.0×10^{-6}	Kasner and Biondi (1968)
$O_2^+ \cdot H_2O$	5.0×10^{-6}	(estimate)
H_3O^+	3.6×10^{-6}	(see text)
$H_3O^+ \cdot OH$	8.0×10^{-6}	(estimate)
$H_3O^+ \cdot H_2O$	8.0×10^{-6}	(see text)
$H_3O^+ \cdot OH \cdot O_2$	1.0×10^{-5}	(estimate)
$H_3O^+ \cdot OH \cdot H_2O$	1.0×10^{-5}	(estimate)
$H_3O^+ \cdot (H_2O)_2$	1.3×10^{-5}	(see text)
$H_3O^+ \cdot (H_2O)_3$	3.0×10^{-5}	(estimate)
$NO^+ \cdot H_2O$	5.0×10^{-6}	(estimate)
$NO^+ \cdot (H_2O)_2$	7.0×10^{-6}	(estimate)
$NO^+ \cdot (H_2O)_3$	1.0×10^{-5}	(estimate)
$NO^+ \cdot CO_2$	1.0×10^{-5}	(estimate)

ninety-nine percent of the ambient ion composition, so that the precise values of the recombination coefficients of many of the species are irrelevant. The major uncertain coefficients are thus those of the NO^+ hydrates.

The remaining unknown quantities needed are the primary production rates of the three species that are directly ionized, i.e., N_2^+, O_2^+, and NO^+. As mentioned above, although appreciable quantities of N_2^+ are produced, the rapidity of its conversion to O_2^+ by charge exchange means that for practical purposes we can consider O_2^+ and NO^+ as the only primary ions. At 80 km, the major source of O_+^2 is ionization of metastable $O_2(^1\Delta_g)$ by solar ultraviolet radiation (Hunten and McElroy, 1968), with a much smaller contribution from X-ray and cosmic-ray ionization of O_2 and N_2. The primary source of NO^+ is ionization of NO by solar Lyman-α radiation. For a solar zenith angle of 60°, corresponding to the electron-density profiles of Figure 1, and for an altitude of 80 km, estimates in an earlier paper (Reid, 1970) gave values of $q(O_2^+)$ of about 1.6 $cm^{-3}\ s^{-1}$, and for $q(NO^+)$ of 6 $cm^{-3}\ s^{-1}$ (the former value was based on a calculation by Hunten and McElroy (1968), and the latter on the NO density quoted by Barth (1966) at 85 km and the assumption of a constant mixing ratio).

Having adopted values for the various recombination coefficients and for the neutral-constituent number densities and ion production rates, we can set up 16 equations for the 16 unknown positive-ion densities, with the condition of charge neutrality providing a 17th equation yielding the electron density. Since the equations are non-linear through the appearance of a recombination term in each, the solution was carried out by initially assuming a value for the electron density, solving the equations for the positive-ion densities, using the charge neutrality requirement to calcuate a new electron density, and iterating until successive values of electron density differed by less than 0.1 percent.

The results of the calculation employing these production rates are shown in the left-hand column of Table III. The electron density of 2.06×10^3 cm^{-3} is much larger than the range of electron densities observed at 80 km as shown in Figure 1, i.e., from about 100 to about 500 cm^{-3}, and the reason for this is evident from the ion composition, since NO$^+$, with its relatively small recombination coefficient, accounts for about eighty percent of the ions. If we define the effective recombination coefficient, α_{eff}, through the relation $q = \alpha_{eff} \, n_e^2$, then we find that α_{eff} is about 1.8×10^{-6} cm^3 s^{-1}, which is an order of magnitude smaller than the recombination coefficients adopted for the water-cluster ions.

TABLE III
Positive ion composition at 80 km

$q(O_2^+)$	(cm^{-3} s^{-1}):	1.6		1.6	
$q(NO^+)$	(cm^{-3} s^{-1}):	6.0		0.92	
n_e	(cm^{-3}):	2093		776	
Mass	Species	Percentage		Percentage	
30	NO$^+$	79.8		45.1	
48	NO$^+ \cdot$H$_2$O	15.5	95.8	22.2	68.9
66	NO$^+ \cdot$(H$_2$O)$_2$	0.5		1.6	
19	H$_3$O$^+$	0.6		3.5	
37	H$_3$O$^+ \cdot$H$_2$O	2.7	4.0	20.7	30.0
55	H$_3$O$^+ \cdot$(H$_2$O)$_2$	0.5		5.4	
73	H$_3$O$^+ \cdot$(H$_2$O)$_3$	0.2		0.4	
32	O$_2^+$	0.2		0.6	
	Total	100.0		99.5	

A more recent determination of the neutral NO profile in the upper atmosphere has been made by Meira (1970), who found a pronounced minimum in the NO density near 80 km. Using this measurement, together with the solar Lyman-α flux measured by Smith et al. (1965) for a solar zenith angle near 60°, gives a value for $q(NO^+)$ of about 0.92 cm^{-3} s^{-1} at 80 km. The effect of using this value, and retaining the value of 1.6 cm^{-3} s^{-1} for $q(O_2^+)$, is shown in the right-hand column of Table III. The electron density of 780 cm^{-3} is now much closer to the observed range, but still well outside it, and the water-cluster ions still account for only about thirty percent of the total ion density. The effective recombination coefficient has now increased to $4.1 \times \times 10^{-6}$ cm^3 s^{-1}.

We see, then, that current estimates of ion production rate run counter to observation on two scores: first, they predict that the dominant ion species should be NO$^+$ and its hydrates, rather than the water-cluster ions that have consistently been observed to be dominant at 80 km, and secondly, they predict electron densities that are much larger than the observations allow. The problem of NO$^+$ production rates in the

D-region is, of course, not a new one, and has been discussed by several authors in the past (Aikin *et al.*, 1964; Donahue, 1966; Reid, 1966). The major emphasis in these discussions, however, was placed on the problem of explaining the relatively low electron densities in the *D*-region in view of the relatively large rate of production of electrons by Lyman-α radiation. It now appears that the low electron densities can be explained by the large recombination coefficients of the water-cluster ions, which are observed to be dominant below the ledge, but the NO problem remains in a rather different form, in that it is very difficult to explain this observed dominance of the water-cluster ions in view of the large production rate of NO^+, which can apparently only be converted to the water-cluster ions very slowly.

By using the same type of calculation, it is possible to choose a certain electron density and find the range of separate production rates of O_2^+ and NO^+ that will yield that electron density. In this way limits can be set on the acceptable production rates in the *D*-region. An example of the results of this kind of calculation is shown in Figure 2, again for an altitude of 80 km. The curves are drawn for various constant electron densities ranging from 100 to 2000 cm^{-3}, thus including the range of values observed as shown in Figure 1. The abscissa and ordinate scales are respectively the production rates of O_2^+ and NO^+ (once again the N_2^+ production rate has simply been added to that of O_2^+), and the individual curves show the range of combined production rates needed to produce that particular ambient electron density. Each curve displays a horizontal portion at the left-hand side, where the small O_2^+ production rate

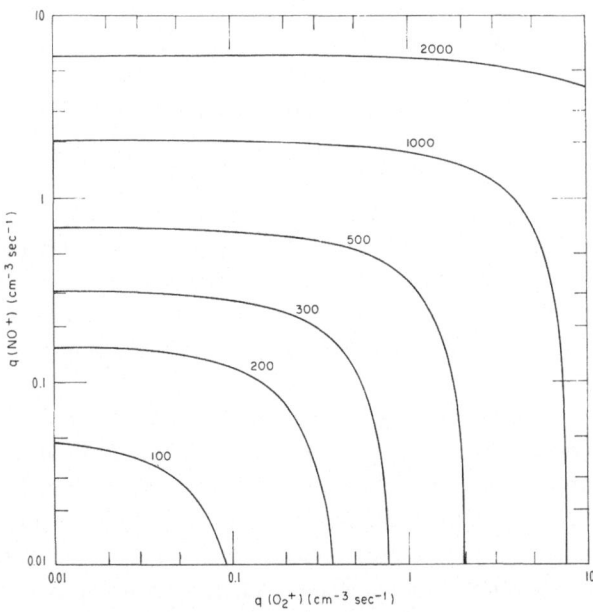

Fig. 2. Ambient electron density at an altitude of 80 km for a solar zenith angle of 60° as a function of the production rates of $O_2{}^+$ and NO^+. The individual curves show the production rates needed to maintain the electron density indicated on the curve (in cm^{-3}).

contributes very little to the electron density, and essentially all the electrons come from NO ionization; in the corresponding vertical regions at the right-hand side practically all the electrons are being produced by O_2 ionization, and the electron density is essentially independent of the NO^+ production rate.

Certain conclusions can immediately be drawn from these curves. If we choose $300 \, \text{cm}^{-3}$ as a typical electron density at 80 km for a solar zenith angle of $60°$, in accordance with the data of Figure 1, we see that even in the complete absence of O_2^+ production the rate of NO^+ production cannot exceed about $0.3 \, \text{cm}^{-3} \, \text{s}^{-1}$, compared to the value of $0.92 \, \text{cm}^{-3} \, \text{s}^{-1}$ that can be predicted from Meira's (1970) NO measurements. As the O_2^+ production is increased, of course, the permissible upper limit of NO^+ production decreases, and at the opposite extreme we find that the O_2^+ production rate cannot exceed about $0.8 \, \text{cm}^{-3} \, \text{s}^{-1}$ if NO^+ production is completely absent. As mentioned above, the rate of production of O_2^+ by ionization of $O_2(^1\Delta_g)$ predicted by Hunten and McElroy (1968) at 80 km for a solar zenith angle of $60°$ is about $1.6 \, \text{cm}^{-3} \, \text{s}^{-1}$. This is only a factor of two larger than the upper limit found above for an electron density of $300 \, \text{cm}^{-3}$, but it must be remembered that this upper limit allows for no NO^+ production at all, and the actual O_2^+ production rate must be considerably lower. A problem therefore exists with both production rates, in that the estimated values for both O_2^+ and NO^+ are individually too large to account for the observed electron densities even if they operate alone.

A partial solution has come from recent calculations of the photoionization cross-section of $O_2(^1\Delta_g)$ carried out by Cairns (1970) and by Huffman et al. (1970). These authors have considered absorption of solar ultraviolet radiation by both O_2 and CO_2 in calculating photoionization rates, and have found that CO_2 absorption greatly reduces the ionization rates in the 80 km region. They find a value of about $0.23 \, \text{cm}^{-3} \, \text{s}^{-1}$ for $q(O_2^+)$ for a solar zenith angle of $60°$ at 80 km, and it can be seen from Figure 2 that this value permits electron densities as low as about $150 \, \text{cm}^{-3}$.

If we again adopt an electron density of $300 \, \text{cm}^{-3}$, together with this estimate of $q(O_2^+)$, we require an NO^+ production rate of about $0.22 \, \text{cm}^{-3} \, \text{s}^{-1}$, and the left-hand column of Table IV shows the results of the calculation using these values. Obviously we now have an electron density that is well within the observed range, but we still have a predominantly NO^+-derived ion composition, despite having reduced the rate of production of NO^+ by about a factor of four from that predicted by the NO observations.

Since all the calculations so far have employed a water-vapor mixing ratio of 3 parts per million, it is worth while at this point to investigate the effect of changing this value, since it presumably has a direct influence on the ion composition. The two right-hand columns of Table IV show the results of carrying out the calculation with water-vapor mixing ratios respectively decreased and increased by a factor of two (i.e., 1.5 and 6 ppm). It can be seen that the effect is quite small, both in the electron density and in the relative ion composition.

It appears, then, that the only way in which we can achieve simultaneously electron densities that are comparable with those observed, and an ion spectrum that is predom-

TABLE IV
Positive ion composition at 80 km

$q(O_2^+)$ (cm^{-3} s^{-1}):		0.23	0.23	0.23
$q(NO^+)$ (cm^{-3} s^{-1}):		0.225	0.225	0.225
H$_2$O mixing ratio (ppm):		3	1.5	6
n_e (cm^{-3}):		298	308	277
Mass	*Species*	*Percentage*	*Percentage*	*Percentage*
30	NO$^+$	31.9 ⎫	32.3 ⎫	31.3 ⎫
48	NO$^+ \cdot$H$_2$O	34.8 ⎬ 71.4	36.8 ⎬ 72.2	32.0 ⎬ 69.7
66	NO$^+ \cdot$(H$_2$O)$_2$	4.7 ⎭	3.1 ⎭	6.4 ⎭
19	H$_3$O$^+$	2.1 ⎫	3.0 ⎫	1.3 ⎫
37	H$_3$O$^+ \cdot$H$_2$O	16.8 ⎬ 28.1	16.3 ⎬ 27.0	15.0 ⎬ 29.9
55	H$_3$O$^+ \cdot$(H$_2$O)$_2$	7.6 ⎭	7.0 ⎭	9.3 ⎭
73	H$_3$O$^+ \cdot$(H$_2$O)$_3$	1.6 ⎭	0.7 ⎭	4.3 ⎭
32	O$_2^+$	0.2	0.2	0.3
	Total	99.7	99.4	99.9

inantly formed of water-cluster ions, is to increase $q(O_2^+)$ to a value somewhere between those predicted by the two calculations presently available, and to reduce $q(NO^+)$ to a value well below even that adopted in the calculations of Table IV. There are obviously an infinite number of such possible solutions available from Figure 2, but as an example we can choose $q(O_2^+)=0.48$ cm^{-3} s^{-1} and $q(NO^+)=0.128$ cm^{-3} s^{-1} (lying on the $n_e = 300$ cm^{-3} curve of Figure 2). The results are shown in Table V, and as expected show a predominance of the water-cluster ions, with the major single species being H$_3$O $\cdot$ H$_2$O (mass 37) in agreement with the observations of Narcisi and Bailey (1965).

TABLE V
Positive ion composition at 80 km

$q(O_2^+)$ (cm^{-3} s^{-1}):		0.48
$q(NO^+)$ (cm^{-3} s^{-1}):		0.128
n_e(cm^{-3}):		293
Mass	*Species*	*Percentage*
30	NO$^+$	18.6 ⎫
48	NO$^+ \cdot$H$_2$O	20.6 ⎬ 42.0
66	NO$^+ \cdot$(H$_2$O)$_2$	2.8 ⎭
19	H$_3$O$^+$	4.4 ⎫
37	H$_3$O$^+ \cdot$H$_2$O	35.3 ⎬ 57.1
55	H$_3$O$^+ \cdot$(H$_2$O)$_2$	14.3 ⎭
73	H$_3$O$^+ \cdot$(H$_2$O)$_3$	3.1 ⎭
32	O$_2^+$	0.5
	Total	99.6

Although the calculations described in this paper are admittedly based on only a rough approximation to D-region conditions, they force us to the conclusion that present estimates of NO^+ production rate are too high by about an order of magnitude to explain either the electron density or the relative positive-ion composition at 80 km altitude. Refinement of the calculation by way of improving the accuracy of the assumed recombination coefficients or the densities of the various neutral species (other than NO) is unlikely to change this conclusion, since it is really based on the fact that there is no very efficient reaction operating to remove NO^+ and produce water-cluster ions instead, analogous to the clustering suffered by O_2^+ with O_2. The only such clustering known to have a large reaction rate in the case of NO^+ occurs with CO_2 (Ferguson, private communication), and has been included in the calculation. CO_2 is, however, a minor constituent by comparison with O_2, and the effect of this reaction is not great.

Since the source of NO^+ production is ionization of neutral NO by solar Lyman-α radiation, we are led to suspect existing estimates of either Lyman-α flux or NO density. The direct measurement of solar Lyman-α flux as a function of altitude by rocket techniques is relatively simple, and has been carried out a number of times. The results agree well with the intensities expected from satellite measurements of Lyman-α flux above the atmosphere and from the profile of molecular oxygen, which is the main absorber of Lyman-α within the atmosphere. It is therefore important to examine the evidence on which the current estimates of NO density in the mesosphere are based. Unfortunately the only technique that has produced data so far has been the measurement of the intensity of the gamma band system of NO as a function of altitude from a rocket. Since the bulk of the NO in the atmosphere lies well above the region of interest here, this technique is relatively insensitive to the NO density with which we are concerned. It would appear very important to attempt to measure the mesospheric NO density by some independent method. Until a reliable independent check on the NO density is available, our understanding of the production and loss of electrons in the D-region will continue to be marred by a very major uncertainty.

Acknowledgements

The author wishes to acknowledge many fruitful discussions of the work described here with his colleagues, especially Drs. E. E. Ferguson and F. C. Fehsenfeld. The work was carried out with the partial support of the Defense Atomic Support Agency.

References

Aikin, A. C., Kane, J. A., and Troim, J.: 1964, *J. Geophys. Res.* **69**, 4621–28.
Barth, C. A.: 1966, *Ann. Geophys.* **22**, 198–207.
Biondi, M. A.: 1970, informal presentation at Fifty-First Annual Meeting, American Geophysical Union, Washington.
Cairns, R. B.: 1970, 'The Photoionization of $O_2(^1\Delta_g)$', paper presented at Fifty-First Annual Meeting, American Geophysical Union, Washington.
Donahue, T. M.: 1966, *J. Geophys. Res.* **71**, 2237–42.

Durden, D. A., Kebarle, P., and Good, A.: 1969, *J. Chem. Phys.* **50**, 805–13.

Fehsenfeld, F. C. and Ferguson, E. E.: 1969, *J. Geophys. Res.* **74**, 2217–22.

Ferguson, E. E.: 1967, *Rev. Geophys.* **5**, 305–27.

Ferguson, E. E.: 1971, this volume, pp. 188–97.

Ferguson, E. E. and Fehsenfeld, F. C.: 1969, *J. Geophys. Res.* **74**, 5743–51.

Ferguson, E. E., Fehsenfeld, F. C., Goldan, P. D., and Schmeltekopf, A. L.: 1965, *J. Geophys. Res.* **70**, 4323–9.

Good, A., Durden, D. A., and Kebarle, P.: 1970, *J. Chem. Phys.* **52**, 222–9.

Gunton, R. C. and Shaw, T. M.: 1965, *Phys. Rev.* **140A**, 756–63.

Huffman, R. E., Larrabee, J. C., and Paulsen, D. E.: 1970, 'New Spectroscopic Data for Oxygen with Application to Metastable Oxygen Ionization Rates', paper presented at Fifty-First Annual Meeting, American Geophysical Union, Washington.

Hunten, D. M. and McElroy, M. B.: 1968, *J. Geophys. Res.* **73**, 2421–8.

Kasner, W. H. and Biondi, M. A.: 1965, *Phys. Rev.* **137A**, 317–29.

Kasner, W. H. and Biondi, M. A.: 1968, *Phys. Rev.* **174**, 139–44.

Massey, H. S. W. and Burhop, E. H. S.: 1952, *Electronic and Ionic Impact Phenomena*, Oxford, 1st Edition, p. 415.

Mechtly, E. A. and Smith, L. G.: 1968, *J. Atmospheric Terrest. Phys.* **30**, 1555–61.

Meira, L. G., Jr.: 1970, 'Rocket Measurements of Upper Atmospheric Nitric Oxide and their Consequences to the Lower Ionosphere', Ph.D. Thesis, University of Colorado.

Narcisi, R. S.: 1970, 'Shock Wave and Electric Field Effects in D-Region Water Cluster Ion Measurements', paper presented at Fifty-First Annual Meeting, American Geophysical Union, Washington.

Narcisi, R. S. and Bailey, A. D.: 1965, *J. Geophys. Res.* **70**, 3687–700.

Pearce, J. B.: 1969, *J. Geophys. Res.* **74**, 853–61.

Reid, G. C.: 1966, 'Ionospheric Implications of Minor Mesospheric Constituents', in *Space Res.* **7** (ed. by R. L. Smith-Rose), Amsterdam, pp. 197–211.

Reid, G. C.: 1970, *J. Geophys. Res.* **75**, 2551–62.

Shimazaki, T. and Laird, A. R.: 1970, *J. Geophys. Res.* **75**, 3221–35.

Smith, L. G., Accardo, C. A., Weeks, L. H., and McKinnon, P. J.: 1965, *J. Atmospheric Terrest. Phys.* **27**, 803–29.

DISCUSSION OF THE FORMATION OF MAJOR
POSITIVE AND NEGATIVE IONS UP TO
THE 50 km LEVEL

VOLKER A. MOHNEN

State University of New York at Albany, Albany, N.Y., U.S.A.

1. Introduction

Tremendous progress has been made in recent years in understanding positive ion formation in the *D*-region. The results of direct measurements (Narcisi, 1969) of *D*-region ion composition can now be understood on the basis of a reaction model calculated with measured reaction rate constants (Fehsenfeld and Ferguson, 1969; Ferguson and Fehsenfeld, 1969). Similar progress has been made on explaining the *D*-region negative ion chemistry, and some very successful results have been obtained in recent months (Ferguson, private communication).

The formation of positive (Mohnen, 1969) and negative (Mohnen, 1970) ions near ground level is now also understood, although details, such as the influence of pollutant gases (especially hydrocarbons), are still under experimental investigation. There exists, however, a serious gap in our knowledge of the nature of ions between 10 and 60 km. Direct mass spectrometric analysis has not been yet made and calculations are rather speculative owing to the uncertainty of rate constants for numerous reactions expected to occur. It is the purpose of this paper to discuss the mechanism of formation of these ions and to predict the major 'terminal' ion species at any altitude up to 50 km.

Near the Earth's surface, ionization is produced mainly by the decay of radioactive emanation and by cosmic radiation. Introducing this energy into the atmosphere leads to many types of chemical reactions including three-body attachment, two-body attachment, positive and negative charge transfer, positive and negative ion molecule reactions, ion neutral associations (two-body and three-body) and binary neutral-interchange reactions. At altitudes up to 60 km, cosmic ray background is still the major ionization source; i.e., photoionization, Bremsstrahlung ionization, collisional ionization, etc., can be neglected. This justifies the extension of our schemes derived at low altitudes (Mohnen, 1969, 1970). Some atmospheric constituents, such as water vapor and ozone, do not have a constant mixing ratio with height, and this must be taken into account by recalculating the reaction schemes as a function of neutral atmospheric composition at various altitudes. These reaction schemes are based on the mean lifetime of ionic species against conversion by any one of the processes mentioned above and are thus a result of solving a set of first-order linear differential equations. The reaction rates, unless otherwise mentioned, are taken from *DASA Reaction Rate Handbook*, and the neutral composition at various altitudes, from U.S. Standard Atmosphere, 1962, and Bortner and Kummler (1968).

Fiocco (ed.), Mesospheric Models and Related Experiments, 210–219. *All Rights Reserved.*

2. Positive Ions

At all altitudes considered here, the initial ion concentration is approximately proportional to the number density of the neutral atmospheric species. Ne^+, H_2^+, CH_4^+, CO_2^+, CO^+, Kr^+, O_3^+, SO_2^+, H_2O^+, Xe^+, H_2O^+ will react via charge transfer with oxygen and thus make O_2^+ the major intermediate ion. At ground level, about half of the initially produced N_2^+ and A^+ will transfer their charge to oxygen; the other half reacts with H_2O and produces H_3O^+. Above 5 km, this process is completely interrupted, and O_2^+ is the only ion left following the initial ionization. It will immediately react with water to form $O_2^+ \cdot H_2O$ and, upon impact with another water molecule, will lead to H_3O^+. Rapid clustering (hydration) finally leads to the terminal ions $H_3O^+ \cdot (H_2O)_n$. Since the concentration of neutral NH_3, I_2, NO_2 and NO together (all having a lower ionization potential than oxygen) is still less than the concentration of water molecules, no charge transfer from O_2^+ to these minor constituents occurs at any level considered here; and since preferential ionization of NO at altitudes below 50 km does not take place, the contribution of NO^+ to the concentration of $H_3O^+ \cdot (H_2O)_n$ is negligible. Figure 1 shows the sequence of positive ion formation valid up to 50 km; the computation shows that the same terminal positive ions are present throughout the atmosphere. This statement holds even up to 85 km, as proven by Fehsenfeld and Ferguson (1969).

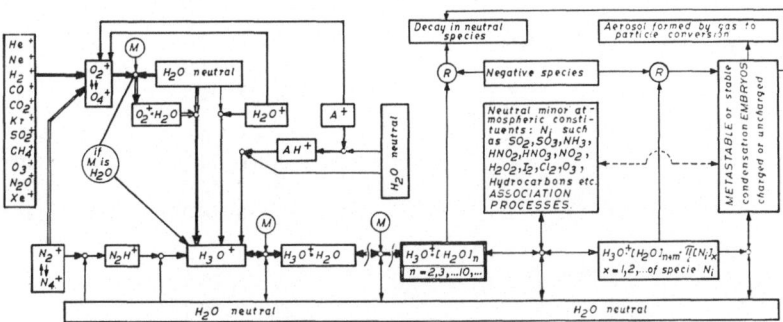

Fig. 1. Reaction scheme for the formation of positive ions (0 to 50 km) and for the formation of aerosols by 'gas-to-particle conversion'.

The next problem to consider is the degree of hydration in the complex $H_3O^+ \cdot (H_2O)_n$. We certainly expect that n follows a statistical distribution function determined by temperature and amount of water vapor present. Kebarle et al. (1967) derived equilibrium constants for $H_2O + H^+ \cdot (H_2O)_{n-1} \rightarrow H^+ \cdot (H_2O)_n$ for $n = 2$–8, in the form

$$\log (K_{n-1, n}) = A/T - B \tag{1}$$

with

$$K_{n-1, n} = \frac{[H^+ \cdot (H_2O)_n]}{[H^+ \cdot (H_2O)_{n-1}] [H_2O]}. \tag{2}$$

In Table I we have summarized equilibrium distribution for some altitudes. Since the water vapor concentration and the temperature undergo quite large changes with altitude and time, it seems advisable to recalculate the $K_{n-1,n}$ values.

In Figure 1 we have indicated a further reaction chain starting with $H_3O^+ \cdot (H_2O)_n$, involving minor atmospheric constituents and described by the term 'associative attachment process'. It has been known for tens of years that the mobility of ions changes 'slowly' with time. The formation of the terminal ions at ground level is achieved in a time less than 10 μsec, whereas this 'ageing effect' has been observed to occur in a millisecond or sometimes a much longer period of time, extending even up to the lifetime of an ion. This effect was not observable in carefully purified gases. Mass spectrometric evidence for this associative attachment process is not available in the literature; however, enough indirect proof seems to exist. Since this leads us to

TABLE I

Equilibrium distribution for $H^+ \cdot (H_2O)_n$ at various altitudes (0 to 50 km)

Height (km)	Temperature (K)	Partial pressure[a] e_{H_2O} (Torr [b])	Distribution of $H^+ \cdot (H_2O)_n$ in %	Mass number
0	288	7.51	$n = 5$: 0.8	91
			$n = 6$: 15.8	109
			$n = 7$: 45.8	127
			$n = 8$: 37.6[c]	145
5	256	0.60	$n = 6$: 7.5	109
			$n = 7$: 43.0	127
			$n = 8$: 49.5	145
10	223	2×10^{-2}	$n = 5$: 0.7	91
			$n = 6$: 24.4	109
			$n = 7$: 56.0	127
			$n = 8$: 18.9	145
20	217	3.9×10^{-4}	$n = 4$: 0.4	73
			$n = 5$: 34.9	91
			$n = 6$: 60.7	109
			$n = 7$: 4.0	127
25	226	2.5×10^{-4}	$n = 4$: 4.5	73
			$n = 5$: 70.5	91
			$n = 6$: 24.4	109
			$n = 7$: 0.6	127
30	250	4.4×10^{-5}	$n = 3$: 3.9	55
			$n = 4$: 87.1	73
			$n = 5$: 8.9	91
50	271	7.0×10^{-6}	$n = 3$: 82.1	55
			$n = 4$: 17.9	73

[a] From Sissenwine et al. (1968).
[b] 1 Torr $= 1.3 \times 10^2$ Nm^{-2}.
[c] Higher hydrates are probable.

the formation of 'condensation embryos', we shall postpone further discussion to Section 4. The associative attachment of molecules (which can be temporary or permanent), such as NH_3, HNO_2, HNO_3, SO_2, NO_2, O_3, etc., is again expected to take place at any altitude wherever their concentration is 'high' enough to allow impact with the ions during their lifetime.

3. Negative Ions

Again, we start from calculations made for ground level conditions (Mohnen, 1970) and investigate extension to higher altitudes by basically the same procedure as in Section 1. We first determine that at any altitude below 50 km dissociative attachment of electrons can be excluded. This makes O_2^- the only precursor for all reaction sequences and leads immediately to the cluster O_4^-. As long as the concentration of H_2O is larger than the concentration of CO_2, the reaction scheme will chain in $O_2^- \cdot (H_2O)_n$ as the dominant ion and in $CO_4^- \cdot (H_2O)_n$ via the binary switching reactions $O_4^- + H_2O \rightarrow O_2^- \cdot (H_2O) + O_2$ and $O_2^- \cdot H_2O + CO_2 \rightarrow CO_4^- + H_2O$ experimentally established in Adams *et al.* (1969), or via $O_2^- \cdot (H_2O)_2 + CO_2 \rightarrow CO_4^- \cdot H_2O + H_2O$. We are unable to predict a time-dependent equilibrium concentration of the two types of clusters present in the lower troposphere.

At altitudes above 8 km, the reaction sequences change to $O_2^- \rightarrow O_4^- \rightarrow CO_4^- \rightarrow$ $\rightarrow CO_4^- \cdot (H_2O)_n$, the latter being the only type of ion present. Above 15 km the increasing concentration of ozone influences the reaction scheme markedly. The following reaction chain is thought to be valid up to 50 km.

$$CO_4^- \cdot H_2O$$

$$O_2^- \xrightarrow[s]{\tau=1/8} O_4^- \xrightarrow{\tau=2 \times 10^{-3}} CO_4^- \overset{\tau=1/10}{\underset{\tau=1/30}{\diagup\diagdown}}$$

$$O_3^- \xrightarrow{\tau=1/40} CO_3^-$$

$$\downarrow \tau=1$$

$$NO_3^- \cdot (H_2O)_n \ldots \leftarrow NO_3^- \cdot H_2O \overset{\tau=0.8}{\leftarrow} NO_2^- \cdot H_2O \overset{\tau=1/3}{\leftarrow} CO_3^- \cdot H_2O$$

$$\downarrow \tau=1/4$$

$$NO_2^- (H_2O)_2$$

The τ-values are estimates for the mean lifetime of each ion species calculated for an altitude of 50 km based on the following concentrations for the neutral atmosphere.

$$N_2 : 1.7 \times 10^{16} \text{ cm}^{-3}$$
$$O_2 : 4.5 \times 10^{15} \text{ cm}^{-3}$$
$$CO_2 : 9.7 \times 10^{12} \text{ cm}^{-3}$$
$$H_2O : 1.2 \times 10^{11} \text{ cm}^{-3}$$
$$O_3 : \ 6 \times 10^{10} \text{ cm}^{-3}$$
$$O_2 (^1\Delta_g) : 3.7 \times 10^{10} \text{ cm}^{-3}$$
$$NO : \ 6 \times 10^{9} \text{ cm}^{-3}.$$

The reactions involved, together with the associated rate constants, are either listed in Mohnen (1970) or taken from Adams *et al.* (1969). This scheme includes various uncertainties due to the lack of information about the possible reactivity of O_3. It is assumed here that $CO_4^- + O_3 \rightarrow O_3^- +$ products, in accordance with the higher electron affinity of ozone (estimated rate constant being 5×10^{-10} cm^3/molec. s). O_3^- will react only with CO_2 leading to CO_3^-, which immediately hydrates to $CO_3^- \cdot H_2O$. A fast binary switching reaction (k estimated 5×10^{-10} cm^3/molec s)

$$CO_3^- \cdot H_2O + NO \rightarrow NO_2^- \cdot H_2O + CO_2$$

is suggested here rather than the commonly used reaction

$$CO_3^- + NO \rightarrow NO_2^- + CO_2,$$

which is extremely slow and would allow the formation of higher hydrates of CO_3^-.

Once $NO_2^- \cdot H_2O$ is formed, the following two possible reactions can occur and are equally likely if the assumed rate constants are true.

$$NO_2^- \cdot H_2O + H_2O + M \rightarrow NO_2^- \cdot (H_2O)_2 + M$$
(k estimated 5×10^{-28} cm^6/molec. s)
$$NO_2^- \cdot H_2O + O_3 \rightarrow NO_3^- \cdot H_2O + O_2$$
(k estimated 2×10^{-11} cm^3/molec. s).

This leads us then to the terminal negative ion, $NO_3^- \cdot (H_2O)_n$. Note that the reaction Chain after CO_3^- is rather speculative. Equilibrium ratios for the hydrated negative ions are not available, but it is obvious that a similar relationship as Equation (1) will describe the temperature dependence of $K_{n-1, n}$.

As in Section 2, we expect the same type of associative attachment processes to

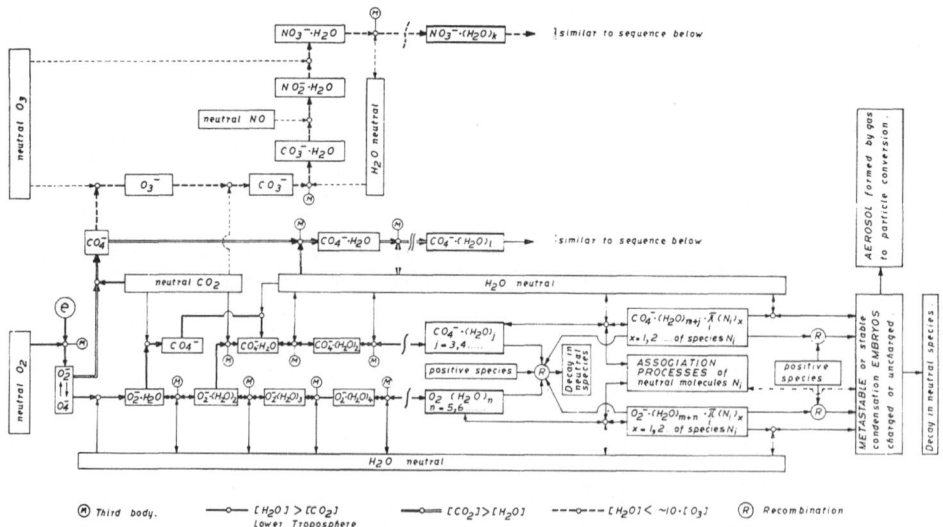

Fig. 2.   Reaction scheme for the formation of negative ions (0 to 50 km) and for the formation of aerosols by 'gas-to-particle conversion'.

occur upon the terminal negative ions. Figure 2 shows the complete negative ion sequence. In summarizing, the nature of negative ions at altitudes between 15 km and 50 km cannot be established theoretically with certainty. For lower altitudes the picture derived here is mostly based on accepted reactions and measured rate constants and therefore believed to be valid, although the degree of hydration is not known.

4. Aerosol

We include the aerosol in our discussion for various reasons. First, if present, it provides at any altitude a surface for ion-pair annihilation. Second, it can act as condensation nuclei under favorable conditions and thus decrease appreciably the partial pressure of water, which determines the degree of hydration of both positive and negative ions (see Equations (1) and (2)). Third, the aerosol could have been produced originally at any altitude by positive and negative ions themselves, involving a so-called 'gas-to-particle conversion' mechanism. Various atmospheric constituents (ionized or neutral) are transferred from a molecular dispersed state into a colloidal state. This aerosol generation is well known at tropospheric level and a general survey has been presented in Mohnen and Lodge (1969). It also occurs in the stratosphere and accounts for the 'Junge layer' (Junge *et al.*, 1961; Scott *et al.*, 1969) at around 25 km height.

Enough experimental proof is available for the existence of aerosol particles throughout the atmosphere, and they need not originate from gas-to-particle conversion. The sizes of particles found in the stratosphere and mesosphere range from less than 50 Å to more than 10 μ. The ultra-small particles appear to be about 100000 times more numerous than particles 0.2 μ in size or larger, as found by Hemenway *et al.* (1961) at an altitude of 26 km. These particles were always found within a larger particle, indicating coagulation processes at work. More recently, Lodge (1969) could demonstrate from rocket samples taken at altitudes between 40 km and 60 km the presence of great numbers of uniform submicroscopic particles 75 Å and less in diameter, which were not incorporated in larger particles. Indirect proof of the existence of great numbers of aerosols with radius less than 0.1 μ has been derived from scattered radiation measurements from infrared and Lidar techniques taken at ground level or aloft.

It seems worth while to recall some basic characteristics of the mechanics of aerosols, such as the sedimentation velocity, the electrical mobility of charged particles and the resulting velocity obtained in an electric field. These parameters have been calculated and are presented in Figures 3 and 4. The velocity of charged particles is derived from the basic relation

$$V = B \cdot E . \tag{3}$$

(B: mobility in cm^2 V^{-1} s^{-1} (Figure 4), E: electric field in V cm^{-1} (Figure 5)).

It is assumed that the aerosol is singly charged. It is, however, easy to take multiple charges on the aerosol into account. Doubling the charge will double the mobility and, therefore, double the velocity obtained in the field.

The calculations have been extended to 70 km to better demonstrate the capability of the Earth's electric field to suspend negatively charged small particles carrying one elementary charge. At altitudes above 60 km, free electrons will strongly bias the charge distribution on the aerosol to the negative fraction and allow, for example, a 30 Å particle to overcome gravity settling up to 68 km. It is obvious from Figure 3 that even a small vertical updraft is much more effective in keeping particles suspended at various altitudes. For example, a typical vertical updraft of the order of 0.2 cm/s would cancel the effect of sedimentation for all particles with radius smaller than 0.1 μ below 50 km. This updraft could, furthermore, transport and keep for any length of time all particles with radius smaller than 100 Å up to at least 70 km.

We now return to the special problem under consideration, aerosol formation initiated by ions or, more generally, by gas-to-particle conversion. Both processes are not well understood theoretically and are part of an extensive experimental investigation being carried out by the author, by Bricard *et al.* (1968) and by Vohra *et al.* (1969). Figures 1 and 2 illustrate how we believe aerosol formation is initiated. By means of dipole attraction, ion induced dipole interaction, or a hydrogen bond, various molecules other than H_2O are being attached (temporarily or permanently) to the terminal positive or negative ions. These complex ions then furnish 'surfaces' for the absorption

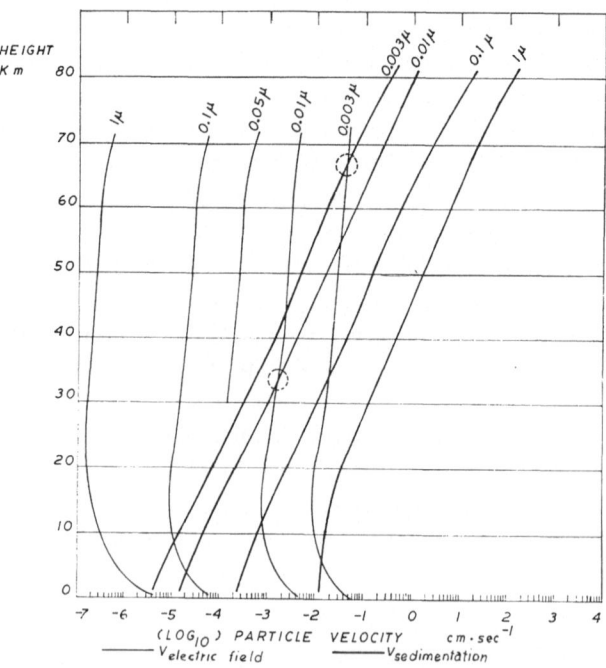

Fig. 3.   Sedimentation velocity V_{sed} and electrical velocity $V_{electr. field}$ of aerosols (parameter: particle radius) as function of altitude (0 to 70 km). Mass density 1.0 g/cm³. Necessary atmospheric parameters are taken from U.S. Standard Atmosphere, 1962.

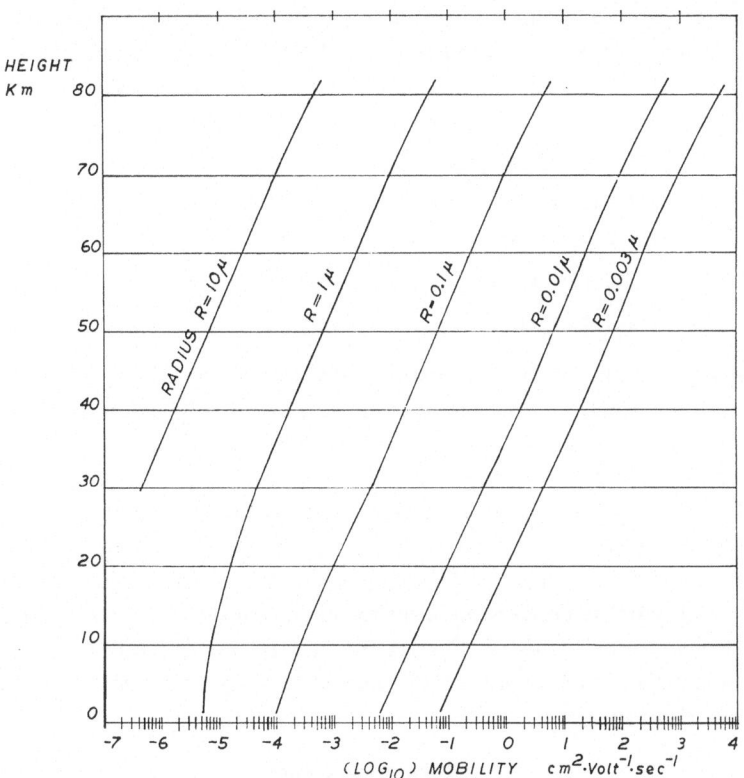

Fig. 4.   Electrical mobilities of aerosol particles carrying one elementary charge as a function of altitude (0 to 70 km). Necessary atmospheric parameters are taken from U.S. Standard Atmosphere, 1962. (Parameter: particle radius.)

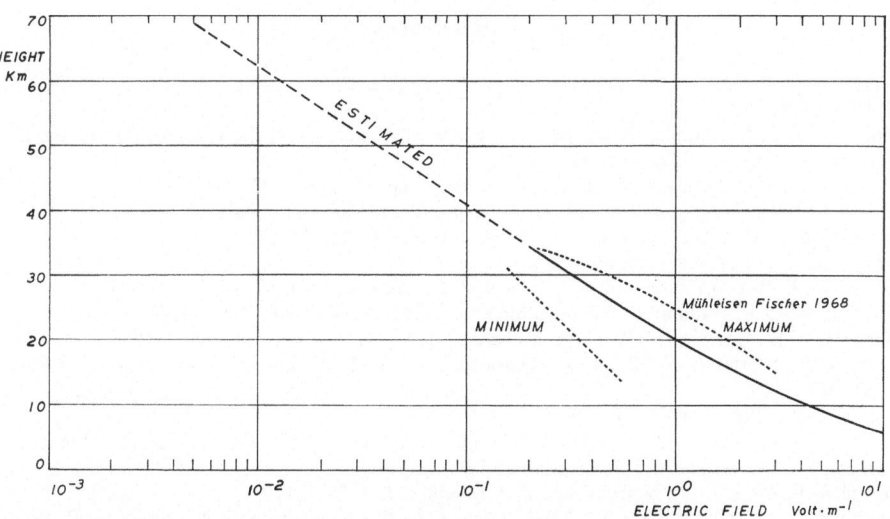

Fig. 5.   Electric field of the Earth as a function of altitude (0 to 70 km). Values above 35 km are estimated by extrapolation.

of excess energy and angular momentum in chemical reactions, enabling the complex to grow and to pick up more impurities. This leads to metastable or stable condensation embryos, as pointed out in the reaction schemes. Of special interest is the recently discovered abundance of HNO_3 in the upper atmosphere (20 to 40 km; Murcray et al., 1969). Complex ions of the form

$$O_2^- \cdot (H_2O)_n \cdot HNO_3$$
$$CO_4^- \cdot (H_2O)_n \cdot HNO_3$$
$$NO_3^- \cdot (H_2O)_n \cdot HNO_3$$
$$H^+ \cdot (H_2O)_n \cdot HNO_3$$

will form stable embryos even after neutralization, since HNO_3 itself is known to initiate aerosol formation although at much slower rates. The aerosol thus formed is very stable against evaporation even under extreme conditions.

Another associative attachment represented by $\pi_i(N_i)_x = NH_3 \cdot SO_2$ has been suggested (Scott et al., 1969) as causing condensation at lower temperatures even without the presence of ions.

In summarizing, associative attachment processes and their possible role in ion induced gas-to-particle conversion seem to play a considerable role in stratospheric and mesospheric ion chemistry. Further investigation is urgently needed, especially relating to neutral minor constituents formed at stratospheric and mesospheric levels, and to their concentration.

Acknowledgments

This work was supported by the Office of Naval Research under contract number N00014-69-C-0043.

References

Adams, N. G., Bohme, D. K., Dunkin, D. B., Fehsenfeld, F. C., and Ferguson, E. E.: 1970, J. Chem. Phys. 52, 3133.
Bortner, M. and Kummler, R.: 1968, 'The Chemical Kinetics and the Composition of the Earth's Atmosphere', G.E.-9500-ECS-SR-1, July.
Bricard, J., Billard, F., and Madelaine, G.: 1968, J. Geophys. Res. 73, 4487–96.
DASA Reaction Rate Handbook, M. H. Bortner (ed.), DASA 1948, 1967.
Fehsenfeld, F. C. and Ferguson, E. E.: 1969, J. Geophys. Res. 74, 2217.
Ferguson, E. E.: private communication.
Ferguson, E. E. and Fehsenfeld, F. C.: 1969, J. Geophys. Res. 74, 5743.
Hemenway, C. L., Fullam, E. F., and Phillips, L. 1961, Nature 190, 4779, 897–8.
Junge, Chr. E., Chagnon, C. W., and Manson, J. E.: 1961, J. Meteorol. 18, 81–108.
Kebarle, P., Searles, S. K., Zolla, A., Scarborough, J., and Arshadi, M.: 1967, J. Amer. Chem. Soc. 89, 6393.
Lodge, J. P.: 1969, Proc. 7th ICCN, Prague.
Mohnen, V. A.: 1969, 'On the nature of tropospheric ions', in S. Coroniti and J. Hughes (eds.). Planetary Electrodynamics, 1, Gordon and Breach.
Mohnen, V. A.: 1970, to be published in J. Geophys. Res. 75, 1717.
Mohnen, V. A. and Lodge, J. P.: 1969, 'General survey of gas-to-particle conversions', Proc. 7th ICCN, Prague.
Murcray, D. G., Kyle, T. G., Murcray, F. H., and Williams, W. J.: 1969, J. Opt. Soc. Amer. 59, 1131.

Narcisi, R. S.: 1969, 'On water cluster ions in the ionospheric D-region', in S. Coroniti and J. Hughes (eds.), *Planetary Electrodynamics*, **2**, Gordon and Breach.
Scott, W. D., Lamb, D., and Duffy, D.: 1969, *J. Atmospheric Sci.* **26**, 727.
Sissenwine, N., Grantham, D. D., and Salmela, H. A.: 1968, 'Humidity up to the Mesopause', AFCRL-68-0050, Oct., p. 46.
U.S. Standard Atmosphere, 1962.
Vohra, K. G., Subba Ramu, M. C., and Vasudevan, K. N.: 1969, 'Studies on the Nucleation of Water Cluster Ions', Proc. 7th ICCN, Prague.

ELECTRONIC COLLISIONS

J. B. HASTED

Birkbeck College, University of London, England

The electronic collision processes of greatest importance to aeronomy are those of excitation, deactivation, attachment and electron-ion recombination. We report resonance electron spectrometry experiments relevant to the first three processes.

Excitation of O_2 to Vibrational and $^1\Delta_g$ Levels

Absolute excitation cross-sections are in some confusion owing to the very rapid energy variation of cross-section which is implied by the reconciliation of the trapped-electron beam experiments of Schulz and Dowell (1962) with the swarm experiments of Hake and Phelps (1967). The former reported very small cross-sections, 3.8×10^{-19} cm^2 for $v'=1$, 8×10^{-20} for $v'=2$, 4×10^{-20} for $v'=3$, 2×10^{-20} for $v'=4$, 3×10^{-20} for $^1\Delta_g$ and 6×10^{-21} for $^1\Sigma_g^+$, but all these were mean values integrated over an energy range 0–0.16 eV above the respective thresholds. The latter interpreted their electron drift velocities and characteristic energies by means of an

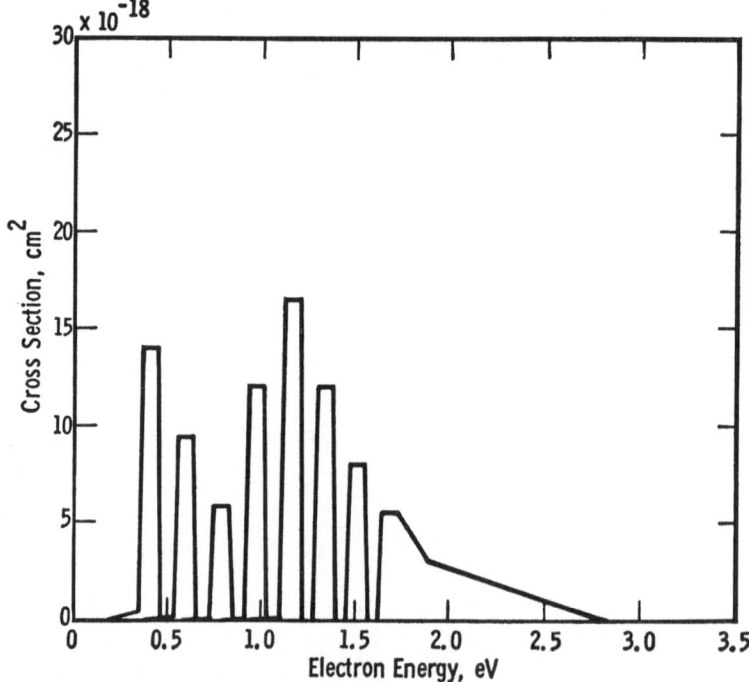

Fig. 1. Individual vibrational excitation cross-sections for electrons in O₂ derived by Hake and Phelps (1967) from transport coefficients.

Fiocco (ed.), Mesospheric Models and Related Experiments, 220–230. All Rights Reserved.
Copyright © 1971 by D. Reidel Publishing Company, Dordrecht-Holland.

excitation function (Figure 1), which contains very much larger cross-sections. Their contention was that, since the electron energy loss cross-sections required by the drift experiments were large, but nevertheless only small cross-sections had been found by Schulz and Dowell at each vibrational threshold, the only consistent function would be comb-shaped. The peaks were taken to commence rising 0.16 eV above each vibrational level of the molecule; each peak corresponds to a loss of energy equal to the vibrational level above which the peak lies.

Although this is from a theoretical point of view an unlikely form of excitation function, some support was lent by the electron spectrometer measurement of Hasted and Awan (1969), who claimed that for each vibrational level a small forward-scattered electron peak could be observed just above threshold. However, extensive operation and improvement of this spectrometer in our laboratory by Mr. Larkin has shown that these peaks are instrumental in origin; some improved vibrational excitation functions are now presented.

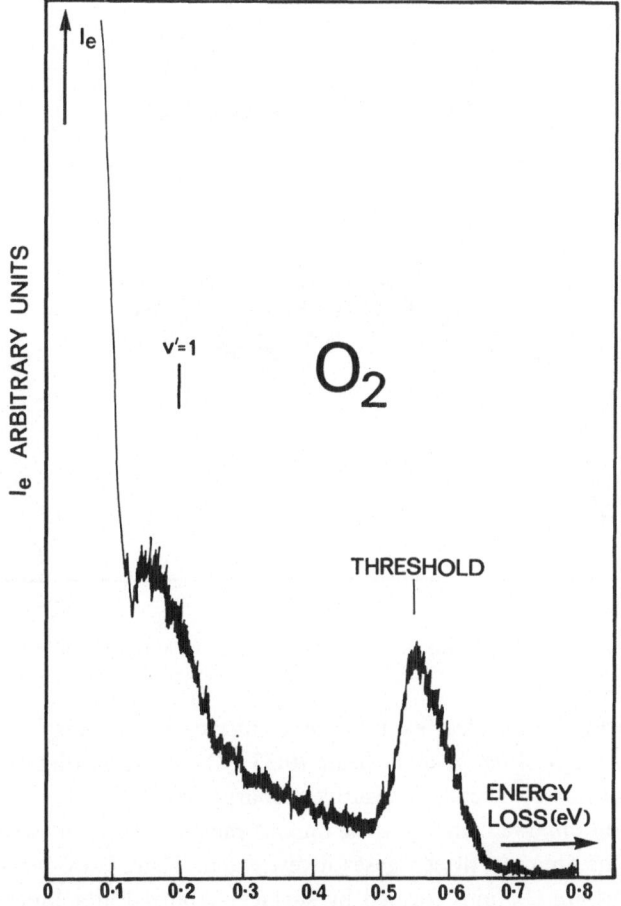

Fig. 2.   Energy loss scan at $\theta = 0°$ for electrons in O_2, taken at impact energy 0.54 eV.

An electron spectrometer consists of a primary monochromated beam of electrons, accelerated to a measured kinetic energy and passed through a collision chamber containing gas; the scattered electrons are monitored at a certain polar scattering angle θ (in this case $0°$) and energy-analyzed before detection. Figure 2 shows the current received in oxygen when the difference between the initial beam energy and the post-collision beam energy is scanned. A peak corresponds to the detection of those electrons which have been scattered after exciting a level of energy equal to this difference. Thus the peak at 0.16 eV, though poorly resolved, corresponds to the $v'=1$ level of oxygen (0.194 eV). The analyzer is held onto this level and the primary beam energy is scanned, thus providing a differential excitation cross-section function, which is shown in Figure 3 for the $v'=1, 2, 3$ levels.

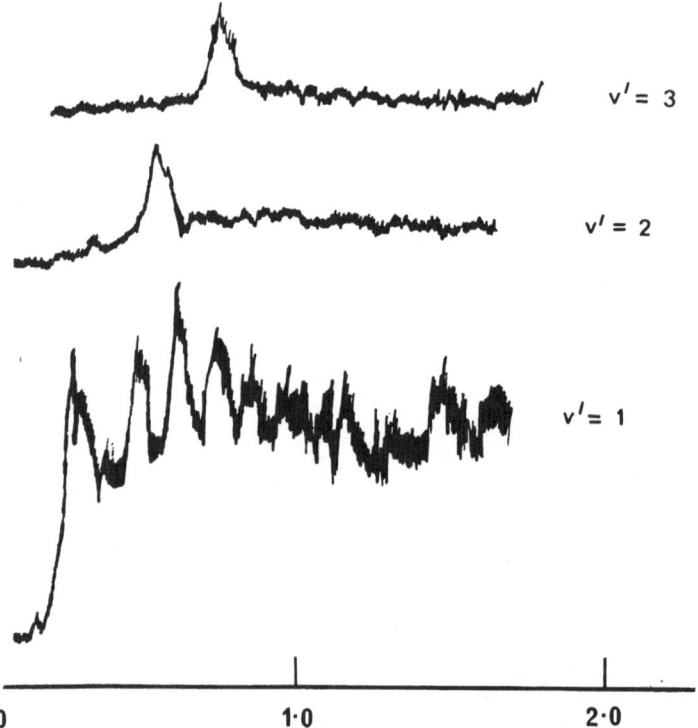

Fig. 3.   Excitation function signals for electrons in oxygen, $v' = 1, 2, 3$.

However in Figure 2 a clear example of an instrumental energy loss peak appears at exactly the energy of the primary beam (0.53 eV). At this setting of the instrument, the analyzer and detector collect negatively charged particles of zero energy. Similar peaks have been observed whatever the impact energy of the primary beam, but only in oxygen, alone amongst fifteen gases investigated. There is evidence that the zero-energy particles are O_2^- ions formed by surface-stabilized attachment, and they are found up to impact energies of 2.5 eV.

The form of our measured differential excitation functions of the O_2 vibrational levels, shown in Figure 3, can therefore be criticized. We believe that part or all of the first peak in the $v' = 1$ function is instrumental. For $v' = 1$ and $v' = 3$ the peaks are entirely instrumental; the real excitation functions can hardly be observed: they must be $\leqslant 10\%$ of that for $v' = 1$.

Our experiments have shown that no cross-section for any level higher than $v' = 1$, or for $^1\Delta_g$ or $^1\Sigma_g^+$, is observable, i.e. is greater than 10% of the cross-section for $v = 1$. Absolute excitation functions can be deduced from measurements with our variable scattering angle spectrometer, but in the meantime we have normalized the $v' = 1$ function (corrected for spectrometer characteristic) against the peaks in the data of Hake and Phelps. The normalized and corrected data are presented in Figure 4. Consistency with the Schulz and Dowell cross-section is still possible provided that the peaks are sufficiently narrow. The width of our observed peaks is probably mostly instrumental (30 meV full width half height resolution of the spectrometer). The real functions fall much closer to zero between the peaks, as does the function we have calculated from Equation (1) below, which is shown as a broken line in Figure 4. This is consistent with the data of Schulz and Dowell. Owing to the possible presence of potential scattering we cannot accurately estimate the line-width of O_2^- $^2\Pi_g$, but a value $\Gamma = 0.005$ eV gives reasonable consistency.

Vibrational excitation of oxygen by electrons takes place by resonance scattering

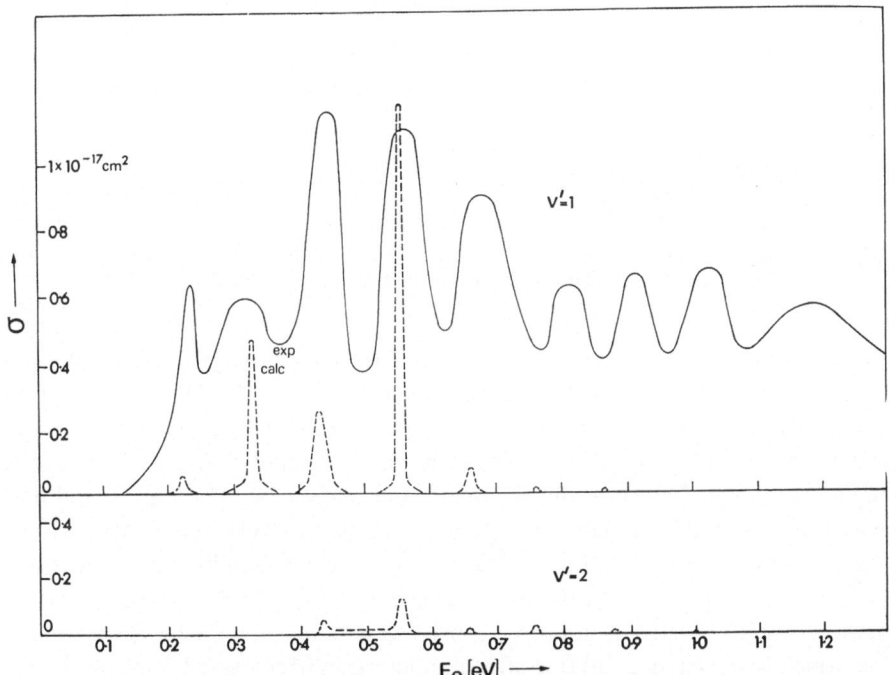

Fig. 4. Normalized excitation functions for electrons in oxygen. Full line, $v' = 1$. Broken lines, calculations from Equation 1, for $v' = 1$ and $v' = 2$.

via temporary states of O_2^-, possibly the higher vibrational levels of the ground $(2p\,\pi_u)^4\,(2p\,\pi_g)^3\,{}^2\Pi_g$ state of O_2^-. According to the compound state theory (Herzenberg, 1967) the cross-section σ is dominated by interference between waves scattered from the different vibrational levels of the negative ion, and (Chen, 1966) at impact energy E_e is proportional to

$$\sigma \propto \left| \sum_{\lambda} \frac{\langle v' \mid \lambda \rangle \langle \lambda \mid 0 \rangle}{E_e - E_a - \lambda h\omega_- c - \tfrac{1}{2}i\Gamma} \right|^2 \tag{1}$$

where the Franck-Condon overlap is calculated between $O_2\,{}^3\Sigma_g^-\,(v=0)$ and $O_2^-\,{}^2\Pi_g\,(v=\lambda)$, and again between $O_2^-\,{}^2\Pi_g\,(v=\lambda)$ and $O_2\,{}^3\Sigma_g^-\,(v=v')$. The electron affinity is E_a and the vibrational constant of O_2^- is ω_-. The line-width of the resonance is Γ, and although it varies with nuclear separation, approximate calculations can be made with single-valued Γ. Successful interpretations of cross-section functions of

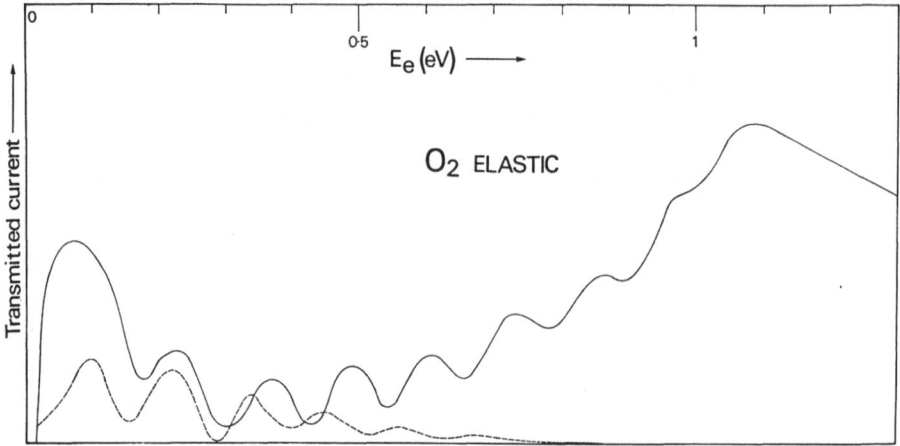

Fig. 5.   Transmission function for elastic scattering of electrons in oxygen (full line), with normalized calculations from Equation 1 (broken line).

other diatomic molecules (Hasted and Awan, 1969) have prompted us to make similar calculations for oxygen, using $E_a = 0.44$ eV (Phelps and Pack, 1961) and $hc\omega_- = 0.11$ eV deduced from the observed oscillations in the elastic scattering function ($v' = 0$). The small cross-sections for $v' = 2$, contrasted with the structured $v' = 0$ and $v' = 1$ functions, are an exceptional feature of this molecule, and many trial values of E_a and nuclear separation r_e for $O_2^-\,{}^2\Pi_g$ have been taken. The best calculations for the $v' = 1$ and 2 functions are illustrated in Figure 4. They are made with equilibrium nuclear separation for $O_2^-\,{}^2\Pi_g$, $r_e = 1.40$ Å; the potential energy curves for the system are shown in Figure 6. The $v' = 1$ curve is not too inconsistent with our observed cross-section function, which is shown as a full line after computerized background removal. The $v' = 2$ cross-sections are about ten times smaller and are considered to be unobservable in the present experiments. The calculated $v' = 3$ excitation functions are smaller than those for $v'=2$ by a factor of 10, and those for higher v' states and for $O_2\,^1\!\Delta_g$ are smaller still. Direct transitions between O_2^- $v=0$ and $O_2\,^3\!\Sigma_g^-$ are Franck-Condon violated, and this explains why the O_2^- photodetachment data (Burch *et al.*, 1958) are dominated by transitions from O_2^- $v=3$, yielding the anomalous electron affinity 0.15 eV. This is of relevance to the sunlight photodetachment rate (Woo *et al.*, 1969).

It is significant that we have been quite unable to fit the scattering data with O_2 electron affinity greater than about 0.6 eV. The sensitivity of the functions to large variations of E_a makes it unlikely that some recent determinations (Stockdale, 1969; Vogt, 1970; Bailey, 1970) could be reconciled with the present data.

The apparent width of the oscillations in the elastic scattering function, shown in Figure 4, is attributable not to the line-width but to a phase shift arising from potential scattering. Although we have fitted these data using an artificial line-width $\Gamma=0.07$ eV in Equation 1, this was undertaken merely to see whether the Franck-Condon factor

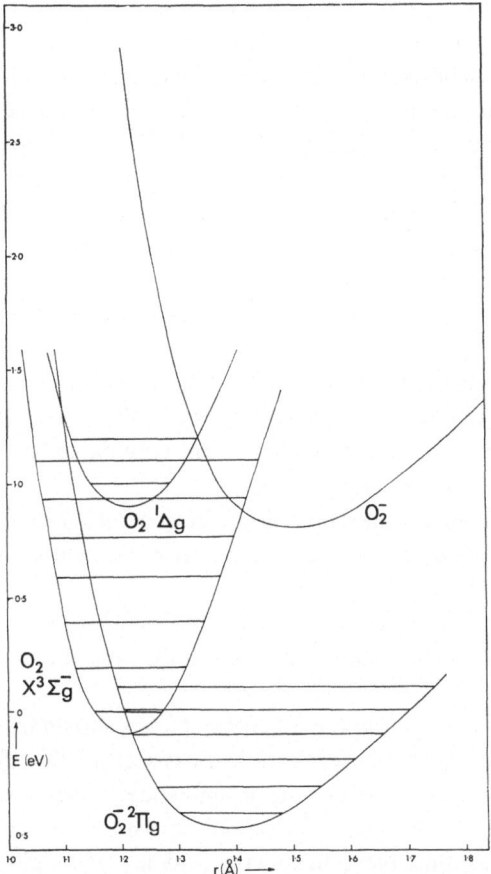

Fig. 6.   Potential energy curves for $O_2\,^3\!\Sigma_g^-$, $O_2\,^1\!\Delta_g$, and $O_2^-\,^2\!\Pi_g$, the latter calculated from parameters giving best fit to the $v'=0$, 1, 2, data. Also included is the unidentified state required to give large $^1\!\Delta_g$ excitation functions from $O_2\,^3\!\Sigma_g^-$ $v=1$.

was capable of extending the oscillations into the 1–2 eV region; there is no appreciable extension, so that contributions from another O_2^- state may be involved. In reality
the function could be fitted by a more complicated equation

$$\sigma = \frac{4\pi}{k^2} \sum_l \sin^2 \delta_{lnr} \left| 1 - (\cot \delta_r + i) \sum_\lambda \frac{\langle \lambda \mid 0 \rangle^2 \frac{1}{2}\Gamma}{E_e - E_a - \lambda hc\omega_- + \frac{1}{2}i\Gamma} \right|^2 \tag{2}$$

where k is the impact electron momentum, δ_{lnr} is the potential (non-resonant) phase
shift for wave of angular quantum number l, and δ_r is the resonant scattering phase-
shift.

Inspection of Figure 6 makes it clear that the formation of the higher vibrational
levels of O_2^- $^2\Pi_g$ from O_2 $v=0$ is precluded by Franck-Condon violation. These
levels are necessary for the formation of higher vibrational channels (states of
O_2 $^3\Sigma_g^-$) and of O_2 $^1\Delta_g$; therefore the excitation of $^1\Delta_g$ from O_2 $v=0$ is unlikely.

Megill et al. (1970) recently made rocket measurements between 70 and 150 km of
the relative intensities (ratio R) of N_2^+ 3914 Å radiation and O_2 $^1\Delta_g$ 1.27 μm radiation
in similar aurorae. Consistent differences of volumes of emission were observed,
implying different mechanisms of excitation. Direct auroral electron excitation would
lead to $R<10$, whereas values between 50 and 1000 are observed. It was proposed by
these authors that there is a mechanism of O_2 $^1\Delta_g$ excitation by ambient electrons in
the aurora, held at a high temperature. Using laboratory cross-sections for N_2^+ 3914 Å
excitation, and measured energies of ion pair creation, Megill et al. calculated that the
efficiencies of converting auroral particle energy into 1.27 μ photons could actually
exceed 100%. It was tentatively proposed that heating of ambient electrons by electric
fields contributed to the radiation energy.

To explain the measured intensities of 1.27 μ, it was necessary to assume a cross-
section for O_2 $^1\Delta_g$ excitation as high as 10^{-17} cm^2. This is consistent with the analysis
of the Hake and Phelps (1967) experiments, but very much larger than the 10^{-21} cm^2
cross-section reported by Schulz and Dowell.

A further experiment on the observation from a rocket of 1.27 μ emission from a
region of the atmosphere exposed to high power radar heating confirms the proposed
mechanism of excitation.

Megill et al. believe that as much as 20% of the O_2 is in the $v=1$ level at 150 km.
These large proportions have been observed in the laboratory by Bader and Ogryzlo
(1964).

On the basis of the oxygen potential energy curves shown in Figure 6, we have calculated from Equation (1) the excitation cross-section functions for excitation of
O_2 $^1\Delta_g$ from $v=0$ and $v=1$ levels of the ground state. It is likely that the cross-section
from $v=0$ will be small, since both the present measurements and those of Schulz and
Dowell would suggest this. Nevertheless the evidence of Megill et al. would suggest a
large cross-section, possibly originating from $v=1$.

However the calculations (normalized in the same way) yield maximum O_2 $^1\Delta_g$
excitation cross-sections 3.3×10^{-20} cm^2 from $v=1$, and 8×10^{-21} cm^2 from $v=0$.

Clearly the Franck-Condon violation is too serious for such a mechanism. Therefore we have repeated the calculations over a wide range of r_e until the Franck-Condon violation is minimized (O_2^- $r_e = 1.52$ Å) and the cross-section maximizes to the values

$$
\begin{array}{cccc}
 & {}^1\Delta_g & {}^1\Delta_g & {}^1\Delta_g \\
 & v' = 0 & v' = 1 & v' = 2 \\
\end{array}
$$

$$
O_2 \ {}^3\Sigma_g^- \quad
\begin{array}{cccc}
v = 0 & 0.29 & 0.004 & 0.016 \times 10^{-17} \ \text{cm}^2 \\
v = 1 & 1.62 & 0.85 & 0.19 \ \times 10^{-17} \ \text{cm}^2.
\end{array}
$$

It is seen that these values are sufficiently large to satisfy the auroral requirements; at the same time the excitation of $O_2 X^3\Sigma_g^-$ $v=0$ is sufficiently unlikely for it to be unobservable in the present measurements. Nevertheless, scattering of electrons from oxygen via O_2^- $r_e = 1.52$ Å, $E_a = 0.44$ eV would involve considerable vibrational excitation of oxygen ($v' = 1, 2, 3 \ldots$). The calculated maximum values of such cross-sections are as follows:

$$v' = 1, \quad 0.79 \times 10^{-17} \ \text{cm}^2$$
$$v' = 2, \quad 0.87 \times 10^{-17} \ \text{cm}^2$$
$$v' = 3, \quad 0.69 \times 10^{-17} \ \text{cm}^2.$$

Therefore it is clear that a state with this nuclear separation and energy, capable though it may be of explaining the auroral measurements, seriously contradicts our laboratory vibrational excitation measurements.

However, such a state, capable of yielding large ${}^1\Delta_g$ cross-sections, could exist at any energy below 0.98 eV. If its energy were, say, 0.8 eV, then it would be responsible for no vibrational excitation of oxygen for electrons below this energy. We have carried out calculations for O_2 ${}^1\Delta_g$ proceeding via a state of O_2^- with $E_a = -0.8$ eV, $\Gamma = 0.07$ eV, $r_e = 1.52$ Å. The maximum excitation cross-section value is $4.0 \times 10^{-17} \ \text{cm}^2$ commencing from $v = 1$, but only $0.4 \times 10^{-17} \ \text{cm}^2$ from $v = 0$. Thus our failure to observe O_2 ${}^1\Delta_g$ excitation in the laboratory is explained, but the possibility of large cross-sections in the aurora, by excitation first to $v = 1$, and thence in a second process to O_2 ${}^1\Delta_g$, is by no means excluded.

Calculations of vibrational excitation via this state show that there is some excitation of $v' = 1$ at energies between 1 and 2 eV. The calculated maximum vibrational excitation functions are

$$v' = 1, \quad 0.19 \times 10^{-17} \ \text{cm}^2 \quad \text{at} \quad 1.45 \ \text{eV}.$$
$$v' = 2, \quad 0.11 \times 10^{-17} \ \text{cm}^2 \quad \text{at} \quad 1.47 \ \text{eV}.$$
$$v' = 3, \quad 0.08 \times 10^{-17} \ \text{cm}^2 \quad \text{at} \quad 1.47 \ \text{eV}.$$

This is not inconsistent with the data in Figures 3 and 4.

The existence of two maxima in the high cross-sections required by the transport data of Hake and Phelps, illustrated in Figure 1, is attributable to the existence of two states of O_2^-, each contributing to vibrational excitation of $v' = 1$.

The probable configuration of the higher state of O_2^- cannot of course be deduced from scattering data without measurements over a range of polar scattering angles. The $(2p\,\pi_u)^3\,(2p\,\pi_g)^4\,^2\Pi_u$ state predicted by Mulliken (1959) is probably at rather higher energy (3.5 eV above $^2\Pi_g$), transitions having been observed in O_2^- in solid alkali halides (Holzer et al., 1968; Rolfe, 1964, 1968). Since presenting this paper at the ESRIN-ESLAB Symposium, a report has been received by the author of molecular orbital calculations by Krauss (1970) of a level $^2\Sigma^+$ lying at 0.6 eV above $O_2\,^3\Sigma_g^-$, and with $r_e = 1.7$ Å. This is sufficiently close to what we have deduced from the auroral, transport and scattering evidence to be regarded as the most probable assignment of the level shown in Figure 6.

Establishment of the O_2^- potential energy curves is important for the discussion of gas temperature variation of the three-body attachment coefficient, and also of deactivation $^1\Delta_g$ by electrons. Calculations for the latter are in progress; a not insignificant cross-section is expected. The electron excitation of oxygen to $^1\Sigma_g^+$ should contain an important contribution from resonance scattering in the region of 2 eV.

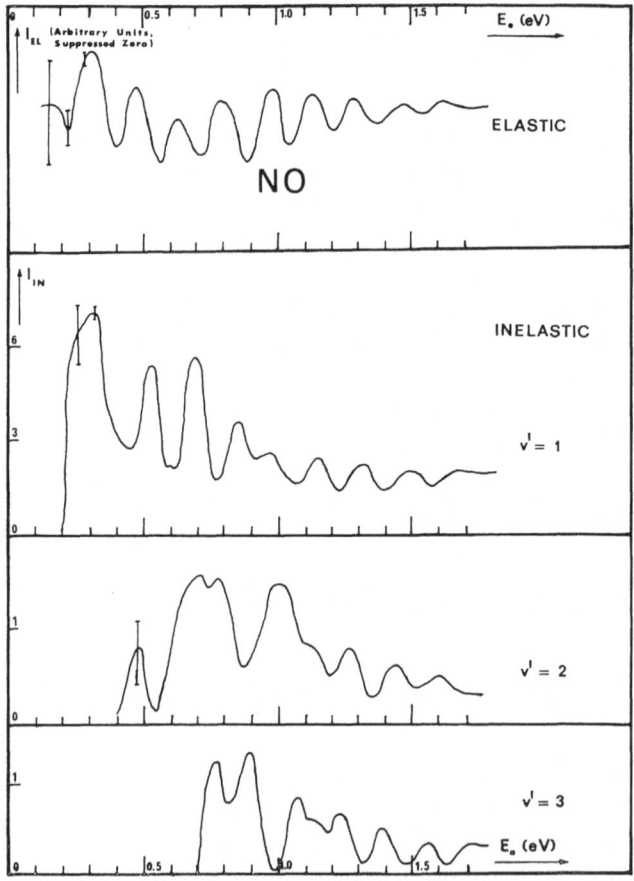

Fig. 7.   Zero angle electron elastic and inelastic scattering functions for NO.

The great sensitivity of all the calculations to vibrational state v of $O_2 X^3\Sigma_g^-$ makes it imperative that gas temperature be varied in electron scattering experiments. This is being undertaken in our laboratory.

Further Resonance Electron Spectrometry Experiments

Resonance electron scattering from N_2 and NO has been reported previously (Hasted and Awan, 1969). New improved data for NO taken by Larkin are presented in Figure 7. They should not result in any serious change in the previously reported values of E_a and r_e for NO^-. Data for N_2O and CO_2 have recently been reported (Larkin and Hasted, 1970). New improved data for NO_2 (previously reported by Boness $et\ al.$ (1968) and presumably arising from the $\tilde{A}$ state of NO_2^- isoelectronic with $O_3\ \tilde{A}$) are presented in Figure 8. These are of importance for the calculation of temperature dependence of the associative detachment processes:

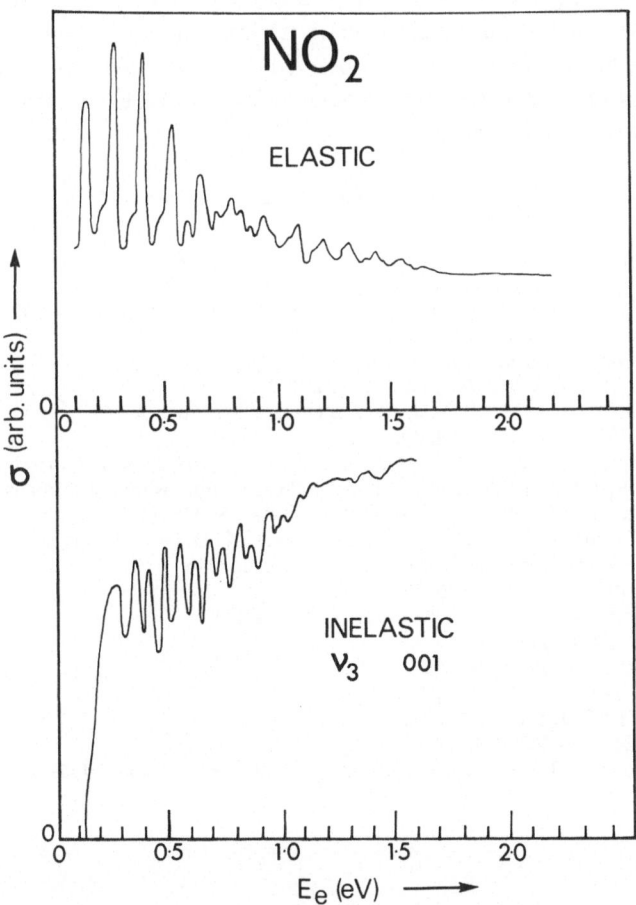

Fig. 8.   Zero angle electron elastic and inelastic scattering functions for NO_2.

$$O^- + N \rightarrow NO + e$$
$$O^- + N_2 \rightarrow N_2O + e$$
$$O^- + CO \rightarrow CO_2 + e$$
$$O^- + NO \rightarrow NO_2 + e$$
$$O_2^- + N \rightarrow NO_2 + e.$$

Similar observed processes

$$O^- + H_2 \rightarrow H_2O + e$$
$$OH^- + H \rightarrow H_2O + e$$
$$H^- + OH \rightarrow H_2O + e$$

and the important

$$O_2^- + O \rightarrow O_3 + e$$

must proceed via negative ion states of H_2O and O_3, and therefore we have searched extensively for structure in the electron scattering functions from these molecules; considerable vibrational excitation is observed, but the absence of structure suggests that the widths of the negative ion levels are in excess of 2 eV, so that they would be observed only with great difficulty.

References

Bader, L. W. and Ogryzlo, E. A.: 1964, *Disc. Faraday Soc.* **37**, 46.
Bailey, T. L.: 1970, *J. Chem. Phys.* **52**, 179.
Boness, M. J. W., Hasted, J. B., and Larkin, I. W.: 1968, *Proc. Roy. Soc.* **A305**, 493.
Burch, D. S., Smith, S. J., and Branscomb, L. M.: 1958, *Phys. Rev.* **112** 171.
Chen, J. C. Y.: 1966, *J. Chem. Phys.* **45**, 2710.
Hake, R. D. and Phelps, A. V.: 1967, *Phys. Rev.* **158**, 70.
Hasted, J. B. and Awan, A. M.: 1969, *J. Phys. B (Proc. Phys. Soc.)* **2**, 367.
Herzenberg, A.: 1967, *Phys. Rev.* **160**, 80.
Holzer, W., Murphy, W. F., Bernstein, H. J., and Rolfe, J.: 1968, *J. Mol. Spectry.* **26**, 543.
Krauss, M. O.: 1970, National Bureau of Standards, Washington. Reported verbally by P. Krupenie.
Larkin, I. W. and Hasted, J. B.: 1970, *Chem. Phys. Letters* **5**, 325.
Megill, L. R., Despain, A. M., Baker, D. J., and Baker, K. D.: 1970, 'Oxygen Atmospheric and Infrared Atmospheric Bands in the Aurora', ESSA, Colorado.
Mulliken, R. S.: 1959, *Phys. Rev.* **115**, 1225.
Phelps, A. V. and Pack, J. L.: 1961, *Phys. Rev. Letters* **6**, 111.
Rolfe, J.: 1964, *J. Chem. Phys.* **40**, 1964.
Rolfe, J.: 1968, *J. Chem. Phys.* **49**, 963.
Schulz, G. J. and Dowell, J. T.: 1962, *Phys. Rev.* **128**, 174.
Stockdale, J. A.: 1969, *J. Chem. Phys.* **50**, 2176.
Vogt, D.: 1970, *Z. Phys.* **232**, 439.
Woo, S. B., Branscomb, L. M., and Beaty, E. C.: 1969, *J. Geophys. Res.* **74**, 2933.

LABORATORY STUDIES OF ELEMENTARY REACTIONS OF
NEUTRAL SPECIES CONTAINING O OR H

B. A. THRUSH

Dept. of Physical Chemistry, University of Cambridge, Cambridge, England

Much information about the reactions of ground state oxygen, hydrogen and nitrogen atoms, of simple radicals such as OH, HO_2 and of metastable molecules such as $O_2(^1\Delta_g)$ and $O_2(^1\Sigma_g^+)$ comes from the study of reactions in discharge-flow systems (Kaufman, 1961; Campbell and Thrush, 1965). In this type of apparatus free atoms are produced by passing the parent molecules through an electrodeless radio-frequency or microwave discharge; typically total pressures between 0.1 Torr and 10 Torr are used to minimise difficulties associated with diffusional effects or pressure drop and the concentrations of active species are usually *ca.* 1% of the total flow. Reactions can be studied at temperatures from 1000 K down to 200 K or below, provided that no species is condensed on the flow tube wall and that the reaction proceeds at a measurable rate. At longer times measurements are limited to about 10 s by losses of atoms to the wall where the fraction of collisions effective in recombination (γ) is typically 10^{-4} to 10^{-6}, and by three-body recombination processes, such as

$$N + N + M = N_2 + M \tag{1}$$

$$O + O_2 + M = O_3 + M. \tag{2}$$

The time required for diffusional mixing places a limit near 5 ms on the shortest reaction times that can be used.

The fastest rate constants that can be measured are governed not only by this factor but also by the sensitivity with which active species can be detected. The use of such sensitive techniques as resonance absorption or fluorescence in the vacuum ultra-violet, in the study of free atoms or gas phase E.S.R. for OH and other linear radicals with dipole moments, makes it possible to study reactions so rapid that they occur at every gas kinetic collision. However, the rate constants often used in aeronomy for such processes as (3), (4) and (5)

$$N + NO = N_2 + O \quad (k_3 = 4 \times 10^{-11} \text{ cm}^3 \text{ s}^{-1}) \tag{3}$$

$$O + OH = O_2 + H \quad (k_4 = 5 \times 10^{-11} \text{ cm}^3 \text{ s}^{-1}) \tag{4}$$

$$N + OH = NO + H \quad (k_5 = 7 \times 10^{-11} \text{ cm}^3 \text{ s}^{-1}) \tag{5}$$

were obtained by competitive methods using the less sensitive technique of chemiluminescence to measure atomic concentrations (Clyne and Thrush, 1961, 1963; Campbell and Thrush, 1968). For example, k_3 was found by competition with reaction (6)

$$N + O_2 = NO + O \quad (k_6 = 1.4 \times 10^{-11} \exp(-7200/RT) \text{ cm}^3 \text{ s}^{-1}) \tag{6}$$

Fiocco (ed.), Mesospheric Models and Related Experiments, 231–239. All Rights Reserved.
Copyright © 1971 by D. Reidel Publishing Company, Dordrecht-Holland.

and k_5 from the competition between reactions (4) and (5). Thus the experimentally determined quantity is k_3/k_6 or k_4/k_5, and the use in an aeronomical calculation of a different expression for k_6 or k_5 from that assumed by the original authors requires the values of k_3 or k_4 used in the calculation to be changed from their published value in the same proportion.

For slower atom transfer reactions such as (6) which cannot compete effectively with the recombination reaction (2) below ambient temperatures, the required rate constant must be obtained by extrapolation from measurements at higher temperatures. There is no experimental evidence that the form $AT^{1/2} \exp(-E/RT)$ (in which the pre-exponential factor is assumed to be proportional to the collision number) fits the behaviour of simple transfer reactions such as (6) or (7) over a wide temperature range

$$O + H_2 = OH + H \tag{7}$$

any better than the simple Arrhenius form $A \exp(-E/RT)$. Indeed the transition state theory (Glasstone *et al.*, 1941) yields expressions for such processes in which A *decreases* slightly with increasing temperature. For reactions such as (7) in which a proton is transferred, its tunneling through the barrier might contribute significantly at low temperatures, producing an upward curvature in the Arrhenius plot.

Such curvature has proved exceptionally difficult to detect, as points near the low temperature limit of experiments are subject to the widest possible experimental error. Figure 1 illustrates this point for reaction (7) where there is exceptional overall agreement between the results of various authors. The four dotted lines at high temperatures refer to various studies in shock tubes. The triangles and filled circles together with the dotted line for intermediate temperatures come from various studies of combustion systems. The remaining data were obtained using discharge flow systems: the crosses by Westenberg and de Haas (1969), the dotted line at low temperatures by Hoyermann *et al.* (1967) using gas phase E.S.R., the open circles by Clyne and Thrush (1963) and Campbell and Thrush (1968) using chemiluminescence. Wong and Potter (1965), whose data are marked by half-filled circles, used a stirred flow reactor with mass spectrometric detection. The lowest temperature point was obtained by Kaufman (1958) under conditions where secondary reactions are known to introduce complications. The line given corresponds to $k_7 = 3 \times 10^{-11}$ $\exp(-9450/RT)$ cm^3 s^{-1} as recommended by Baulch *et al.* (1968) who give a detailed assessment of all but the most recent data on this system. It is clear from Figure 1 that the points at the lowest temperatures, where the tunnel effect might be observed, are the least certain, and that accurate measurements over the wide range in $(1/T)$ which can be obtained in discharge flow systems provide the best hope for achieving a realistic extrapolation over the range of temperatures important in aeronomy.

Apart from reactions in which there is an energy barrier to recombination (such as the spin forbidden combinations of $O(^3P) + CO$ and $O(^3P) + N_2$ where the barrier heights are *ca.* 4 and 20 kcal/mole respectively), three-body combination reactions show negative temperature coefficients. Although a number of theoretical treatments predict this property and give overall rates of recombination similar to those observed, the experimental temperature dependences are generally greater than the theoretical ones, and no theory has yet satisfactorily explained the wide range of third-body efficiencies encountered. Both these factors depend on the nature of the interaction with the third body. Indeed there is no sound theoretical preference for representing the data by either an AT^{-n} or an $A \exp(+E/RT)$ form. For iodine atom recombination Russell and Simons (1953) noted an approximately linear relation between the

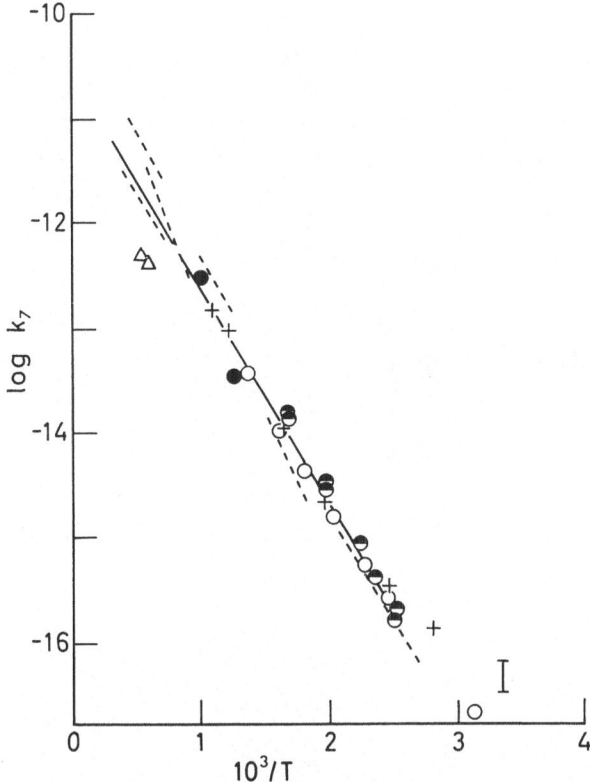

Fig. 1.   Arrhenius plot for the reaction $O + H_2 = OH + H$.

logarithm of the rate constant with a given third body and its boiling point, stressing the importance of the intermolecular potential, while Porter and Smith (1961) identified the negative activation energy with the energy of interaction between an iodine atom and the third body which was attributed to charge transfer effects. Such effects should not be important for O, H or N.

Fortunately atmospheric gases, with the exception of the minor constituent water, do not show a wide range of three-body efficiencies, and there is no real evidence that the relative efficiencies change significantly with temperature.

For the recombination reactions (1) and (8)

$$O + O + M = O_2 + M \tag{8}$$

the experimental evidence clearly supports the use of Arrhenius forms over wide temperature ranges, the recommended form being $k_1 = 1.4 \times 10^{-33} \exp(1000/RT)$ cm^6 s^{-1} and $k_8 = 2.8 \times 10^{-34} \exp(+1400/RT)$ cm^6 s^{-1} for M $=$ N$_2$; the former holds well between 100 K and 6000 K (Campbell and Thrush, 1967; Clyne and Stedman, 1967), the latter between 200 K and 3500 K although it fails at higher temperatures (Campbell and Thrush, 1967).

The combination of ground state oxygen atoms with oxygen molecules (reaction (2)) has been studied over a much smaller temperature range, since the data on these combination reactions at high temperatures is deduced from shock tube studies of the reverse dissociation reaction and ozone is rapidly dissociated at much lower temperatures than are oxygen or nitrogen. For this reason both $\exp(E/RT)$ and T^n dependence of k_2 are shown in Figures (2) and (3), the data being denoted by the initial letters of the workers' names. Only data for which the authors considered such problems as the dissociation of ozone by O$_2(^1\Delta_g)$ and O$_2(^1\Sigma_g^+)$ and role of atomic hydrogen in catalysing the combination reaction:

$$H + O_2 + M = HO_2 + M$$

$$O + HO_2 \quad = OH + O_2$$

$$O + OH \quad = H + O_2$$

have been included.

Error limits are however not given, and the vertical bars on the shock tube studies of Jones and Davidson (1962) and the pyrolysis work of Benson and Axworthy

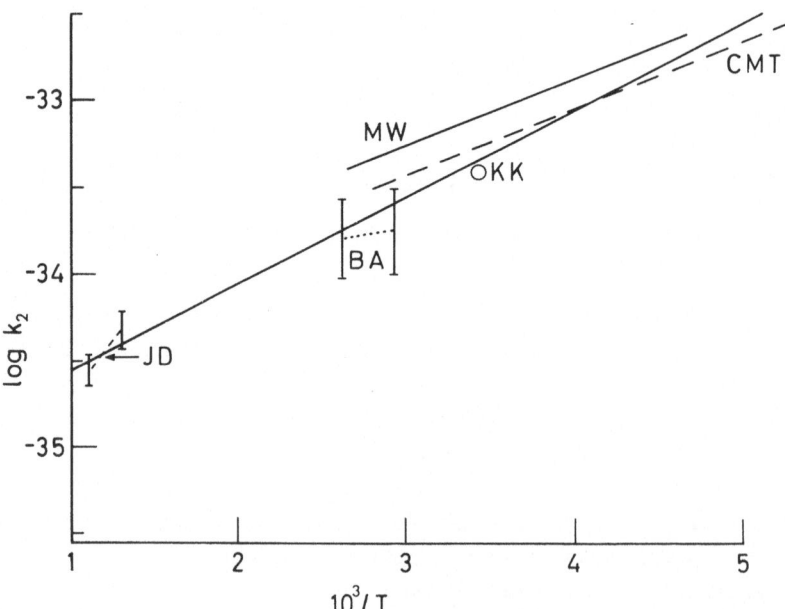

Fig. 2.   Experimental data on the reaction O $+$ O$_2$ $+$ Ar $=$ O$_3$ $+$ Ar in Arrhenius form.

(1957, 1965) correspond to the range of values of k_2 deduced from studies of the dissociation of ozone due to an uncertainty of ± 400 cal/mole in its heat of formation. This would contribute a factor of two at room temperature. The two studies using flow systems by Clyne *et al.* (1965) and Mulcahy and Williams (1968) give remarkably similar temperature dependences, but differ by 50% for no obvious reason; the former is closer to the room temperature point of Kaufman and Kelso (1967). The solid lines correspond to $k_2 = 9 \times 10^{-36} \exp(+2300/RT)$ cm^6 s^{-1} and $k_2 = 6.4 \times 10^{-34}$ $(T/273)^{-2.6}$ cm^6 s^{-1} respectively for argon as third body. This was chosen as providing the best direct comparison between the various experiments. It is generally agreed that the values for $M = N_2$ and $M = O_2$ are factors of 1.5 and 1.6 greater respectively than for argon.

This temperature coefficient is similar to those found for other combination reactions of atoms with diatomic molecules, e.g. $H + O_2 + M$, $O + NO + M$, $H + NO + M$. The presence of a negative temperature coefficient, and the increase in concentration of the reacting species at low temperatures, for a given total pressure mean that the study of recombination reactions in discharge flow systems is more favourable at temperatures below ambient where there is less competition from bimolecular and heterogeneous processes. It is fortunate that this temperature range is the most useful for comparison with atmospheric data.

We now describe some experiments carried out by R. G. Derwent in the author's laboratory. In this work, oxygen containing about 5% of $O_2(^1\Delta_g)$ is obtained by

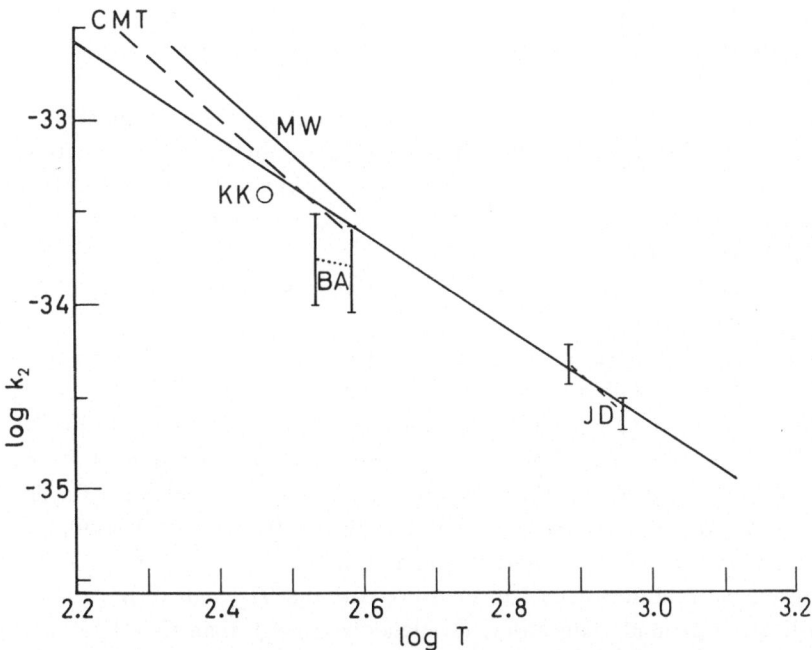

Fig. 3.   Experimental data on the temperature dependence of $O + O_2 + Ar = O_3 + Ar$ in log-log form.

passing the products of a discharge in molecular oxygen at a few Torr pressure through a tube coated with mercuric oxide, which removes virtually all the atomic oxygen. Arnold *et al.* (1966) observed that added nitrogen dioxide or iodine are excited electronically in these systems but the maximum energy of the emitted photons is more than twice the excitation energy of $O_2(^1\Delta_g)$. As much information about atmospheric processes has come from study of light emission, and $O_2(^1\Delta_g)$ is by no means a negligible constituent of the upper atmosphere, a more detailed investigation of this process appeared worth while.

At relatively high concentrations of $O_2(^1\Delta_g)$, the higher excited state $O_2(^1\Sigma_g^+)$ is populated by the 'energy pooling' process

$$O_2(^1\Delta_g) + O_2(^1\Delta_g) = O_2(^1\Sigma_g^+) + O_2(^3\Sigma_g^-). \tag{9}$$

Our measured value of $k_9 = 2.2 \times 10^{-17}$ cm^3 s^{-1} exceeds that found for this reaction by Arnold and Ogryzlo (1967). We also find that this reaction populates level $v' = 1$ of the $^1\Sigma_g^+$ state to the extent of 5%, this being much greater than the 2.3% predicted by Franck-Condon factors for the $^1\Delta_g$–$^1\Sigma_g^+$ transition assuming a vertical transition. The closer energy resonance of the level $v' = 1$ ($\Delta E = -1235$ cm^{-1} as against $\Delta E = -2643$ cm^{-1}) is presumably responsible. Unlike $O_2(^1\Delta_g)$, $O_2(^1\Sigma_g^+)$ is destroyed quite rapidly at the glass walls of the flow tube (collisional efficiency, $\gamma = 10^{-2}$) and is also rapidly removed by added water, for which we find rate constants of $k = 4.7 \times 10^{-12}$ and 6.7×10^{-12} cm^3 s^{-1} for $v' = 0$ and $v' = 1$ respectively. The rate constant for quenching $O_2(^1\Delta_g)$ by water is only about 10^{-20} cm^3 s^{-1}.

Using these criteria it is possible to distinguish and analyse the reactions of $O_2(^1\Delta_g)$ and $O_2(^1\Sigma_g^+)$ in our system.

When iodine is added to discharged oxygen in our apparatus its molecular emission from all vibrational levels of the $^3\Pi_{o+u}$ state up to convergence limit to $I(^2P_{1/2}) + I(^2P_{3/2})$ at 20044 cm^{-1} is observed. There is clear kinetic evidence that in our system this emission is due to excitation of molecular iodine and not to the recombination of $I(^2P_{3/2}) + I(^2P_{1/2}) + M$.

Energetically the excitation must involve at least three molecules of $O_2(^1\Delta_g)$ or $O_2(^1\Sigma_g^+)$ plus $O_2(^1\Delta_g)$.

Quenching experiments and kinetic studies show that the latter explanation is correct and the most obvious excitation mechanism is for $O_2(^1\Sigma_g^+)$ to excite I_2 from its ground state to the metastable $^3\Pi_{1u}$ state, followed by excitation to the $^3\Pi_{o+u}$ state by the much more populous $O_2(^1\Delta_g)$. Near-vertical transition from the ground state of I_2 would also reach the rising limb of the $^3\Pi_{1u}$ potential above the dissociation limit; such a process would account for the efficient dissociation of iodine observed.

With bromine, very little light emission or dissociation is observed. This is entirely consistent with the above mechanism, since both the $^3\Pi_{1u}$ state of Br_2 and its dissociation limit to ground state atoms lie higher in energy than $O_2(^1\Sigma_g^+)$, although the $^3\Pi_{o+u}$ states of Br_2 and I_2 lie in closely similar energy regions (Table I).

The emission by added nitrogen dioxide is also much weaker than that of iodine;

in this case, $O_2(^1\Sigma_g^+)$ has sufficient energy to produce electronic excitation of NO_2. The difference in intensity is consistent with the known efficient quenching of electronically excited NO_2 (Myers et al., 1966).

TABLE I

Energy levels of Br_2, I_2, O_2

	Br_2 (cm^{-1})	I_2 (cm^{-1})
$^3\Pi_{1u}$	13737	11803
$^3\Pi_{0^+u}$	15814	15725
$X_2 = X + X$	15900	12441
$X_2 = X + X$	19585	20044

$O_2(^1\Delta_g) = 7882$ cm^{-1} $O_2(^1\Sigma_g^+) = 13121$ cm^{-1}.

Another factor enhancing the emission with I_2 is the iodine atom catalysed population of the $^1\Sigma_g^+$ state of O_2, which proceeds by the mechanism

$$O_2(^1\Delta_g) + I(^2P_{3/2}) = O_2(^3\Sigma_g^-) + I(^2P_{1/2}) \tag{10}$$

$$O_2(^1\Delta_g) + I(^2P_{1/2}) = O_2(^1\Sigma_g^+) + I(^2P_{3/2}) \tag{11}$$

$$O_2(^3\Sigma_g^-) + I(^2P_{1/2}) = O_2(?) + I(^2P_{3/2}). \tag{12}$$

We find that $k_{10} = 2.6 \times 10^{-11}$ cm^3 s^{-1} and $k_{11} = 3 \times 10^{-14}$ cm^3 s^{-1} using $k_{12} = 9 \times 10^{-12}$ cm^3 s^{-1} (Donovan and Husain, 1966) and statistical mechanics to calculate the equilibrium constant for reaction (10). It can be deduced that all of the quenching collisions of $I(^2P_{1/2})$ in reaction (12) populate $O_2(^1\Delta_g)$, although 20% of all collisions have enough energy to do so. The low rate constants found here for spin allowed reactions involving $O_2(^1\Delta_g)$, except when the process is nearly resonant, stress the lack of reactivity of this species reported by other workers. In the present case, this is certainly not due to restrictions placed by correlation rules, and the low reactivity of $O_2(^1\Delta_g)$ relative to $O_2(^1\Sigma_g^+)$ is in striking contrast to the high reactivity of $O(^1D)$ compared with $O(^1S)$. Our results do however show clearly that the 'energy pooling' processes in which an emitting species acquires the energy of several $O_2(^1\Delta_g)$ molecules occurs by stepwise electronic excitation; $O_2(^1\Sigma_g^+)$ provides a convenient 'stepping stone'. No evidence has been found of the participation of vibrationally excited species, despite the evidence from shock tube studies of the strong coupling between vibrational energy (e.g. in N_2) and electronic excitation of atoms, and of the formation of vibrationally excited N_2, CO, etc. in the quenching of Na, Hg, etc. It is unlikely that 'energy pooling' reactions of $O_2(^1\Delta_g)$ are responsible for any significant atmospheric emission.

A study of vibrationally excited OH by gas phase E.S.R. is being carried out in collaboration with P. N. Clough and A. H. Curran.

The reaction

$$H + O_3 = OH + O_2 + 77 \text{ kcal/mole}$$

which yields vibrationally excited OH radicals up to $v' = 9$ is of considerable importance in the upper atmosphere. Hitherto this excitation has been studied in the laboratory using infra-red chemiluminescence (McKinley et al., 1955), a method which has now been developed so that emission can be observed at pressures around 1 mTorr where vibrational relaxation is negligible (Anlauf et al., 1968). This technique does not yield direct information about the rate of population of level $v' = 0$, and the dipole moment parameters needed to determine the vibrational energy distribution have to be found from relative band intensities. Studies of this reaction by gas phase E.S.R. have the advantage that level $v' = 0$ can be observed directly and that accurate values of the dipole moment in different vibrational states can be obtained by measurement of the Stark effect, although we have not yet been able to make these latter measurements.

Experimentally, hydrogen atoms in an argon carrier are produced by flowing ca. 5% hydrogen in argon through a microwave discharge. The quartz flow tube of 20 mm i.d. is expanded to a pill box form which fits snugly into a Varian X-band 'large access' cavity operated in a $TE_{0,1,1}$ or $TE_{0,1,2}$ mode. Highly concentrated ozone gas is introduced via a movable axial probe whose tip is positioned close to or inside the cavity; total pressures around 50 mTorr give the best compromise between sensitivity and line resolution in our system. A conventional Varian E.S.R. spectrometer with 12 in. magnet and 100 kHz modulation is used. Transitions involving the $J = \frac{5}{2}$ and $J = \frac{3}{2}$ levels of the $^2\Pi_{3/2}$ state of OH have been observed and measured up to $v' = 4$, and no measurements on higher levels are in progress.

In the E.S.R. spectrum the electric dipole transitions, which are some 10^4 times stronger than magnetic dipole ones, occur between $+$ and $-$ components of the Λ-doublets; they are therefore displaced by an amount corresponding to the Λ-doubling frequency on either side of the expected resonance. For the $J = \frac{3}{2}$ level of $^2\Pi_{3/2}$, the resonance for Hund's case (a) would occur near $g = 0.8$, the actual values decreasing from 0.936 to 0.922 from $v' = 0$ to $v' = 4$. The decrease in Λ-doubling with increasing vibrational energy ensures that the E.S.R. spectra corresponding to different vibrational energies are well resolved. These Λ-doubling frequencies deduced from the $J = \frac{3}{2}$ transitions are given in Table II. Although a full interpretation must await accurate measurements on the $J = \frac{5}{2}$ level, this table shows that the simple assumption that the Λ-doubling is proportional to B_v^2 expected for Hund's case (b) holds remarkably well, with a value only 4.2% less than for the case of 'pure precession' with $l = 1$

TABLE II

Λ-doubling frequencies in OH($^2\Pi_{3/2}$) in MHz

v'	Observed	Calculated
0	1666.29 ± 0.2	(1666.29)
1	1537.92 ± 0.3	1541.29
2	1413.37 ± 0.3	1422.66
3	1293.30 ± 0.3	1309.58
4	1176.43 ± 0.3	1202.56

(van Vleck, 1929). Although OH is generally regarded as being closer to Hund's case (a) for slow rotation (Dieke and Crosswhite, 1948), the observed behaviour is not unexpected since the Λ-doubling of a $^2\Pi_{3/2}$ state obeying Hund's case (a) is very small.

The dimensions of the X-band microwave cavity used in this work mean that the average diffusion distance to the walls for OH radicals is 1.5 cm, giving diffusion times of 1 to 10 ms. Since rapid vibrational relaxation occurs at the walls, this system is not ideal for the study of initial vibrational distribution. It is however a good system for the study of the reactions of ground state and vibrationally excited OH, for instance with ozone, since OH radicals can be detected at concentrations where they undergo on the average only one collision with another OH radical during their residence time in the cavity.

References

Anlauf, K. G., Macdonald, R. G., and Polanyi, J. C.: 1968, *Chem. Phys. Letters* **1**, 619–22.
Arnold, S. J., Finlayson, N., and Ogryzlo, E. A.: 1966, *J. Chem. Phys.* **44**, 2529–30.
Arnold, S. J. and Ogryzlo, E. A.: 1967, *Can. J. Phys.* **45**, 2053–61.
Baulch, D. L., Drysdale, D. D., and Lloyd, A. C.: 1968, *High Temperature Reaction Rate Data No. 2* Leeds University.
Benson, S. W. and Axworthy, A. E.: 1957, *J. Chem. Phys.* **26**, 1718–26.
Benson, S. W. and Axworthy, A. E.: 1965, *J. Chem. Phys.* **42**, 2614–15.
Bunker, D. L.: 1966, *Theory of Elementary Gas Reaction Rates*, Pergamon, Oxford.
Campbell, I. M. and Thrush, B. A.: 1965, *Ann. Rep. Chem. Soc. (London)* **62**, 17–38.
Campbell, I. M. and Thrush, B. A.: 1967, *Proc. Roy. Soc.* **A296**, 201–32.
Campbell, I. M. and Thrush, B. A.: 1968, *Trans. Faraday Soc.* **64**, 1265–74.
Clyne, M. A. A., McKenney, D. J., and Thrush, B. A.: 1965, *Trans. Faraday Soc.* **61**, 2701–9.
Clyne, M. A. A. and Stedman, D. H.: 1967, *J. Phys. Chem.* **71**, 3071–2.
Clyne, M. A. A. and Thrush, B. A.: 1961, *Proc. Roy. Soc.* **A261**, 259–73.
Clyne, M. A. A. and Thrush, B. A.: 1963, *Proc. Roy. Soc.* **A275**, 544–58.
Dieke, G. H. and Crosswhite, H. M.: 1948, *Bumblebee Report No. 87*, Johns Hopkins University, Baltimore.
Donovan, R. J. and Husain, D.: 1966, *Trans. Faraday Soc.* **62**, 2023–9.
Glasstone, S., Eyring, H., and Laidler, K. J.: 1941, *The Theory of Rate Processes*, McGraw Hill, New York.
Hoyermann, K., Wagner, H. G., and Wolfrum, J.: 1967, *Ber. Bunsengesell. Physik. Chem.* **71**, 599–602.
Jones, W. M. and Davidson, N.: 1962, *J. Amer. Chem. Soc.* **84**, 2868–78.
Kaufman, F.: 1958, *Proc. Roy. Soc. A*, **243**, 123–39.
Kaufman, F.: 1961, *Prog. Reaction Kinetics* **1**, 1–39.
Kaufman, F. and Kelso, J. R.: 1967, *J. Chem. Phys.* **46**, 4541–42.
McKinley, J. D., Garvin, D., and Boudart, M. J.: 1955, *J. Chem. Phys.* **23**, 784–6.
Myers, G. H., Silver, D. M., and Kaufman, F.: 1966, *J. Chem. Phys.* **44**, 718–23.
Mulcahy, M. F. R. and Williams, D. J.: 1968, *Trans. Faraday Soc.* **64**, 59–70.
Porter, G. and Smith, J. A.: 1961, *Proc. Roy. Soc.* **A261**, 28–37.
Russell, K. E. and Simons, J.: 1953, *Proc. Roy. Soc.* **A217**, 271–9.
Van Vleck, J. H.: 1929, *Phys. Rev.* **33**, 467–506.
Westenberg, A. A. and De Haas, N.: 1969, *J. Chem. Phys.* **50**, 2512–16.
Wong, E. L. and Potter, A. E.: 1965, *J. Chem. Phys.* **43**, 3371–82.

THE PHOTOCHEMISTRY OF OZONE AND SINGLET
MOLECULAR OXYGEN IN THE ATMOSPHERE

R. P. WAYNE

Physical Chemistry Laboratory, Oxford University, South Parks Road, Oxford, England

Hypothetical reaction schemes are frequently used in the interpretation of atmospheric photochemistry, and rate parameters chosen to give a good fit to the observed data. Although the close agreement between the experimental and calculated results gives considerable confidence in the proposed mechanisms, it is clearly more satisfactory if laboratory investigations yield information about reactions and their rates which predicts with some degree of accuracy the atmospheric observations. In this paper, some laboratory studies on the photochemistry of the O_3–O_2–O system are presented, with particular reference to the participation of $O_2({}^1\Delta_g)$ in the various processes. These studies have permitted the evaluation of atmospheric models, and the relevance of the studies to atmospheric photochemistry will be discussed. Detailed descriptions of the experimental techniques are to be found in the papers to which reference is made.

Optical emission from singlet molecular oxygen is a major component of the day and night airglows (see, for example, Wayne, 1967; Evans and Llewellyn, 1970), and $O_2({}^1\Delta_g)$ reaches concentrations higher than 10^{10} molecule cm^{-3} during the day. It has frequently been supposed that $O_2({}^1\Delta_g)$ is produced in the primary step of ozone photolysis by ultraviolet radiation:

$$O_3 + h\nu \rightarrow O_2({}^1\Delta_g) + O({}^1D) \tag{1a}$$

and, indeed, Houghton (1965) has shown on the basis of energy considerations that absorption of solar radiation by ozone must be the precursor to $O_2({}^1\Delta_g)$ formation. Laboratory studies of ozone photolysis are presented, and the nature of the primary process is discussed in some detail. Quenching of $O_2({}^1\Delta_g)$ by atmospheric gases, and reaction of $O_2({}^1\Delta_g)$ with ozone, are both potentially important processes whose efficiency must be known before a satisfactory atmospheric model may be set up from the laboratory data. Laboratory investigations of these processes are described next and the implications of the results for atmospheric photochemistry is then considered. The relatively high concentrations of $O_2({}^1\Delta_g)$ suggest that this species could play a significant role in the ion chemistry of the atmospheric D-region; finally, two possible reactions which could lead to ion formation are discussed.

Photolysis of ozone in the ultraviolet region yields $O({}^1D)$ as one of the primary products (McGrath and Norrish 1957, 1960; Norrish and Wayne, 1965a, b; De More and Raper, 1962, 1966). In the strong absorption region of the Hartley band it is reasonable to suppose that the primary process is spin-conserved

$$O_3({}^1A) + h\nu_{uv} \rightarrow O_3 \text{ (excited singlet)} \rightarrow \text{two triplet or}$$
$$\text{two singlet products}. \tag{2}$$

Fiocco (ed.), Mesospheric Models and Related Experiments, 240–252. All Rights Reserved.
Copyright © 1971 by D. Reidel Publishing Company, Dordrecht-Holland.

Table I shows the wavelengths at which the formation of atomic and molecular fragments in various electronic states becomes energetically possible. It is seen that $O_2(^1\Delta_g)$ production

$$O_3 + h\nu \rightarrow O_2(^1\Delta_g) + O(^1D) \tag{1a}$$

can occur isothermally for wavelengths shorter than about 3100 Å. For $\lambda < 2660$ Å, the formation of $O_2(^1\Sigma_g^+)$ is also possible

$$O_3 + h\nu \rightarrow O_2(^1\Sigma_g^+) + O(^1D). \tag{1b}$$

De More and Raper (1962, 1966) have, in fact, shown that in the condensed phase the efficiency for $O(^1D)$ production drops markedly with increasing wavelength at $\lambda \sim 3100$ Å, which suggests that $O(^1D)$ is not formed when the spin-conserved reaction (1a) is no longer energetically possible.

TABLE I
Long wavelength limits (Å) for the production of various electronic states of O_2 and O in the primary step of ozone photolysis

O	O_2				
	$(^3\Sigma_g^-)$	$(^1\Delta_g)$	$(^1\Sigma_g^+)$	$(^3\Sigma_u^+)$	$(^3\Sigma_u^-)$
(^3P)	11 800	6110	4630	2300	1700
(^1D)	4110	3100	2660	1670	1500
(^1S)	2340	1960	1790	1290	1080

Kinetic evidence also indicated that singlet molecular and atomic products are formed in the photolysis at $\lambda = 2537$ Å (Norrish and Wayne, 1965a). The quantum yield for ozone decomposition can reach values as high as 16 or 17 as a result of an energy chain initiated by $O(^1D)$

$$O(^1D) + O_3 \rightarrow O_2^* + O_2 \tag{3}$$

$$O_2^* + O_3 \quad \rightarrow 2O_2 + O(^1D), \tag{4}$$

where O_2^* is some energy rich state of O_2. The quantum yield is a linear function of pressure, as shown in Figure 1. The quantum yield extrapolated to zero ozone pressure is four within experimental error, and this result can reasonably be explained only by the production of singlet molecular oxygen (referred to below as $O_2^\dagger$) in the primary step. At low pressures, the chain propagating step (4) is dominated by deactivation of O_2^*, and the limiting sequence of steps is

$$O_3 + h\nu \quad \rightarrow O(^1D) + O_2^\dagger \tag{1c}$$

$$O(^1D) + O_3 \rightarrow O_2^{(*)} + O_2 \tag{3}$$

$$O_2^\dagger + O_3 \quad \rightarrow 2O_2 + O(^3P) \tag{4}$$

$$O(^3P) + O_3 \rightarrow 2O_2 \tag{5}$$

giving the observed quantum yield of four.

In order to see whether the quantum yield remains four at lower pressures than those originally studied, we have investigated the photolysis of ozone at $\lambda = 2537$ Å in a low-pressure flow system (Jones *et al.*, 1970). Figure 2 shows that the quantum

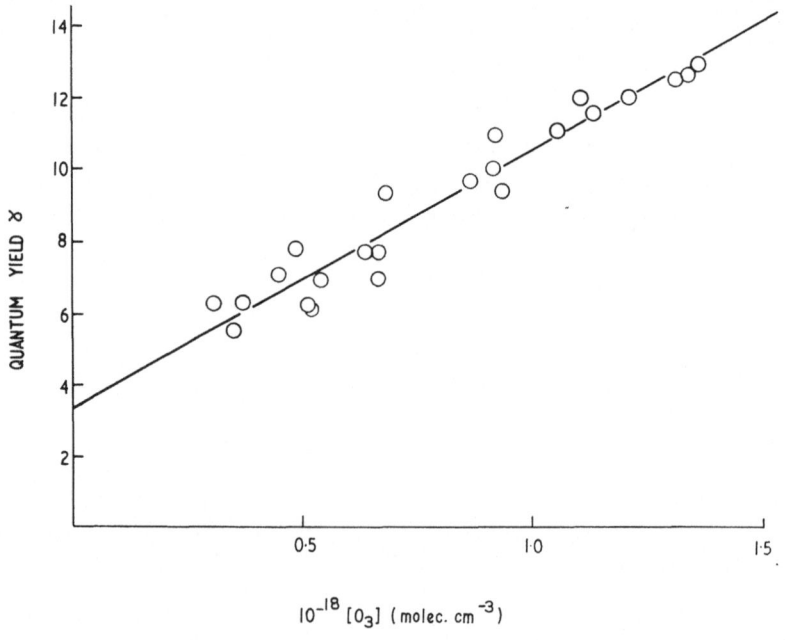

Fig. 1.   Variation in quantum yield with [O₃] for the photolysis of ozone at $\lambda = 2537$ Å.

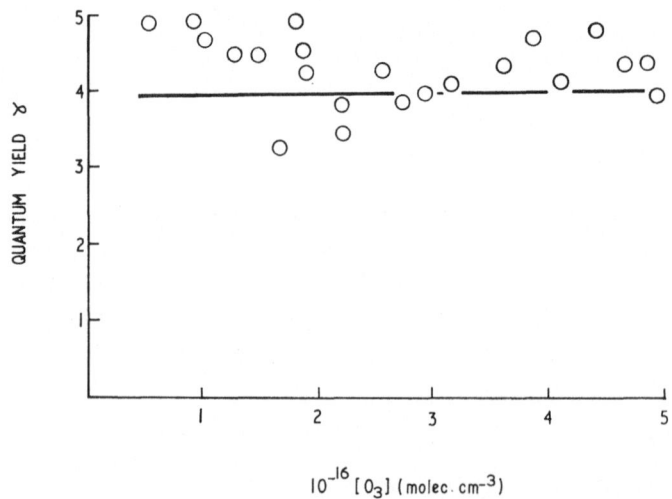

Fig. 2.   Quantum yields for photolysis at $\lambda = 2537$ Å of ozone-oxygen mixtures containing 90% ozone.

yield remains close to four down to pressures of less than 5×10^{-15} molec cm^{-3}, or about 0.2 Torr.

It now became of interest to investigate the effect of wavelength of photolysis on the slope and intercept of the quantum yield *vs.* [O$_3$] plot. The slope is expected to decrease towards zero as the ratio of [O(^1D)] : [O(^3P)], produced in the primary step,

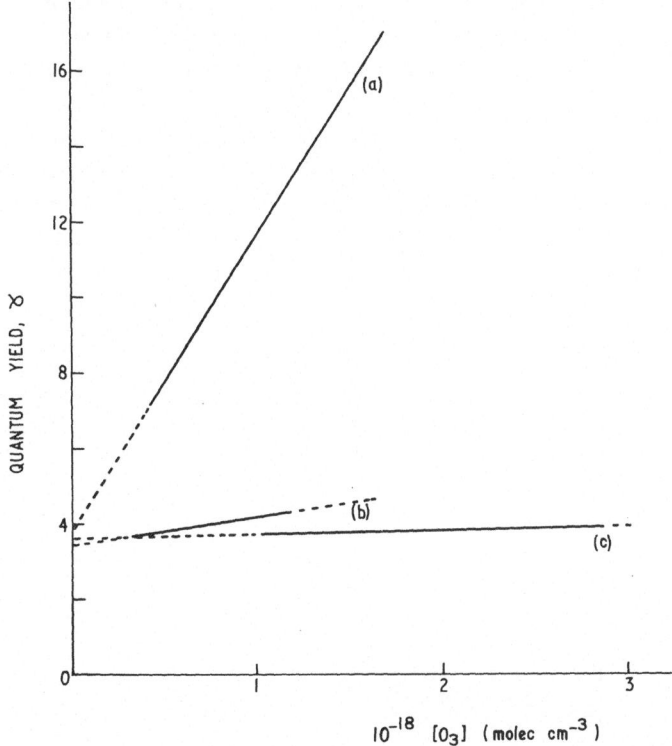

Fig. 3. Least squares lines showing the variation in quantum yield with [O$_3$] for the photolysis of ozone. (a) $\lambda = 2890, 2920$ Å; (b) $\lambda = 3130$ Å; (c) $\lambda = 3340$ Å. The solid portions of the lines show the concentration regions over which quantum yield determinations were made.

decreases, while the intercept should decrease from four to two when O$_2^{\dagger}$ is no longer a primary product. We have now studied the photolysis at six wavelengths in the wavelength region 2480 to 3340 Å (Jones and Wayne, 1969, 1970). The results for $\lambda = 2890/2920$, 3130 and 3340 Å are shown in Figure 3. The slopes do, in fact, decrease with increasing wavelength, and if the primary quantum yield for O(^1D), $\phi_{O(^1D)}$, is taken to be unity at $\lambda = 2537$ Å, the slopes provide a measure of $\phi_{O(^1D)}$ at other wavelengths. Figure 4 shows $\phi_{O(^1D)}$ as a function of wavelength.

Contrary to expectation, however, the limiting low pressure quantum yield remains four, even out to $\lambda = 3340$ Å. The implication is that O$_2^{\dagger}$ is produced in the primary step with a quantum efficiency of unity at wavelengths longer than those at which the spin-conserved process can occur. Table I shows that there is sufficient energy for the formation of $O_2(^1\Delta_g)$ (or of $O_2(^1\Sigma_g^+)$) at $\lambda = 3340$ Å if the atomic fragment is in the ground, ^3P, state

$$O_3 + h\nu \rightarrow O_2^\dagger + O(^3P). \tag{1d}$$

That spin is not conserved at $\lambda > 3100$ Å is probably a consequence of the relatively weak absorption in the long wavelength tail of the Huggins band. ΔS for the optical absorption may no longer be zero, and the excited O_3 from which dissociation occurs may thus be a triplet; alternatively, intramolecular energy transfer from an excited singlet first formed may populate a dissociating triplet of ozone in this wavelength region.

The limit of $\lambda \sim 3100$ Å for production of $O(^1D)$ still stands, since there is insufficient energy for formation of *both* $O_2(^1\Delta_g)$ and $O(^1D)$ at longer wavelengths, and the experimental results confirm this limit. However, the original argument that $O_2(^1\Delta_g)$ production also ceases at $\lambda \sim 3100$ Å must now be modified. Some consequences in atmospheric photochemistry of $O_2(^1\Delta_g)$ formation at $\lambda > 3100$ Å are considered later.

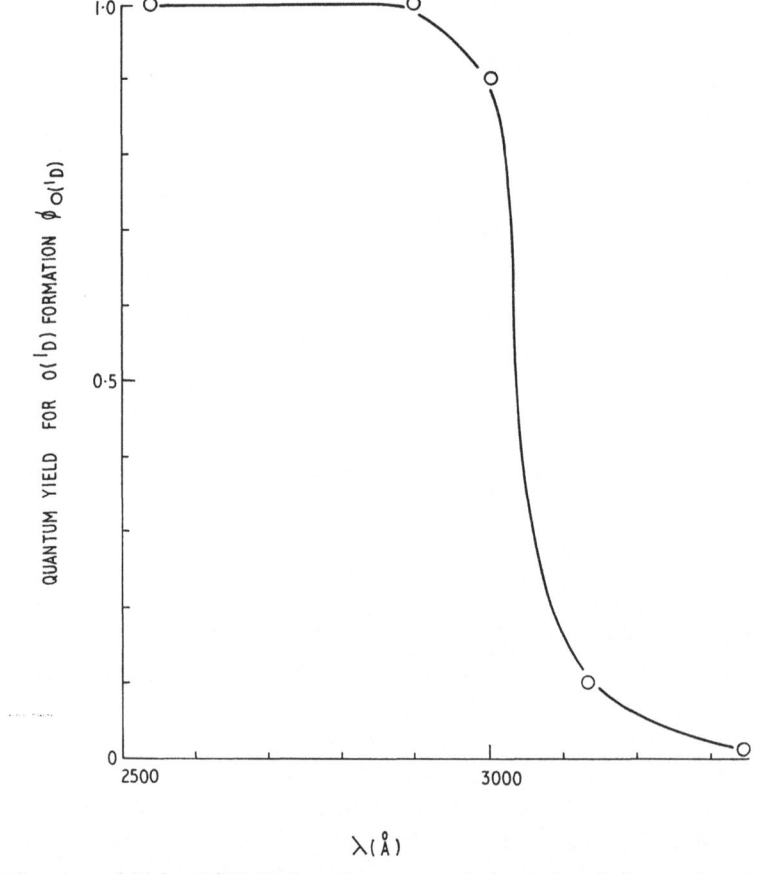

Fig. 4.   Quantum yield for $O(^1D)$ in the primary step of ozone photolysis as a function of wavelength. The absolute measurements assume that $\phi_{O(^1D)} = 1$ at $\lambda = 2537$ Å.

The evidence presented so far for $O_2(^1\Delta_g)$ formation in the primary step of ozone photolysis is circumstantial. For some time it has been clear that the production of $O_2(^1\Delta_g)$ must be demonstrated directly. We were first able to show, by observation of the $O_2(^1\Delta_g \leftrightarrow {}^3\Sigma_g^-)$ emission, that $O_2(^1\Delta_g)$ is formed on photolysis of ozone some years ago (Izod and Wayne, 1968) by use of a low pressure fast flow system in which photolysis at $\lambda = 2537$ Å took place upstream of the detectors. At that time, we were unable to detect $O_2(^1\Delta_g)$ except when molecular oxygen was present in excess, and we were forced to the conclusion that $O_2(^1\Delta_g)$ was formed in a secondary energy transfer process.

$$O(^1D) + O_2 \rightarrow O_2(^1\Delta_g) + O(^3P). \tag{6}$$

Since the kinetic evidence points to the formation of one of the singlet states in the primary step at $\lambda = 2537$ Å, we further concluded that $O_2(^1\Sigma_g^+)$ is the primary molecular product at that wavelength. Measurements of the infrared atmospheric band intensity in the airglow left little doubt that at longer wavelengths ($\lambda > 2660$ Å: see Table I), $O_2(^1\Delta_g)$ is the primary product.

These preliminary experiments suffered from the lack of a good detector for the infrared atmospheric band at $\lambda = 1.27$ μm, which required us to perform to experiments under kinetically unsatisfactory conditions. Recently, however, a sensitive germanium photodiode has become available commercially, and an intrinsic germanium photoconductive device has also been developed in this laboratory. Using a germanium detector with a low pressure photolysis system without time resolution, Gauthier and Snelling (1970) have obtained results consistent with $O_2(^1\Delta_g)$ formation in the primary step at $\lambda = 2537$ Å. Absence of time resolution can lead to possible dangers of interpretation, and we have accordingly re-examined the photolysis in the time-resolved system using the high sensitivity detectors (Jones and Wayne, 1971). The new data leave little doubt that $O_2(^1\Delta_g)$ is a primary product of photolysis, and that the earlier results are in error. Our experimental system enables us to measure the absolute efficiency for $O_2(^1\Delta_g)$ formation, and it is shown that the value is close to unity (Jones, this symposium).

Loss processes for $O_2(^1\Delta_g)$ in the atmosphere include radiative decay and reaction with, or deactivation by, the atmospheric constituents. Badger et al. (1965) give the Einstein A factor for radiation as 2.58×10^{-4} s^{-1}. Laboratory studies have now been made of the rates of reaction of $O_2(^1\Delta_g)$ with ozone (Clark et al., 1970),

$$O_2(^1\Delta_g) + O_3 \rightarrow 2O_2 + O(^3P), \tag{4a}$$

with atomic oxygen (Clark and Wayne, 1969a), and with atomic nitrogen (Clark and Wayne, 1970a); and of the deactivation of $O_2(^1\Delta_g)$

$$O_2(^1\Delta_g) + M \rightarrow O_2 + M \tag{7}$$

where $M = O_2$, N_2, Ar, CO_2 and H_2O (Clark and Wayne, 1969b).

Early measurements of the rate constant for reaction (4a) were open to doubts about whether $[O_3]$ decayed in the system used (Wayne and Pitts, 1969) or whether vibrationally excited molecular oxygen played a part in the reaction (McNeal and Cook,

1967). The development of kinetic photoionization of $O_2(^1\Delta_g)$ by Ar-resonance radiation as a technique for the determination of $[O_2(^1\Delta_g)]$ (see Clark and Wayne, 1969b) has made it possible to obtain kinetic data under conditions where $[O_2(^1\Delta_g)] \ll [O_3]$ and where vibrationally excited oxygen cannot persist. Reaction (4a) may be expected to possess an energy of activation, and the temperature dependence of rate was determined so as to allow computation of the rate of reaction at any altitude in the atmosphere. An Arrhenius plot of the experimental data is given in Figure 5. The Arrhenius parameters $[k = A \exp(-E/RT)]$ are

$$A = 6.7 \pm 2.5 \times 10^{-13} \text{ cm}^3 \text{ molec}^{-1} \text{ s}^{-1}$$

$$E = 13.0 \pm 1.6 \text{ kJ mole}^{-1} (0.13 \pm 0.02 \text{ eV}).$$

If the rate constant for the process

$$O + O_2(^1\Delta_g) \rightarrow \text{products} \tag{8}$$

were sufficiently great, then appreciable loss of $O_2(^1\Delta_g)$ by this mechanism would occur in the atmosphere. However, the measured rate constant at room temperature (Clark and Wayne, 1969a) is less than 1.3×10^{-16} cm^3 molec^{-1} s^{-1}, and the process is therefore probably unimportant in the atmosphere. The analogous loss process involving atomic nitrogen is discussed later.

Laboratory investigations of the physical quenching of $O_2(^1\Delta_g)$ in reaction (7) were at first hampered by the domination of reaction (4a) over (7). Ozone is produced in the laboratory system by recombination of atomic oxygen, from the electric discharge used to generate $O_2(^1\Delta_g)$,

$$O + O_2 + M \rightarrow O_3 + M \tag{9}$$

and much of the observed quenching may be due to the ozone. The high sensitivity of the photoionization technique allows the use of a diluted-discharge flow technique in which the ozone concentration is sufficiently small for reaction (4a) not to interfere.

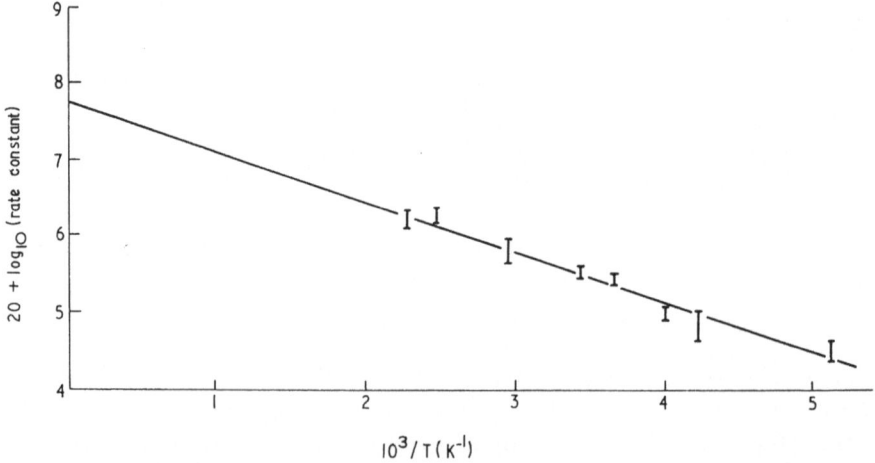

Fig. 5.   Arrhenius plot of rate constants for the reaction $O_2(^1\Delta_g) + O_3 \rightarrow 2O_2 + O$.

Room temperature rate constants for quenching of $O_2(^1\Delta_g)$ by several atmospheric gases, as well as for the reactions with O_3, O and N, are given in Table II.

TABLE II

Rate constants at room temperature for the chemical and physical deactivation of $O_2(^1\Delta_g)$

Quenching species	Rate constant $(cm^3 \ molec^{-1} \ s^{-1})$
O	$\leqslant 1.3 \times 10^{-16}$
O_2	$2.4 \pm 0.2 \times 10^{-18}$
O_3	$3.4 \pm 0.5 \times 10^{-15}$
N	$2.7 \pm 1.0 \times 10^{-15}$
N_2	$\leqslant 1.1 \times 10^{-19}$
Ar	$\leqslant 2.1 \times 10^{-19}$
CO_2	$3.9 \pm 0.8 \times 10^{-18}$
H_2O	$1.5 \pm 0.5 \times 10^{-17}$
dry air	$4.3 \pm 0.7 \times 10^{-19}$
dry air *calculated* from N_2, O_2, Ar and CO_2 rates	$5.3 \pm 1.0 \times 10^{-19}$

The laboratory data described in the last two sections make it possible to assess the rates of excitation and deactivation of $O_2(^1\Delta_g)$ in the atmosphere without the use of any variable fitting parameters. Data are needed, of course, for solar intensities and for the atmospheric ozone profile. In the latter case, the experimental information may not be entirely reliable, but for an initial calculation of atmospheric $[O_2(^1\Delta_g)]$ profiles, an ozone profile closely following that of Johnson *et al.* (1952) has been adopted: Table III gives the concentrations as a function of altitude.

TABLE III

Ozone profile adopted in calculations of atmospheric $[O_2(^1\Delta_g)]$

Altitude (km)	$[O_3]$ (molec cm^{-3})
70	1.6×10^9
65	3.2×10^9
60	8.0×10^9
55	2.3×10^{10}
50	8.0×10^{10}
45	1.8×10^{11}
40	6×10^{11}
35	1.8×10^{12}
30	3.0×10^{12}
25	4.0×10^{12}
20	3.0×10^{12}
15	1.0×10^{12}
10	5.0×10^{11}
5	3.0×10^{11}
0	1.5×10^{11}

The reactions which must be considered are

$$O_3 + h\nu \qquad \rightarrow O_2\,(^1\varDelta_g) + O\,(^3P \text{ or } ^1D) \tag{1e}$$

$$O_2\,(^1\varDelta_g) \qquad \rightarrow O_2 + h\nu_{\lambda=1.27\mu m} \tag{10}$$

$$O_2\,(^1\varDelta_g) + \text{air} \rightarrow O_2 + \text{air}. \tag{9a}$$

The mixing ratios of atomic oxygen, or of ozone, with air are never high enough for reactions (8) or even (4a) to compete with (9a), according to the rate data presented earlier.

Profiles for $[O_2(^1\varDelta_g)]$ have been calculated at a number of different solar zenith angles (Crutzen *et al.*, 1971), and allowing for $O_2(^1\varDelta_g)$ formation in (1e) up to $\lambda = 3100$ Å and up to $\lambda = 3500$ Å. The effect of $O_2(^1\varDelta_g)$ formation at $\lambda > 3100$ Å is most marked at large zenith angles and at low altitudes, where the increased solar intensity offsets the decrease in ozone absorption cross section with increasing wavelength.

Figures 6–8 show some typical results of calculations for secants of the zenith angle equal to 1, 4 and 20. The solid lines represent $O_2\,(^1\varDelta_g)$ formation in the primary photolysis of ozone throughout the ultraviolet absorption region, while the broken lines indicate the concentrations expected if there were a long wavelength limit at $\lambda = 3100$ Å.

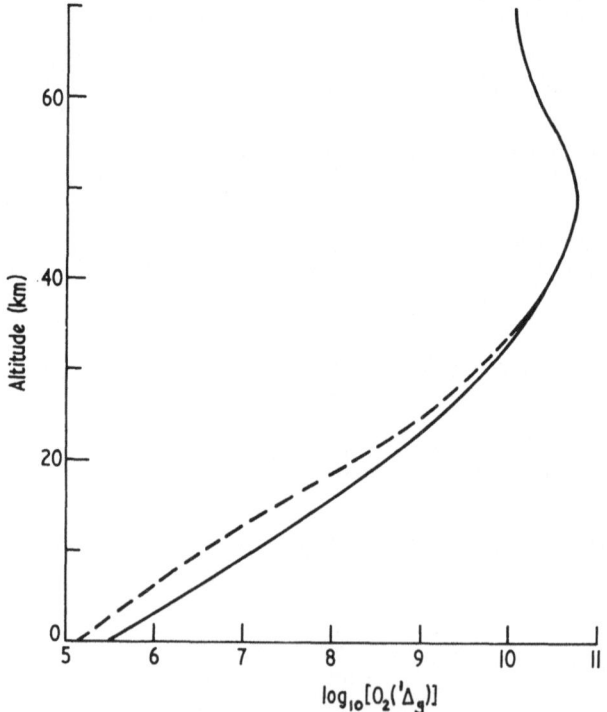

Fig. 6.   Calculated $[O_2(^1\varDelta_g)]$ – altitude profile for $\sec\theta = 1$. Solid line allows for $O_2(^1\varDelta_g)$ formation at $\lambda < 3500$ Å, broken line at $\lambda < 3100$ Å.

The airglow data of Evans *et al.* (1968) correspond to a zenith angle for which $\sec\theta = 4$, and it may be compared with the predictions of Figure 7: the observations, as normalized by Evans and Llewellyn (1970), are shown as crosses at 5 km altitude points. It is not profitable at this stage to attempt to distinguish between the two wavelength limits for $O_2(^1\Delta_g)$ formation on the basis of these airglow data because of the uncertainties in the ozone profile. Evans and Llewellyn (1969) have shown clearly that a very close fit to the observations can be obtained by adjustment of the ozone profile and the quenching parameters. Our comparison is provided solely to illustrate that quite good agreement between observed and predicted $O_2(^1\Delta_g)$ profiles is obtained using purely experimental data for the various processes. It may be that the most satisfactory procedure is now to derive an ozone profile from the airglow measurements and the laboratory kinetic data.

The higher concentrations of $O_2(^1\Delta_g)$ predicted if $O_2(^1\Delta_g)$ is produced at $\lambda > 3100$ Å may be of importance in connection with photochemical pollution. Suggestions have been made (Pitts *et al.*, 1969) that $O_2(^1\Delta_g)$ plays a part in the photochemistry of polluted urban environments, particularly with respect to processes leading to the oxidation of nitric oxide. It has previously been supposed that ozone photolysis

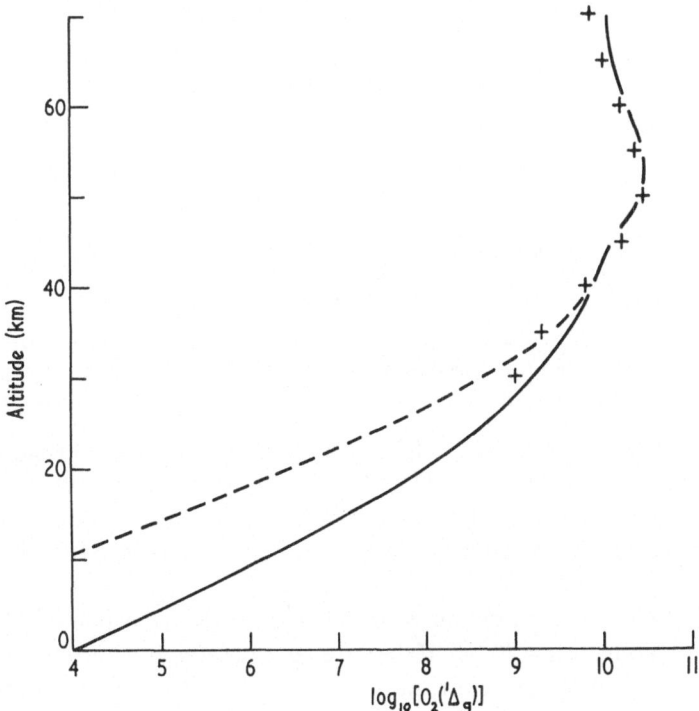

Fig. 7. Calculated $[O_2(^1\Delta_g)]$ – altitude profile for $\sec\theta = 4$. Solid line allows for $O_2(^1\Delta_g)$ formation at $\lambda < 3500$ Å, broken line at $\lambda < 3100$ Å. $+$: points taken at 5 km intervals from the experimental data of Evans and Llewellyn (1970); these data are derived from normalised rocket observations of Evans *et al.* (1968) at a solar elevation of 14.5°.

cannot be a major source of $O_2(^1\Delta_g)$ in the lower atmosphere because of the virtual absence of radiation at $\lambda < 3100$ Å. However, this conclusion may no longer be valid if photolysis at longer wavelengths is effective in producing the singlet species. It should further be noted that the $O_2(^1\Delta_g)$ concentrations given in Figures 6–8 are lower limits, since $[O_3]$ in a polluted atmosphere is likely to be much higher than the values used in the calculations.

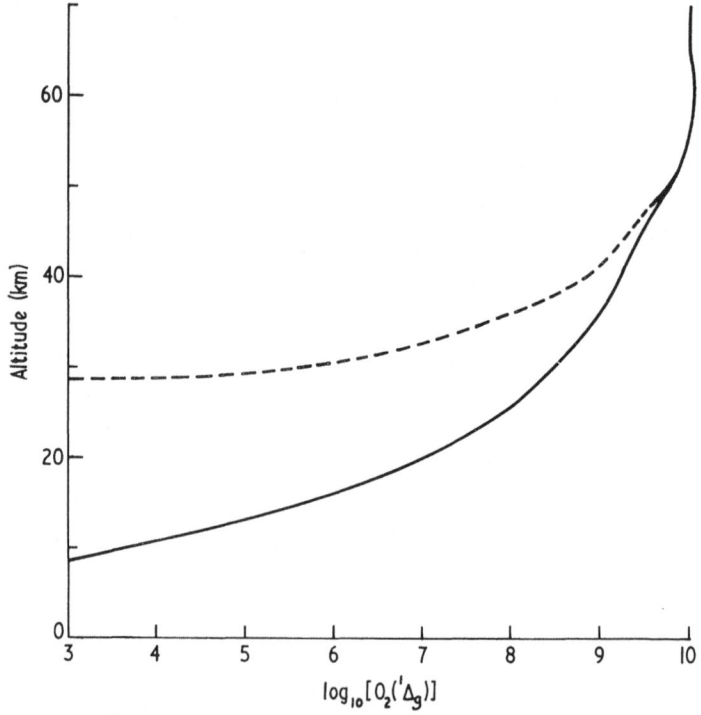

Fig. 8.   Calculated $[O_2(^1\Delta_g)]$ – altitude profile for $\sec\theta = 20$. Solid line allows for $O_2(^1\Delta_g)$ formation at $\lambda < 3500$ Å, broken line at $\lambda < 3100$ Å.

The relative abundance of $O_2(^1\Delta_g)$ in the atmospheric D region suggests that any ionization phenomena involving $O_2(^1\Delta_g)$ may be major contributors to the total ion formation. Hunten and McElroy (1968) have suggested two possible processes, and in this section we describe laboratory experiments designed to test these suggestions.

First, $O_2(^1\Delta_g)$ can be photoionized by the wavelength band 1027–1118 Å, parts of which can penetrate below 70 km. If the cross section for photoionization were around 3×10^{-18} cm^2 in the wavelength region below 1118 Å, then the rate of production of O_2^+ would be comparable with that of NO^+ formation from NO (Hunten and McElroy, 1968).* We have measured photoionization currents at $\lambda = 1067$ Å produced in known

* However, this calculation appears not to allow for absorption by atmospheric carbon dioxide.

concentrations of $O_2(^1\Delta_g)$, and compared these with similar ionization currents due to nitric oxide, whose cross section for photoionization is known (Clark and Wayne, 1970b). The cross section at $\lambda = 1067$ Å can thus be calculated to be

$$3.2 \pm 0.8 \times 10^{-18} \text{ cm}^2.$$

Cairns and Samson (1965) have published curves which permit the ionization cross section to be calculated in the wavelength region 1037–1118 Å using our absolute value at $\lambda = 1067$ Å. Figure 9 shows the cross section as a function of wavelength. It is seen that the cross section exceeds 3×10^{-18} cm^2 at four of the five O_2 transmission windows in the wavelength region, and that it is about 6×10^{-18} cm^2 at the important O VI lines (1032, 1038 Å). Thus the estimates of Hunten and McElroy stand, and are, if anything, slightly low.

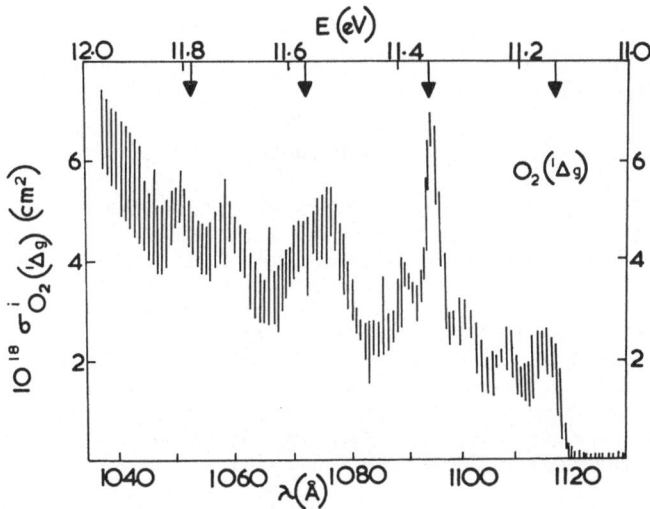

Fig. 9.   Photoionization cross section for $O_2(^1\Delta_g)$ as a function of wavelength.

A second suggestion of Hunten and McElroy (1968) is that the reaction

$$N + O_2(^1\Delta_g) \rightarrow NO + O \tag{11a}$$

could be sufficiently rapid for it to be a major source of nitric oxide which could subsequently undergo photoionization. The corresponding reaction of atomic nitrogen with ground state O_2

$$N + O_2(^3\Sigma_g^-) \rightarrow NO + O \tag{11b}$$

is relatively slow at room temperature, but this is largely a result of the activation energy for the reaction (~ 30 kJ mole^{-1}) and the pre-exponential factor is normal (1.5×10^{-11} cm^3 mole^{-1} s^{-1}). Hunten and McElroy require the rate constant for (11a) to be approximately 3×10^{-13} cm^3 molec^{-1} s^{-1} at 200 K for the reaction to be the dominating nitric oxide production mechanism in the D region. It might be expected that excitation energy of $O_2(^1\Delta_g)$ $(106\,\text{kJ mole}^{-1})$ would overcome virtually all the activation barrier to reaction of the ground state species, so that the rate constant for (11a) at $200\,\text{K}$ might approach the pre-exponential factor for (11b) of $\sim 10^{-11}\,\text{cm}^3\,\text{molec}^{-1}\,\text{s}^{-1}$. In the event, however, this turns out not to be the case. We have investigated the kinetics of reaction (11a) in the laboratory (Clark and Wayne, 1970a), and find the Arrhenius parameters

$$A \leqslant 2 \times 10^{-14}\,\text{cm}^3\,\text{molec}^{-1}\,\text{s}^{-1}$$

$$E < 5.0\,\text{kJ mole}^{-1}.$$

Thus, although the activation barrier in (11b) is largely overcome by the excitation energy of $O_2(^1\Delta_g)$, the pre-exponential factor is abnormally low: we have discussed this result in terms of the transition state for the reaction. At room temperature, the rate constant is $2.7 \pm 1.0 \times 10^{-15}\,\text{cm}^3\,\text{molec}^{-1}\,\text{s}^{-1}$, and, even if the activation energy were zero, reaction (11a) cannot be sufficiently fast at $200\,\text{K}$ for it to be an important source of nitric oxide.

References

Badger, R. M., Wright, A. C., and Whitlock, R. F.: 1965, *J. Chem. Phys.* **43**, 4345–50.
Cairns, R. B. and Samson, J. A. R.: 1965, *Phys. Rev. A*, **139**, 1403–7.
Clark, I. D. and Wayne, R. P.: 1969a, *Chem. Phys. Letters* **3**, 405–7.
Clark, I. D. and Wayne, R. P.: 1969b, *Proc. Roy. Soc.* **A314**, 111–27.
Clark, I. D. and Wayne, R. P.: 1970a, *Proc. Roy. Soc.* **A316**, 539–50.
Clark, I. D. and Wayne, R. P.: 1970b, *Molecular Phys.* **18**, 523–31.
Clark, I. D., Jones, I. T. N., and Wayne, R. P.: 1970, *Proc. Roy. Soc.* **A317**, 407–16.
Crutzen, P. J., Jones, I. T. N., and Wayne, R. P.: 1971, *J. Geophys. Res.* **76**, 1490–97.
De More, W. B. and Raper, O. F.: 1962, *J. Chem. Phys.* **37**, 2048–52.
De More, W. B. and Raper, O. F.: 1966, *J. Chem. Phys.* **44**, 1780–3.
Evans, W. F. J., Hunten, D. M., Llewellyn, E. J., and Vallance Jones, A.: 1968, *J. Geophys. Res.* **73**, 2885–96.
Evans, W. F. J. and Llewellyn, E. J.: 1970, *Ann. Geophys.* **26**, 167–78.
Gauthier, M. and Snelling, D. R.: 1970, *Chem. Phys. Lett.* **5**, 93.
Houghton, J. T.: 1965, *Proc. Roy. Soc.* **A288**, 545–55.
Hunten, D. M. and McElroy, M. B.: 1968, *J. Geophys. Res.* **73**, 2421–8.
Izod, T. P. J. and Wayne, R. P.: 1968, *Nature* **217**, 947–8.
Johnson, F. S., Purcell, J. D., Tousey, R., and Watanabe, K.: 1952, *J. Geophys. Res.* **57**, 157–76.
Jones, I. T. N. and Wayne, R. P.: 1969, *J. Chem. Phys.* **51**, 3617–8.
Jones, I. T. N. and Wayne, R. P.: 1970, *Proc. Roy. Soc.* **A319**, 273–87.
Jones, I. T. N. and Wayne, R. P.: 1971, *Proc. Roy. Soc.* **A321**, 409–24.
Jones, I. T. N., Kaczmar, U. B. and Wayne, R. P.: 1970, *Proc. Roy. Soc.* **A316**, 431–9.
McGrath, W. D. and Norrish, R. G. W.: 1957, *Proc. Roy. Soc.* **A242**, 265–76.
McGrath, W. D. and Norrish, R. G. W.: 1960, *Proc. Roy. Soc.* **A254**, 317–26.
McNeal, R. J. and Cook, G. R.: 1967, *J. Chem. Phys.* **47**, 5385–9.
Norrish, R. G. W. and Wayne, R. P.: 1965a, *Proc. Roy. Soc.* **A288**, 200–11.
Norrish, R. G. W. and Wayne, R. P.: 1965b, *Proc. Roy. Soc.* **A288**, 361–70.
Pitts, J. N., Jr., Khan, A. U., Smith, E. B., and Wayne, R. P.: 1969, *Environ. Sci. Tech.* **3**, 241–4.
Wayne, R. P.: 1967, *Quart. J. Roy. Meteor. Soc.* **93**, 69–78.
Wayne, R. P. and Pitts, J. N., Jr.: 1969, *J. Chem. Phys.* **50**, 3644–5.

LABORATORY INVESTIGATIONS OF THE PRIMARY
PRODUCTS OF OZONE PHOTOLYSIS

I. T. N. JONES*

Physical Chemistry Laboratory, Oxford University, South Parks Road, Oxford, England

The photochemistry of the $O/O_2/O_3$ system has been described by Wayne (this volume, p. 240) and so will not be discussed further here. Measurements of the atmospheric dayglow (Noxon, 1968) have shown that the (0, 0) bands of the $O_2(^1\Delta_g \leftrightarrow {}^3\Sigma_g^-)$ 'infrared atmospheric' system at $\lambda = 1.27\ \mu m$ are the strongest of all atmospheric emissions although they cannot be seen from the ground because of reabsorption in the lower atmosphere. $O_2(^1\Delta_g)$ is thus the metastable species present in the largest concentration in the upper atmosphere. The photolysis of ozone by ultraviolet radiation is the only atmospheric production mechanism which could produce sufficient quantities of $O_2(^1\Delta_g)$ to explain the dayglow measurements. It is thus important from the atmospheric, as well as the fundamental photochemical, point of view to know the primary products of ozone photolysis in the ultraviolet spectral region.

Emission from $O_2(^1\Delta_g)$ at $\lambda = 1.27\ \mu m$ was first observed during ozone photolysis at $\lambda = 2537$ Å by Izod and Wayne (1968). They were able to observe the $\lambda = 1.27\ \mu m$ band only when excess molecular oxygen was present in their system; $O_2(^1\Delta_g)$ was formed by an energy transfer process

$$O(^1D) + O_2 \rightarrow O_2(^1\Delta_g) + O(^3P) \tag{1}$$

while the primary step at $\lambda = 2537$ Å must be

$$O_3 + h\nu \rightarrow O_2(^1\Sigma_g^+) + O(^1D). \tag{2}$$

Reaction (2) can proceed isothermally at wavelengths shorter than 2660 Å and it was necessary for Izod and Wayne (1968) to assume a change in the primary molecular product about this wavelength with $O_2(^1\Sigma_g^+)$ being produced at $\lambda < 2660$ Å and $O_2(^1\Delta_g)$ at $\lambda > 2660$ Å. These early experiments were difficult to perform because of a lack of suitable detectors for the 1.1–1.3 μm region; also the rate constants for the deactivation of $O_2(^1\Delta_g)$ and its reaction with ozone were not known.

However, germanium photodiodes now available commercially are sensitive to radiation from 1 μm to 1.8 μm with a peak response at $\lambda \sim 1.4\ \mu m$. They are thus very suitable for detecting emissions in the $\lambda = 1.27\ \mu m$ region.

Gauthier and Snelling (1970) have used such a germanium detector and obtained results in a non-time-resolved system which suggest that $O_2(^1\Delta_g)$ *is* a primary product of ozone photolysis at $\lambda = 2537$ Å,

$$O_3 + h\nu \rightarrow O_2(^1\Delta_g) + O(^1D) \tag{3}$$

* Present Address: Department of Chemistry, University of California, Los Angeles, California 90024, U.S.A.

Fiocco (ed.), Mesospheric Models and Related Experiments, 253–260. All Rights Reserved.
Copyright © 1971 by D. Reidel Publishing Company, Dordrecht-Holland.

and also that $O_2(^1\Sigma_g^+)$ is formed only on addition of molecular oxygen to their system, by energy transfer from $O(^1D)$:

$$O(^1D) + O_2 \rightarrow O_2(^1\Sigma_g^+) + O(^3P). \tag{4}$$

In the work described in this paper a germanium photodiode (R.C.A. 6904A) and an intrinsic germanium device made in the Oxford laboratories have been used to detect optical emission from the $O_2(^1\Delta_g \leftrightarrow ^3\Sigma_g^-)$ system at $\lambda = 1.27\ \mu m$. A time-resolved flow system was used to study the kinetics of ozone decomposition and $O_2(^1\Delta_g)$ production during photolysis by $\lambda = 2537\ Å$ radiation. Experiments were carried out in the presence of a large excess of inert carrier gas (Ar, N_2 or He) to ensure essentially constant flow conditions.

A diagram of the principal section of the flow system is shown in Figure 1. The irradiation zone consisted of a quartz tube, joined to a gold-plated pyrex glass section with a viewing slit. $O_2(^1\Delta_g)$ emission measurements were made at this slit using an infrared grating monochromator fitted to the front of the germanium diode/metal dewar assembly. Slightly downstream from the gold-plated section two parallel quartz windows were fitted to the flow tube. These were the entrance and exit points for the monochromatic ($\lambda = 2537\ Å$) cross-beam used to measure $[O_3]$ by optical absorption. Full details of the apparatus and experimental procedure will appear elsewhere (Jones and Wayne, 1971).

A scan of the products of ozone photolysis with the monochromator showed that the $\lambda = 1.27\ \mu m$ band was the only detectable emission in the region of spectral response of the germanium photodiode.

It was found that when small increments of $[O_3]$ were added to the carrier gas flow, upstream from the photolysis zone, the observed $\lambda = 1.27\ \mu m$ band intensity gradually increased. However, when the $[O_3]$ was constant, the observed $O_2(^1\Delta_g \leftrightarrow ^3\Sigma_g^-)$ emission intensity was unaffected by addition of large excesses of molecular oxygen. The same effect was noted when He, N_2 and Ar were used as carrier gases.

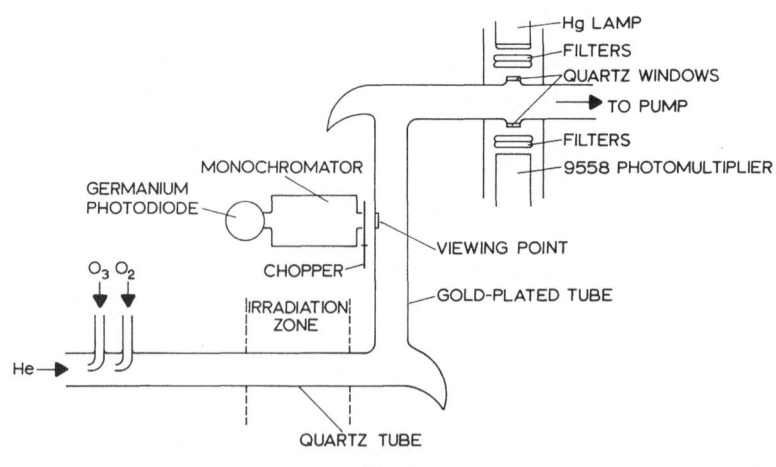

Fig. 1.

These kinetic experiments provide conclusive evidence therefore that $O_2(^1\Delta_g)$ is produced in the primary step of ozone photolysis at $\lambda = 2537$ Å by reaction (3), in agreement with the observations of Gauthier and Snelling (1970).

The kinetics of the reaction of $O_2(^1\Delta_g)$ with ozone have now been investigated (Clark et al., 1970)

$$O_2(^1\Delta_g) + O_3 \rightarrow 2O_2 + O(^3P) \tag{5}$$

and it seems that the early results of Izod and Wayne (1968) may be explained by the removal of $O_2(^1\Delta_g)$ by reaction (5) in their system and by variation in residence time caused by large additions of O_2 used as a carrier gas.

To measure quantitatively the primary quantum efficiency of $O_2(^1\Delta_g)$ formation the germanium detector's response was calibrated at $\lambda = 1.27$ μm in absolute units. The standard technique for optical emission calibrations involves the use of the NO+O continuum as described by Fontijn et al. (1964). However, their calibration data are not very accurate in the infrared region and are only quoted to within an order of magnitude at $\lambda = 1.27$ μm. Hence the calibration method used was the titration of $O_2(^1\Delta_g)$ with ozone.

Reaction (5) results in the removal of an ozone molecule and formation of $O(^3P)$ atom for every $O_2(^1\Delta_g)$ reacting. The titration method depends on knowing the major loss process for the O atoms formed in reaction (5) in order to calculate $[O_2(^1\Delta_g)]$ from measured changes in $[O_3]$.

Consider the reactions

$$O + O_3 \rightarrow 2O_2 \tag{6}$$

$$O + O_2 + O_2 \rightarrow O_3 + O_2 \tag{7}$$

$$O + \text{wall} \rightarrow \tfrac{1}{2}O_2. \tag{8}$$

At low $[O_3]$ and high $[O_2]$ reaction (7) dominates and reaction (5) would produce no observed change in $[O_3]$. At low $[O_3]$ and low $[O_2]$, reaction (8) results in the change in $[O_3]$ approaching the change in $[O_2(^1\Delta_g)]$. However, as $[O_3]$ is increased reaction (6) becomes important and a point is attained where $\delta[O_3]$ tends to twice $\delta[O_2(^1\Delta_g)]$. For a total $[O_3]$ above this value,

$$\frac{\delta[O_2(^1\Delta_g)]}{\delta I} = \frac{1}{2}\left(\frac{\delta[O_3]}{\delta I}\right)_{\text{limit}} \tag{9}$$

where $(\delta[O_3]/\delta I)_{\text{limit}}$ is the limiting value of the ratio of the measured change in $[O_3]$ to the measured change in $\lambda = 1.27$ μm emission intensity due to reaction (5).

For the calibration, the apparatus shown in Figure 1 was slightly modified. $O_2(^1\Delta_g)$ was produced by a microwave discharge through oxygen and O atoms were removed by passing the products of the discharge over HgO powder. This ensured that all O atoms in the system were produced by reaction (5). For full details of experimental procedure and precautions see Jones and Wayne (1971).

A plot of $\delta[O_3]/\delta I$ against $[O_3]$ is given in Figure 2. It clearly shows the limiting plateau values reached at high $[O_3]$ and gives the required calibration factor at $\lambda = 1.27\ \mu m$ for use in Equation (9) at the appropriate amplifier gain setting.

It was now necessary to measure the total amount of light absorbed under the same conditions as the $O_2(^1\Delta_g)$ emission intensity measurements were made in order to find the amount of ozone photolysed to produce a given measured $[O_2(^1\Delta_g)]$. This was done using the optical absorption at $\lambda = 2537$ Å in the cross-beam. Changes in absorption due to O_3, caused by turning the photolysis lamps on and off, gave the percentage of ozone decomposed. The $[O_3]$ used were such that absorptions were always weak and a linear relationship between fractional absorption and $[O_3]$ was always obeyed. (See Jones *et al.*, 1970.)

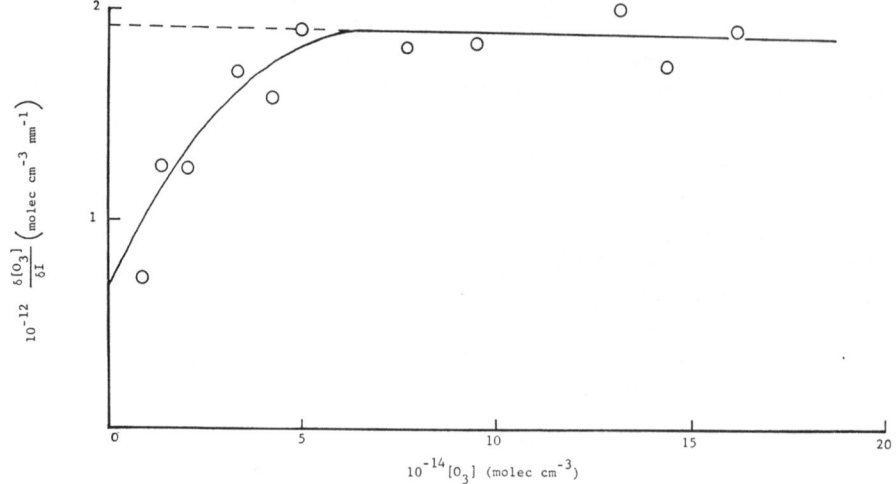

Fig. 2.   Plot of $\delta[O_3]/\delta I$ against $[O_3]$.

The fractional decomposition of ozone was a function of the ozone concentration, the incident light intensity I_0, and the residence times in the irradiated and dark regions of the flow tube.

At all $[O_3]$ used, the processes

$$O_3 + h\nu \quad \xrightarrow{\phi} O_2(^1\Delta_g) + O(^1D) \tag{3}$$

$$O(^1D) + O_3 \rightarrow 2O_2 \tag{10}$$

occurred rapidly (Biedenkapp and Bair, 1970) and the quantum yield for ozone decomposition was therefore at least 2. In addition, the relatively slower (Clark *et al.*, 1970) reactions,

$$O_2(^1\Delta_g) + O_3 \xrightarrow{k_5} 2O_2 + O(^3P) \tag{5}$$

$$O(^3P) + O_3 \quad \rightarrow 2O_2 \tag{6}$$

decomposed ozone in the dark, so that the total effective quantum yield, by the time the gases had reached the observation point, was $(2+Z)$, where Z can take values between 0 and 2. Z depends on $[O_3]$ and on the residence time in the dark portion of the flow tube.

Analysis of the above four Equations (3, 10, 5 and 6) gave a series of simultaneous differential equations for $[O(^3P)]$, $[O_2(^1\Delta_g)]$ $[O_3]$ etc., exact solution of which is impossible by analytical methods. However, a simplification was possible, assuming that, in solving the equations for $d[O_2(^1\Delta_g)]/dt$ and $d[O(^3P)]/dt$, $[O_3]$ was sensibly constant. No great error was introduced by this approximation, since the maximum observed percentage decomposition of ozone was 15–20%. Thus the decomposition in the dark section cannot be $>10\%$ and was probably $\not> 5\%$.

Using the above simplification the expression for Z became

$$Z = \frac{e^{-2\beta t_1}}{\frac{1}{2}\phi(1 - e^{-2\beta t_1})} \ln \frac{[O_3]_{t_1}}{[O_3]_{t_2}} \tag{11}$$

where $t_1 =$ the time the gases are in the irradiation zone, $t_2 =$ the time the gases are in the dark region, $\phi =$ primary efficiency of $O_2(^1\Delta_g)$ production, and $\beta = \phi I_0 \varepsilon d$.

If $2\beta t_1$ is small, Equation (11) gives

$$[O_3]_{t_1} = [O_3]_{t_2} e^{\beta t_1 Z} \tag{12}$$

and $e^{\beta t_1}$ is thus a small correction factor, actually $<4\%$, whose value is not critically dependent on Z or βt_1. To calculate Z the ozone concentration, was taken as $[O_3]_{t_2}$ since it has been shown that this value is probably in error by $\not> 5\%$. Solution of the equations also gives, for a residence time t_1 in the irradiation zone,

$$[O_3]_{t_1} = [O_3]_0 e^{-2\beta t_1} \tag{13}$$

$$[O_2(^1\Delta_g)]_{t_1} = \frac{1}{2}\phi[O_3]_0 (1 - e^{-2\beta t_1}). \tag{14}$$

For the percentage ozone decomposition studies, performed using He as the carrier gas, $[O_3]_0$, the initial ozone concentration before irradiation, was known. $[O_3]_{t_2}$ was calculated from the measured percentage decomposition and $[O_3]_0 \cdot [O_3]_{t_2}$ was then corrected to $[O_3]_{t_1}$ using Equation (12). Figure 3 shows a plot of $[O_3]_{t_1}$ against $[O_3]_0$. From this graph

$$\frac{[O_3]_0}{[O_3]_{t_1}} = e^{2\beta t_1} = 1.12. \tag{13}$$

Hence $2\beta t_1 = 0.11$, and since $t_1 = 0.029$s in these experiments, $\beta = \phi I_0 \varepsilon d = 1.9$ s^{-1}. Thus the percentage decomposition studies gave a value for β, which was, effectively, a measure of the absorbed light intensity.

$[O_2(^1\Delta_g)]_{t_1}$ can now be calculated from the observed emission intensities using Equation (15)

$$[O_2(^1\Delta_g)]_{t_1} = [O_2(^1\Delta_g)]_{obs} \exp[k_5 [O_3]_{t_2} t] \tag{15}$$

where $k_5 = 3.5 \times 10^{-15}$ molec^{-1} cm^3 s^{-1} as it is known that reaction (5) is the only significant loss process for $O_2(^1\Delta_g)$ under the present experimental conditions.

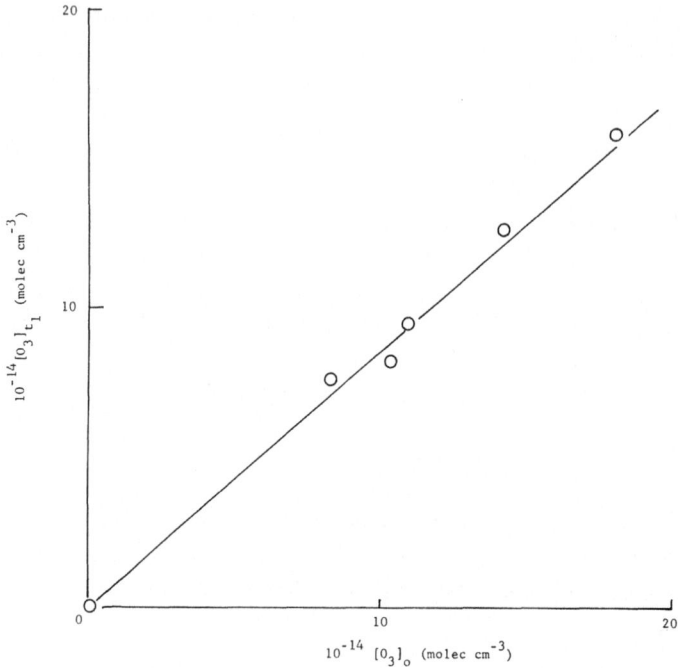

Fig. 3.   Plot of $[O_3]_{t_1}$ against $[O_3]_0$.

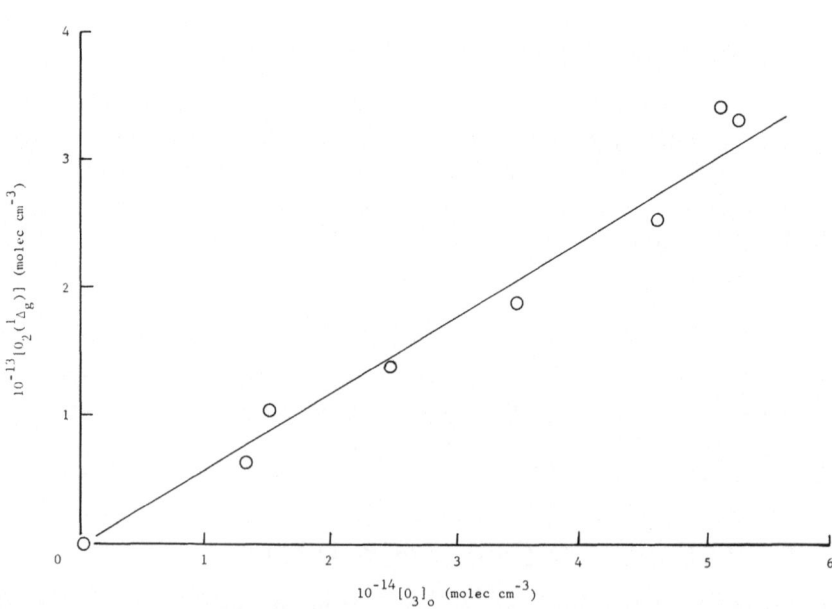

Fig. 4.   Plot of $[O_2(^1\varDelta_g)]$ against $[O_3]_0$ for helium carrier gas, $[He] = 2.24 \times 10^{16}$ molec cm^{-3}.

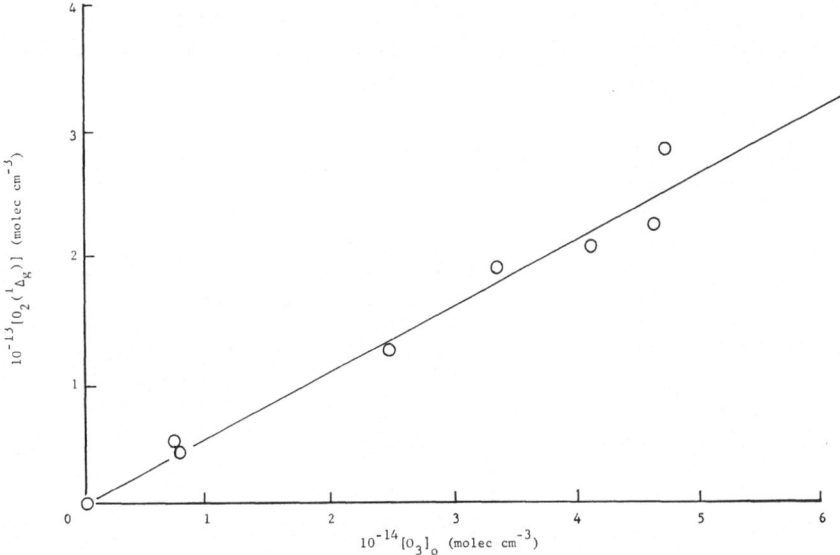

Fig. 5.   Plot of $[O_2(^1\Delta_g)]$ against $[O_3]_0$ for argon carrier gas, $[Ar] = 2.24 \times 10^{16}$ molec cm^{-3}.

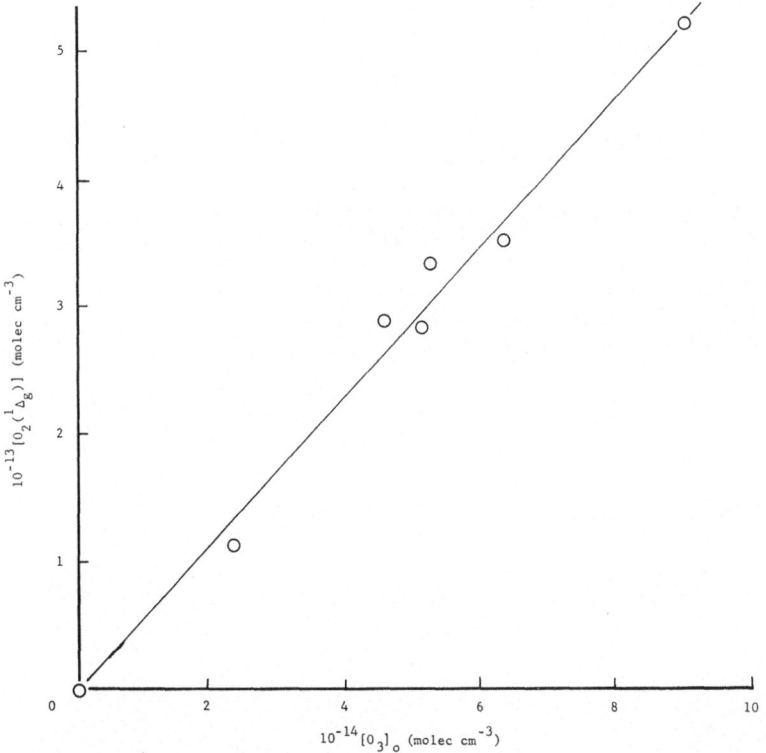

Fig. 6.   Plot of $[O_2(^1\Delta_g)]$ against $[O_3]_0$ for nitrogen carrier gas $[N_2] = 2.74 \times 10^{16}$ molec cm^{-3}.

Figures 4, 5 and 6 give the plots of $[O_2(^1\Delta_g)]_{t_1}$ against $[O_3]_0$ for He, Ar and N_2 respectively as carrier gases. The slopes of these graphs, when combined with the value of β already calculated and the appropriate value of t_1 in each case, enabled ϕ to be evaluated, using Equation (14). This assumed the value of β, i.e. the lamp intensity did not vary between experiments. The results are given in Table I. It will be seen that within experimental error, ϕ is essentially unity and certainly $\not< 75\%$.

TABLE I

Values of ϕ, calculated from Equation (14) for ozone photolysis by $\lambda = 2537$ Å radiation, using various inert carrier gases

Gas	t_1 (s)	$2\beta t_1$	$\frac{1}{2}(1 - e^{-2\beta t_1})$	$\dfrac{[O_2(^1\Delta_g)]_{t_1}}{[O_3]_0}$	ϕ
He	0.029	0.11	0.05	0.053	1.06
Ar	0.036	0.137	0.065	0.059	0.91
N_2	0.038	0.145	0.065	0.055	0.85

The results presented in this paper show that the photolysis of ozone at $\lambda = 2537$ Å produces $O_2(^1\Delta_g)$, as the molecular product in the primary photolytic step, with a quantum efficiency of essentially unity.

References

Biedenkapp, D. and Bair, E. J.: 1970, *J. Chem. Phys.* **52**, 6119–25.
Clark, I. D., Jones, I. T. N., and Wayne, R. P.: 1970, *Proc. Roy. Soc.* **A317**, 407–16.
Fontijn, A., Meyer, C. B., and Schiff, H. I.: 1964, *J. Chem. Phys.* **40**, 64–70.
Gauthier, M. and Snelling, D. R.: 1970, *Chem. Phys. Letters* **5**, 93–6.
Izod, T. P. J. and Wayne, R. P.: 1968, *Nature* **217**, 947–8.
Jones, I. T. N. and Wayne, R. P.: 1971, *Proc. Roy. Soc.* **A321**, 409–24.
Jones, I. T. N., Kaczmar, U. B., and Wayne, R. P.: 1970, *Proc. Roy. Soc.* **A316**, 431–9.
Noxon, J. F.: 1968, *Space Sci. Rev.* **8**, 92–134.
Wayne, R. P.: 1971, this volume, pp. 240–52.

THE AIRGLOW REACTION $NO + O + (M) \rightarrow NO_2^* + (M)$
AT LOW PRESSURES

K. H. BECKER, W. GROTH, and D. THRAN

Institut für Physikalische Chemie der Universität Bonn, West Germany

The increase of the emission intensity of the chemiluminescent reaction $NO + O \rightarrow NO_2^*$ due to a third body M was first shown in 1967 in the pressure region of 3 to 20 mTorr (Becker *et al.*, 1967, 1968a). Although these findings were confirmed independently by two other groups (Jonathan and Petty, 1968; McKenzie and Thrush, 1968), experimental limitations did not permit interpreting these results as definitive evidence against a two-body mechanism for the airglow reaction in the pressure region of 1 to 100 mTorr (Reeves *et al.*, 1964).

The present paper gives some new results about the airglow reaction which definitely show that radiative recombination as well as three-body recombination occurs in the pressure region of 0.1 to 50 mTorr.

The experiments were carried out by flowing a mixture of nitric oxide and active oxygen through a reaction sphere of 2×10^5 liters (Becker *et al.*, 1968b), or by filling this reactor to 0.1 mTorr with NO and active oxygen, followed by a stepwise increase of the total pressure during the decay of the airglow intensity (controlled by the wall recombination of the oxygen atoms). A decay time of 1 to 5 min was observed for the O atoms. By using resonance fluorescence which was tested previously (Becker *et al.*, 1969) as an analytical tool for a relative measure of low atomic concentrations, it was shown that the decay of the oxygen atoms by wall recombination and the decay of the airglow intensity occurred with the same time constant. These decay times correspond to a recombination coefficient for O atoms on stainless steel of $\gamma = 10^{-5}$ to 10^{-4} (the surface might be slightly contaminated by DC-705 pump oil, and the low values for γ could be obtained only after flowing active oxygen through the reaction vessel for a long period of time). Up to pressures of 50 mTorr the gas inlets for different gases such as O_2, N_2, H_2, He, Ne, Ar, CO_2, and SF_6 could be operated fast enough compared to the decay time of the airglow.

The light emission was measured with a photo-multiplier, E.M.I.-6255S, combined with interference filters centered at 4450 Å (halfwidth $\Delta\lambda = 80$ Å) or at 5600 Å (halfwidth $\Delta\lambda = 110$ Å). Without spectral filters the strong emission of the atmospheric (0, 0) band of O_2^*, which originates from the energy pooling reaction of two $O_2(^1\Delta_g)$ molecules (Becker *et al.*, 1970) (always produced with O atoms in a microwave discharge in oxygen), could not be suppressed even though the photocathode had an S-type spectral response. In addition, the spectrum of NO_2^* chemiluminescence was measured by an $f = 0.5$ m Czerny-Turner monochromator equipped with a cooled photomultiplier, E.M.I.-9558QB, allowing single photon counting. The spectral resolution was limited to $\Delta\lambda = 30$ Å by the sensitivity of the detection system.

In contrast to fluorescence excitation at low pressures (Sakurai and Broida, 1969),

Fiocco (ed.), Mesospheric Models and Related Experiments, 261–265. All Rights Reserved.

the spectra from the airglow reaction showed no structure at pressures below 1 mTorr. With a rotatable plate in the centre of the reaction sphere it was shown that the emission at total pressures below 1 mTorr was caused by volume reactions and not by surface effects. With increasing pressure a small red-shift of the chemiluminescence spectrum was observed similar to the results of fluorescence experiments (Keyser *et al.*, 1968; Schwarz and Johnston, 1969).

Figure 1 shows a plot of I/I_0 as a function of the pressure p_M of the added gases at 4450 and 5600 Å, I being the chemiluminescence intensity due to NO_2^* at the pressure p_M, and I_0 the initial intensity at a pressure $p_{M_0} = 0.1$ mTorr of active oxygen and nitric oxide before the stepwise addition of O_2 or CO_2. For all runs the initial oxygen atom concentration and the nitric oxide concentration were kept constant. Most of the

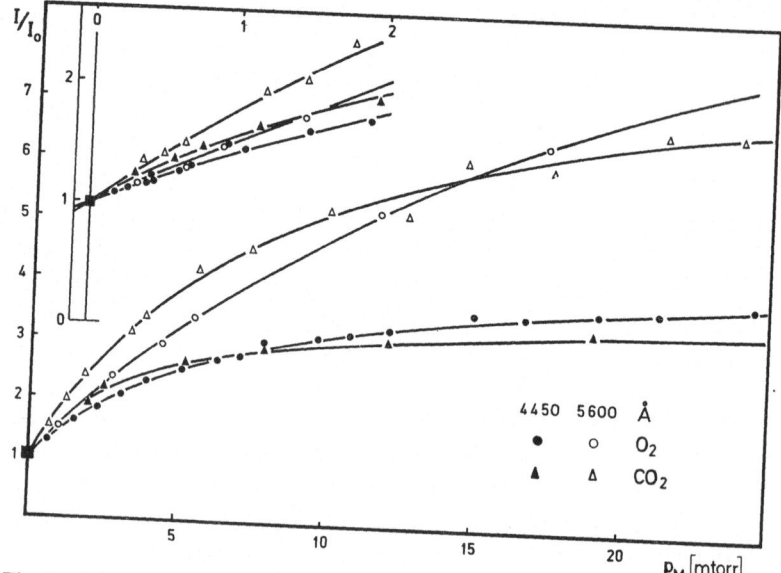

Fig. 1.   The pressure dependence of $I_{NO^*_2}$ for O_2 and CO_2 at 4450 and 5600 Å.

initial pressure p_{M_0} (more than 95%) is given by molecular oxygen. Observing the pressure dependence of the airglow during the decay of atomic oxygen in a closed reaction chamber guaranteed an isotropic distribution of the reacting particles. In addition, under these conditions the formation of NO clusters is prevented (Milne and Greene, 1967; Golomb and Good, 1968).

At low pressures the kinetics of the NO_2^* chemiluminescence follows relation (1):

$$I_{NO^*_2} = (NO) \cdot (O) \left[k_1 + \frac{k_2 \cdot (M)}{1 + k_q/k_r \cdot (M)} \right]. \tag{1}$$

The value k_1 represents that part of the chemiluminescence which originates from a radiative recombination of O and NO in a two-body process, k_2 is the rate constant of the three-body recombination, k_q the rate constant for quenching of NO_2^* by M, and

k_r the radiative transition probability of $NO_2^* \to NO_2 + h\nu$. The quenching constant q is given by k_q/k_r. Because of the short lifetimes of the emitting quasimolecules $[NO_2^*]$ which are always in equilibrium with NO and O because of the high rate of redissociation, the quenching of the emission caused by the two-body reaction can be completely neglected even at medium pressures. The complex relaxation processes of the stable electronically excited states of NO_2 which are known from fluorescence measurements (Sakurai and Broida, 1969; Keyser et al., 1968; Schwartz and Johnston, 1969; Myers et al., 1966) will also be neglected in a first simplified model at pressures in the mTorr region. Especially at the short wavelength side of the NO_2^* chemiluminescence, relaxation and quenching cannot be distinguished from each other. According to Equation (1) the experimental results were computed by Equation (2):

$$\frac{I}{I_0} = \frac{k_1 + \left(k_2(M) + k_2^{O_2}\cdot(O_2)\right)/(1 + q(M) + q_{O_2}(O_2)_0)}{k_1 + \left(k_2^{O_2}\cdot(O_2)_0\right)/(1 + q_{O_2}\cdot(O_2)_0)}. \tag{2}$$

By a 'least-square-fit' program the parameters $d = k_2/k_1$ and q were calculated first for molecular oxygen according to Equation (2) assuming $k_1 \neq 0$. By using these results for $d_{O_2}\cdot(O_2)_0$ and $q_{O_2}\cdot(O_2)_0$ the d and q values for the other gases were obtained as shown in Table 1.

The interpretation of the experimental results depends on whether the pressure dependence of I/I_0 measured from 0.1 mTorr upwards can be extrapolated to zero total pressure. The linear dependences of I/I_0 for $M = O_2$ in Figure 1 make such an extrapolation very likely. The curves in Figure 1 definitely do not go through the origin as p_{O_2} becomes zero which should occur with $k_1 = 0$. This rules out that the chemiluminescence is due only to a three-body reaction with O_2 as third body M. An unusually high three-body efficiency of NO or O atoms $((NO)_0 < 0.01$ mTorr, $(O)_0 < 0.001$ mTorr) in the pressure region of P_{M_0} was excluded by varying $(O)_0$ as well as $(NO)_0$. A linear dependence of $I_{NO^*_2}$ on (O) as well as on (NO) was observed. The linear increase of $I_{NO^*_2}$ with (NO) also rules out a $(NO)_2$ dimer reaction (Milne

TABLE I

Parameters q and d from Equation (2) for several added gases at 4450 Å or for [ ] at 5600 Å

gas	q (Torr^{-1})	d (Torr^{-1})	d/q
O$_2$	140	510	3.6
	[58]	[590]	[10.2]
N$_2$	144	560	3.9
H$_2$	155	611	3.9
He	81	365	4.5
Ne	71	303	4.3
Ar	108	412	3.8
CO$_2$	286	740	2.6
	[115]	[855]	[7.4]
SF$_6$	403	735	1.8
	[231]	[1111]	[4.8]

and Greene, 1967; Golomb and Good, 1968) with O atoms as source of the pressure-independent chemiluminescence. At 4450 Å no interference by the two-body chemi-luminescent reaction $NO + O_3 \rightarrow NO + O_2$ can occur (Clough and Thrush, 1967; Ackerman, 1967).

Therefore, the intercept at zero total pressure strongly indicates $k_1 \neq 0$, and the plot of I/I_0 in Figure 1 shows the intensity I of the airglow emission relative to I_0 which is approximately that part of the emission originating from a radiative recombination of $NO + O$.

The values for d in Table I give the ratios of the three-body to the two-body rate constant for different M. At 4450 Å the two-body and the three-body contributions to the NO_2^* chemiluminescent reaction become equal $(d \cdot (M) = 1)$ between 1.5 and 3.8 mTorr depending on the nature of the third body if quenching of NO_2^* by M is not considered.

The values of d/q, shown in Table 1, give the ratios of the apparent two-body rate constant at pressures in the Torr region to the rate constant of the two-body recombination at low pressures. Except for CO_2 and SF_6, this ratio has about the same value for different M. This means that the three-body and the quenching effi-ciencies depend similarly on the nature of M. The quenching of the NO_2^* emission at a particular wavelength can be caused by several processes such as collisionally induced redissociation, collisionally induced transitions to nonradiative states of NO_2, trans-fer of electronic energy to M, and relaxation processes of vibrational and rotational energy of NO_2^*.

Probably for CO_2 and SF_6, the deactivation of NO_2^* by energy transfer to M is relatively faster than the stabilization of the quasimolecule $[NO_2^*]$ by the third body.

The wavelength dependence of q can be interpreted as the result of relaxation pro-cesses in agreement with the results of recent fluorescence measurements (Schwartz and Johnston, 1969). On the other hand, also a collisionally induced redissociation would explain the wavelength dependence of q which would increase at shorter wavelengths near the dissociation limit.

More experimental results are needed for a quantitative interpretation of the q values by the different processes and for their comparison with q values from fluore-scence experiments.

The wavelength dependence of $d = k_2/k_1$ very likely is a result of different spectral distributions of the chemiluminescence caused by the radiative recombination and that caused by the three-body recombination. Obviously, a stabilisation of the quasi-molecule by M shifts the emission to longer wavelengths. The wavelength dependences of d and of q both let d/q grow with increasing wavelength, e.g.

$$(d/q)_{5600\,\text{Å}} / (d/q)_{4450\,\text{Å}} = 2.8 \quad \text{for} \quad M = O_2.$$

Therefore, only a rough estimate can be made for the absolute value of k_1 compared to $k_2/q = 6.4 \times 10^{-17}$ cm^3 molecule^{-1} s^{-1} which was measured for the whole spectrum at pressures above 1 Torr (Fontijn et al., 1964). The contribution from the radiative recombination to this rate constant lies within the error limit of 30%. From the present results, obtained at 4450 Å, k_1 is roughly $\frac{1}{4}$ of k_2/q. As d/q shows at 5600 Å, an integration over the whole spectrum might lower the rate constant for the radiative recombination to a value between 10^{-18} and 10^{-17} cm³ molecule^{-1} s^{-1}. Such a rate constant for the two-body recombination of NO + O can reasonably well be explained by a lifetime of about 10^{-12} to 10^{-11} s for the quasimolecules [NO$_2^*$] (Gaedtke and Troe, 1970) which emit the recombination continuum at a rate of 1 to 2×10^4 s^{-1} (Schwartz and Johnston, 1969).

Acknowledgement

The support of this work by the Bundesministerium für Bildung und Wissenschaft and by the Landesamt für Forschung des Landes Nordrhein-Westfalen is gratefully acknowledged.

Permission from *Chemical Physics Letters* to reproduce the paper in this volume is gratefully acknowledged.

Note added in proof. By recent measurements on the reaction NO + O + (M)→NO₂ + +(M)+$h\nu$ a rate constant of 3.4×10^{-18} cm³ molecule^{-1} s^{-1} for the radiative recombination and a rate constant of 7×10^{-32} cm³ molecule^{-2} s^{-1} for the three-body chemiluminescent reaction with M = O₂ was determined.

References

Ackerman, M.: 1967, *Bull. Acad. Roy. Belg.* **11**, 1311.
Becker, K. H., Groth, W., and Joo, F.: 1967, Abstracts, Part I, 91, International Conference on Photochemistry, Sept. 6–9, München.
Becker, K. H., Groth, W., and Joo, F.: 1968a, *Bunsengesell.* **72**, 157.
Becker, K. H., Elzer, A., Groth, W., Harteck, P., and Kley, D.: 1968b, Forschungsbericht SHA/1, Institut für Physikal. Chemie der Univ. Bonn.
Becker, K. H., Groth, W., and Jud, W.: 1969, *Z. Naturforsch.* **24a**, 1953.
Becker, K. H., Groth, W., and Schurath, U.: 1970, *Ber. Bunsengesell.* **74**, 930; 1971, *Chem. Phys. Lett.* **8**, 259.
Clough, P. N. and Thrush, B. A.: 1967, *Trans. Faraday Soc.* **63**, 915.
Fontijn, A., Meyer, C. B., and Schiff, H. I.: 1964, *J. Chem. Phys.* **40**, 64.
Gaedtke, H. and Troe, J.: 1970, *Z. Naturforsch.* **25a**, 789.
Golomb, D. and Good, R. E.: 1968, *J. Chem. Phys.* **49**, 4176.
Jonathan, N. and Petty, R.: 1968, *Trans. Faraday Soc.* **64**, 1240.
Keyser, L. F., Kaufman, F., and Zipf, E. C.: 1968, *Chem. Phys. Lett.* **2**, 523.
McKenzie, A. and Thrush, B. A.: 1968, *Chem. Phys. Lett.* **1**, 681.
Milne, T. A. and Greene, F. T.: 1967, *J. Chem. Phys.* **47**, 3668.
Myers, G. H., Silver, D. M., and Kaufman, F.: 1966, *J. Chem. Phys.* **44**, 718.
Reeves, R. R., Harteck, P., and Chace, W. H.: 1964, *J. Chem. Phys.* **41**, 764.
Sakurai, K. and Broida, H. P.: 1969, *J. Chem. Phys.* **50**, 2404.
Schwartz, S. E. and Johnston, H. S.: 1969, *J. Chem. Phys.* **51**, 1268.

ON THE ORIGIN OF NOCTILUCENT CLOUDS*

GIORGIO FIOCCO

European Space Research Institute, 00044 Frascati, Italy

and

GERALD GRAMS

*National Center for Atmospheric Research**, Boulder, Colo. 80302, U.S.A.*

Summary

In previous studies (Fiocco and Grams, 1966, 1969), we presented the results of two series of observations of the vertical distribution of mesospheric aerosols at times when noctilucent clouds (NLC) were expected. Observed enhancements of the dust content near 70 km have led us to consider a model for the formation of noctilucent clouds, dependent on the simultaneous presence of dust particles and water vapour.

To assess the effect of mesospheric motions on the spatial distribution of extra-terrestrial dust, we calculated meridional trajectories of individual dust particles of varying size and density, for a midsummer-midwinter velocity field in accordance with the meridional circulation model of Murgatroyd and Singleton (1961). The altitude range is from 50 km to 100 km; the summer latitudes are from 90° to 0°, and the winter latitudes from 0° to $-90°$. The important parameter is the product $r_1\varrho_1$ of particle radius and density. If we assume $\varrho_1 = 1$ g cm^{-3}, particles smaller than $\sim 0.1\ \mu$ are excluded from regions below approximately 65 km at 60°N latitude in the summer-time. To calculate dust concentration profiles, we assumed a steady influx of extra-terrestrial dust and evaluated the changes in small volume elements originating from various latitudes at 100 km and moving at the particle velocities. The laser-scattering experiments would be expected to detect the presence of dust layers in the 60–70 km altitude interval, since particles having sizes of about 0.1 μ radius are efficient scatterers. On the other hand, smaller particles, typically in the range of 0.01 to 0.05 μ radius, would be concentrated near 80 km. All these particles could act as condensation nuclei to form NLC particles. Thus, if the water content of the region is sufficiently high to permit condensation, a layer of water or ice may form on the particle so that it becomes heavier and falls rapidly to lower altitudes. However, since condensation or sublimation can occur only in a relatively narrow and sharply defined region near the mesopause, the particle soon reaches the boundary of the region where the liquid or solid phase can exist, and the condensed layer evaporates while the solid particle continues its descent. This trapping mechanism could then increase the amount of water vapour in the region of the boundary until the local concentration of water vapour is sufficient to sustain a visible cloud.

In our analysis of this suggested mechanism for the formation of NLC, we estimated

---

* Details of this study will be published in *J. Atmospheric Terrestr. Phys.*
** The National Center for Atmospheric Research is sponsored by the National Science Foundation.

that a local water vapour concentration of about 10^{10} cm^{-3} water vapour molecules is required for NLC particles to grow to a size of radius 10^{-5} cm in falling through a layer of thickness 3×10^4 cm. This requires that the trapping mechanism accumulate a total columnar number of uncondensed water molecules of about 3×10^{14} cm^{-2}, which can accumulate in less than a day, either by a small updraught of the order of 1 cm s or by upward flux produced by eddy diffusion, in accordance with the estimates of Hesstvedt (1968). Thus, the time required to trap a sufficient amount of water to sustain steady-state NLC activity is consistent with the observed durations of NLC displays, and also with the maximum allowable time before photochemical destruction of non-equilibrium amounts of water vapour can appreciably affect the process. Calculations of the water budget for the NLC model showed that the upward and downward flux of water substance balance if the optical thickness of the NLC is about 2×10^{-5}. This is an acceptable value since our optical radar observation of a strong NLC display indicated an optical thickness of approximately 10^{-4}. Finally, we verified that the influx of dust particles implied by the trapping mechanism was not excessive. We calculated that a total influx of extraterrestrial dust of ~ 100 tons/day was required by the model. This influx is compatible with our measurements and with measurements obtained by other means.

References

Fiocco, G. and Grams, G.: 1966, *Tellus* **18**, 34–8.
Fiocco, G. and Grams, G.: 1969, *J. Geophys. Res.* **74**, 2453–8.
Hesstvedt, E.: 1968, *Geofys. Publik.* **27**, No. 4, April.
Murgatroyd, R. J. and Singleton, F.: 1961, *Quart. J. Roy. Meteorol. Soc.* **87**, 125–35.

THE RATE CONSTANTS OF SOME ION-MOLECULE AND ION-ELECTRON REACTIONS, FROM IN SITU IONOSPHERIC MEASUREMENTS

ALAIN GIRAUD

Centre National d'Etude des Télécommunications, 92 – Issy Les Moulineaux, France

There is some generality in the attempt to derive directly reaction rates from in situ measurements, whatever the geophysical importance of the reactions considered, because many upper atmospheric models are based on laboratory-measured rates, and it is legitimate to wonder to what extent upper atmospheric conditions, indeed, relate to laboratory conditions.

For such a comparison, it is necessary to select the simplest type of situation, because it is not easy to properly assess the quantitative validity of models when they involve as many parameters as some tens of reaction rates, the density of several neutral constituents, their absorption and ionization cross-sections, temperature profile, and so on.

From this point of view, the F_2 region of the ionosphere is interesting because:

Chemical equilibrium obtains at all times for molecular ions, whose lifetime is always shorter than a few minutes;

Ion temperature does not vary as a function of altitude, so that ion-neutral reaction rates can be assumed to be constant;

Photoionization rates are small and can be neglected, or, even if they are important, they follow a simple barometric profile.

Still, during the day, many reactions have to be taken into account. In addition, the electronic temperature T_e is larger than the ionic temperature T_i, and increases sharply with altitude. During night time, however, the situation is particularly simple because of the disappearance of N_2^+ and N^+, with the result that there only remain the reactions:

$$O^+ + \begin{cases} O_2 \to O_2^+ + O & \text{with coefficient } K_1 \\ N_2 \to NO^+ + N & \text{with coefficient } K_2 \end{cases}$$

$$e^- + \begin{cases} O_2^+ \to O^* + O^* & \text{with coefficient } \alpha_1 \\ NO^+ \to N^* + O^* & \text{with coefficient } \alpha_2. \end{cases}$$

These reactions are of primary importance because they determine both the chemical loss frequency

$$\beta = K_1 n(O_2) + K_2 n(N_2)$$
$$= \alpha_1 n(O_2^+) + \alpha_2 n(NO^+),$$

in the continuity equations for electrons and O^+ ions, and the equilibrium distribution

Fiocco (ed.), Mesospheric Models and Related Experiments, 268–272. All Rights Reserved.
Copyright © 1971 by D. Reidel Publishing Company, Dordrecht-Holland.

of O_2^+ and NO^+. In addition, the second and fourth are a source of nitrogen atoms, responsible for the production of nitric oxide (cf. Nicolet, this symposium, p. 1).

Since above about 250 km, $n(e^-) = n(O^+)$, it follows that:

$$\frac{n(O_2)}{n(O_2^+)} = \frac{\alpha_1}{K_1}, \qquad \frac{n(N_2)}{n(NO^+)} = \frac{\alpha_2}{K_2}.$$

The validity of the above scheme can be verified by checking that the O_2^+ and NO^+ profiles have a scale height corresponding to O_2 and N_2; this can be expressed in terms of barometric temperatures.

The experimental results from the night-time flight of a rocket-borne ion mass spectrometer, launched in France on 6 February 1969 (Giraud *et al.*, 1970), are shown in Figure 1.

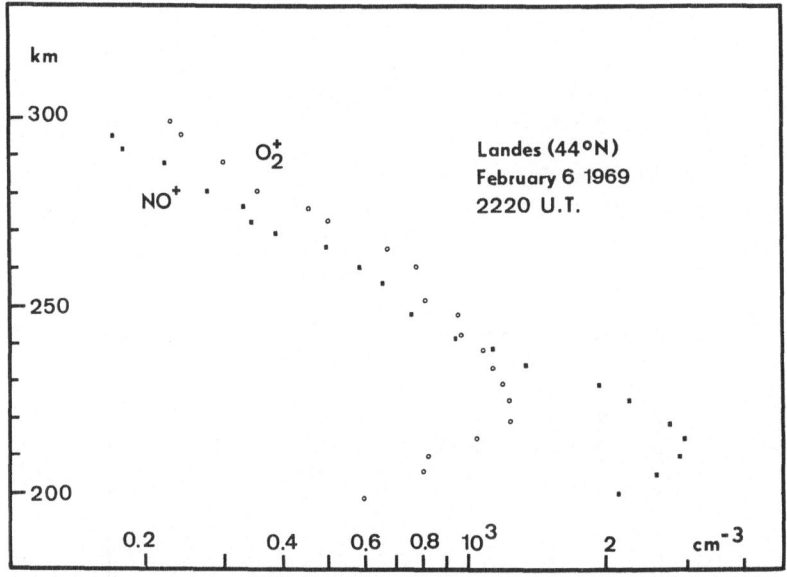

Fig. 1.   Measured profiles of the molecular ions in the night F region.

A least square fit to the data points between 260 and 300 km gives temperatures of 866 and 929 K for NO^+ and O_2^+ respectively. This is within $\pm 4\%$ of the temperature at 300 km measured from the ground by the CNET incoherent scatter sounder during flight. Using CIRA 1965 densities for O_2 and N_2 corresponding to this temperature, we obtain the results given in Table I. An error bar of $\pm 20\%$ on the absolute value of the molecular ion densities can be considered as the experimental uncertainty, while neutral densities will be discussed below.

The α/K ratios obtained can be compared with those derived from laboratory measurements, summarized in Table II. Some extrapolating is necessary, because reliable measurements have not been made near a gas temperature of 1000 K.

The temperature dependences of the dissociative recombination rates α have been taken from Hansen (1968) and Cunningham *et al.* (1969) as:

$$\sqrt{(300/T_e)}\,(1 - e^{-2300/T_i}) \quad \text{for } \alpha_1,$$

and

$$\sqrt{(300/T_e)}\,(1 - e^{-3086/T_i}) \quad \text{for } \alpha_2.$$

The value of the experimental uncertainty for the coefficient at 300 K is taken from Biondi (1969).

TABLE I

Altitude (km)	260	300
Temperature (K) measured	850	900
CIRA (model 4)	864	891
$n(O_2)$ CIRA (cm^{-3})	1.9×10^7	3.9×10^6
$n(O_2^+)$ measured (cm^{-3})	7.5×10^2	2.2×10^2
$\alpha_1/K_1 = n(O_2)/n(O_2^+)$	2.5×10^4	1.8×10^4
$n(N_2)$ CIRA (cm^{-3})	2.6×10^8	6.3×10^7
$n(NO^+)$ measured (cm^{-3})	5.7×10^2	1.5×10^2
$\alpha_2/K_2 = n(N_2)/n(NO^+)$	4.6×10^5	4.2×10^5

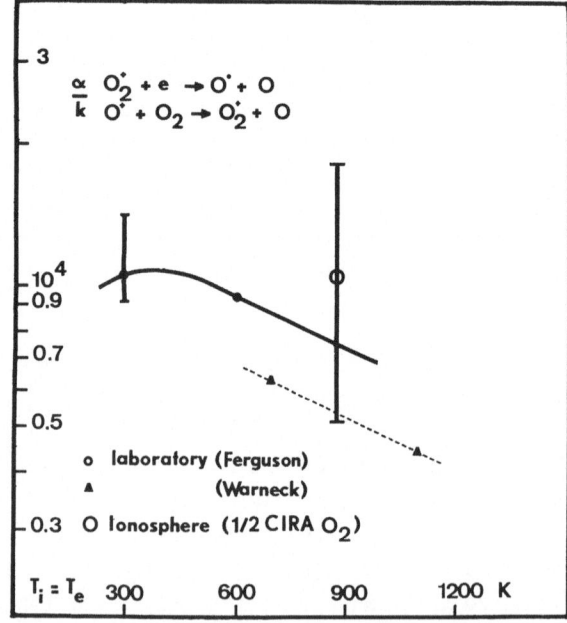

Fig. 2. Comparison between the in situ and laboratory values of α_1/K_1. The laboratory dissociative recombination rate α_1 has an assumed temperature dependence $(300/T_e)^{1/2}$.

TABLE II

$T_i = T_e$ (K)	300	600	900	1200
α_1 (see text)	2×10^{-7}	1.39×10^{-7}	1.07×10^{-7}	0.85×10^{-7}
Kasner and Biondi	$2.2 \pm 0.2 \times 10^{-7}$			
K_1 Ferguson	$1.9 \pm 0.2 \times 10^{-11}$	$1.45 \pm 0.1 \times 10^{-11}$	$\longrightarrow$	$\longrightarrow$
Warneck		$\longleftarrow$ 2×10^{-11}		$\longrightarrow$
α_2 (see text)	4×10^{-7}	2.8×10^{-7}	2.24×10^{-7}	1.85×10^{-7}
Weller and Biondi	$4.1 \, {}^{+0.3}_{-0.2} \times 10^{-7}$			
K_2 Ferguson	$1.2 \pm 0.1 \times 10^{-12}$	0.6×10^{-12}	$\longrightarrow$	$\longrightarrow$
Warneck		$\longleftarrow$ 4.6×10^{-12}		$\longrightarrow$

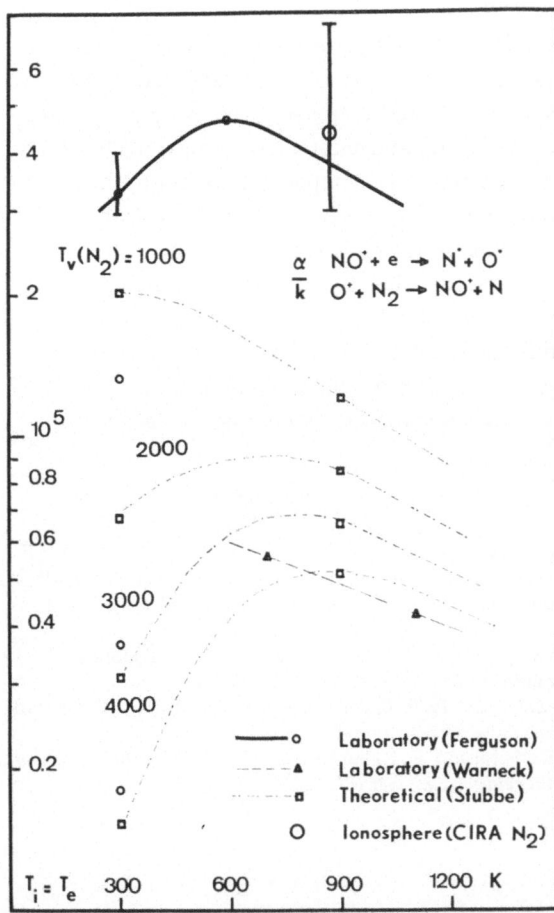

Fig. 3. Comparison between the in situ and laboratory values of α_2/K_2. The laboratory dissociative recombination rate α_2 has an assumed temperature dependence $(300/T_e)^{1/2}$.

The ion-molecule rates have been taken to be constant above 600 K, as suggested by Warneck (1967) and Dunkin et al. (1968). It is known that K_2 strongly depends on the nitrogen vibration temperature (Schmeltekopf et al., 1968). But as this dependence has not been measured for high gas temperature, the theoretical dependence computed by Stubbe (1969) has been used as indicative, even though it does not quite fit the measured rates for low temperatures. It can be seen that the spread of laboratory measured values for K_2 extends over an order of magnitude, a reason for concern as regards their application to ionospheric models.

The data from Table I and Table II are compared in Figures 2 and 3.

Figure 2 shows that, for α_1/K_1, the agreement between in situ and laboratory values is within the bounds of experimental error only if $n(O_2)$ is at least twice smaller than in CIRA 1965. Because of the large amount of data suggesting that O_2 densities in CIRA are indeed overestimated by about this factor (e.g. Gross and von Zahn, 1970), it is felt that the comparison is excellent and not indicative of any significant discrepancy.

Figure 3 shows that the in situ value for α_2/K_2 is in excellent agreement with the lowest laboratory value for K_2, i.e. 6×10^{-13} cm^3 sec^{-1}. Even reducing the N_2 density with respect to CIRA and increasing the laboratory value for α_2 by a factor 2 hardly raises the inferred K_2 above the theoretical value for $T_v(N_2) = 1000$ K. This can mean either that the vibrational temperature of N_2 in the night thermosphere is low, or that K_2 has only a small variation with $T_v(N_2)$ near gas temperature of 1000 K, in contrast to its behavior at ambient temperature.

In conclusion, it appears that uncertainties in our knowledge of the neutral constituents are probably the main factor obscuring a detailed understanding of ionospheric chemical equilibria. It follows that ion composition measurements, together with laboratory values for the reaction rates, should now be considered as a valuable source of information on the neutral atmosphere and its variations.

References

Biondi, M. A.: 1969, Can. J. Chem. 47, 1711.
Cunningham, A., Hobson, R., and O'Malley, T. F.: 1969, Bull. Amer. Phys. Soc. 14, 101.
Dunkin, D. B., Fehsenfeld, F. C., Schmeltekopf, A. L., and Ferguson, E. E.: 1968, J. Chem. Phys. 49, 1365.
Giraud, A., Scialom, G., Pokunkov, A., Poloskov, S., and Toulinov, G.: 1970, Paper presented at COSPAR, Leningrad, to be published in Space Res. 11.
Gross, J. and von Zahn, U.: 1970, Paper presented at COSPAR, Leningrad.
Hansen, C. F.: 1968, Phys. Fluids 11, 904.
Schmeltekopf, A. L., Ferguson, E. E., and Fehsenfeld, F. C.: 1968, J. Chem. Phys. 48, 2966.
Stubbe, P.: 1969, Planetary Space Sci. 17, 1221.

ON THE SEMIANNUAL VARIATION OF THE
UPPER ATMOSPHERE

V. P. BHATNAGAR

Institut für Astrophysik und extraterrestrische Forschung der Universität Bonn,
53 Bonn, West Germany

Abstract. The effects of the seasonal variations of the dynamical properties of the lower thermosphere at and below 120 km, such as the dissociation of the molecular oxygen, the turbopause levels of different constituents and the temperature on the semiannual variation (SAV) in the upper atmosphere, have been studied on the basis of the steady state models. It is concluded that the seasonal variations in the mass density and the temperature as adopted from the US Standard Atmosphere Supplements (1966) in conjunction with the observed dissociation ratio (O/O_2) do not yield the observed semiannual amplitude (SAA) (maximum value in equinoctial months/ minimum value in solstitial months) in the upper atmosphere. In the lower thermosphere, the observed variations of the dissociation ratio lead to lower boundary conditions which are different from those obtained from the observed variations of the temperature.

It is also concluded that the SAV of the dissociation ratio is in phase or out of phase with that of the atmospheric density, depending on whether the SAA in the temperature is less or greater than about 1.1 at 120 km.

The increase of the SAA within the solar cycle, when the solar activity changes from 80 to 150 $\bar{F}$-units, can be explained if the dissociation increases more rapidly with solar activity in equinoctial months as compared to that in solstitial months.

At altitudes where helium predominates, the observed SAA indicates that the turbopause levels of helium lie at about 100 km in winter, 102 km in equinox, and 104 km in summer.

(This version is published with the permission of the publishers of *Planetary and Space Science*.)

Fiocco (ed.), Mesospheric Models and Related Experiments, 273. All Rights Reserved.
Copyright © 1971 by D. Reidel Publishing Company, Dordrecht-Holland.

ROCKET MEASUREMENTS OF ION AND ELECTRON
DENSITIES IN THE *D*-REGION DURING SUNRISE

A. PEDERSEN*

Space Science Department, ESTEC, Noordwijk, Netherlands

and

J. A. KANE

NASA Goddard Space Flight Center, Greenbelt, Md., U.S.A.

This paper discusses the results of two rockets launched near sunrise at White Sands, New Mexico, when positive ion and electron densities were measured in the *D*-region for solar zenith angles $\chi = 91°$ and $\chi = 79°$.

The electron density has been measured during sunrise on many occasions by ground-based techniques and by rockets. However, conclusions about ionisation and detachment of negative ions during sunrise have been limited by poor knowledge concerning ion densities. The combination of positive ion and electron density measurements on the same rocket makes it possible to assess the total charge density and negative ion density and thus provide a better basis for a discussion of these problems.

From observations of VLF waves reflected in the ionospheric *D*-region during morning twilight it has been found that electron densities below about 80 km are increasing when going from night conditions to solar zenith angles $\chi = 98°-99°$ (Thomas and Harrison, 1969). This has been interpreted as detachment of electrons from negative ions induced by radiation at wavelengths larger than 3000 Å. Shorter wavelengths do not penetrate to altitudes below 80 km at these zenith angles. The same authors have found that detachment at altitudes above 80 km seems to occur near or after ground sunrise.

Rocket measurements of electron densities during twilight for solar zenith angles between 108° and 94° (Mechtly and Smith, 1968; Sechrist, 1968) also show an electron increase indicating detachment near or below 80 km for these solar zenith angles.

The measurements reported here cover the height range 80–110 km and complement the VLF observations and the rocket measurements of electron density mentioned above. It will be shown that the majority of negative ions in this height range are detached near to or after ground sunrise. On the basis of these measurements it is possible to discuss electron affinity of negative ions and ion production functions during twilight.

The ion densities were measured with an ion collector, which has certain similarities with the Gerdien condenser probe. A set of parallel plates was connected electrically as indicated in Figure 1a. The advantage of such a configuration is that the stray-field is small, which means that the current collection is not varying strongly with bias on

---

* This work was done while the author was a research associate of the National Academy of Sciences at Goddard Space Flight Center.

Fiocco (ed.), Mesospheric Models and Related Experiments, 274–278. All Rights Reserved.
Copyright © 1971 by D. Reidel Publishing Company, Dordrecht-Holland.

the plates. To demonstrate this, the current, as a function of bias referred to the plasma potential $(V - V_p)$, is compared with two Langmuir probes in Figure 1b for the case of approximately orbital limited current (electron collection). These curves are experimental results from a large chamber with a low density plasma where the Debye length was similar to the probe size, but several times smaller than the chamber.

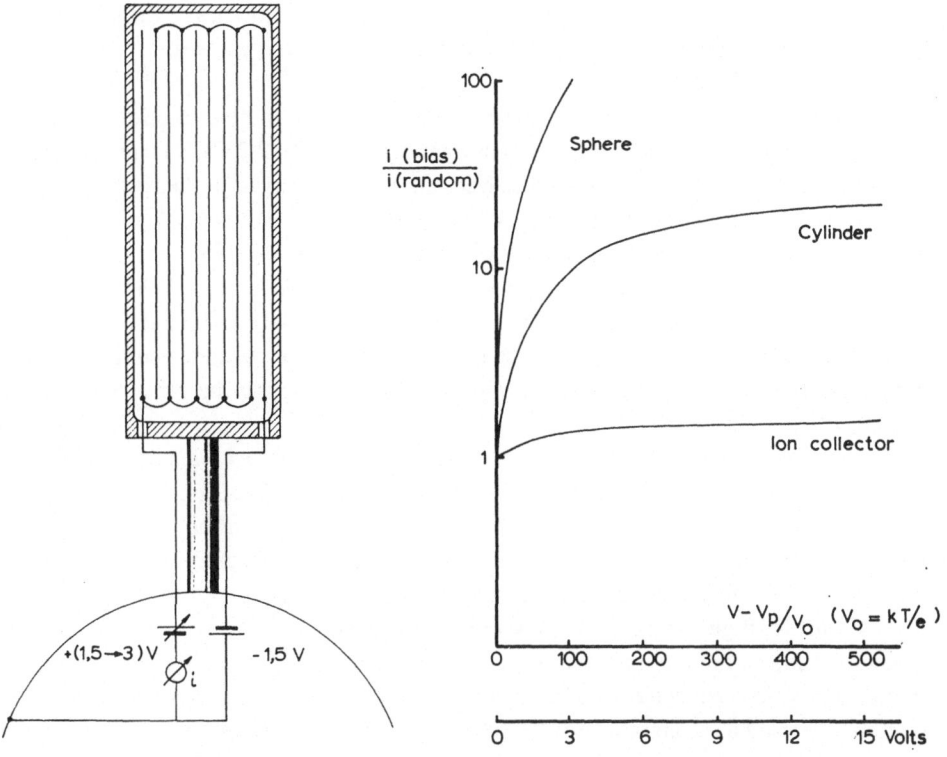

Fig. 1a.   Ion collector.   Fig. 1b.   Current-voltage characteristics for
                                        ion collector and Langmuir probes.

The potential at the middle of the gap between the plates ideally should be at plasma potential in order to have the correct ion collection. This potential, however, is expected to be a volt or so negative with respect to the plasma potential. Therefore the positive bias which was referred to the rocket was increased in a sweep going between 1.5 and 3.0 V so that the potential between the plates would be near to, or pass through, the plasma potential. In this way it was possible to check that deviation from plasma potential had no serious effect on the current collection.

When the probe is moving through the medium at a velocity v_R, which is high compared with the ion thermal velocity v_i, the ram effect of the ions will dominate. For the case of a pure ram current we can write:

$$i_+ = N_+ e v_R A \cos\theta \tag{1}$$

where i_+ = positive current collected, N_+ = positive ion density, v_R = rocket velocity, A = probe area, corresponding to open area of frame in Figure 1a, and θ = angle of attack.

In the collision-dominated region below 80 km the shock wave will tend to sweep some ions past the probe so that i_+ tends to be smaller than the value given by formula (1). Sonin (1967) has discussed this for a probe at the blunt end of a cylinder. It seems that the corrections will be relatively small for the velocities of the rockets and the size of the probes used. The reduction factor will at the most amount to 0.5 at 70 km and be negligible at 80 km.

The electron densities were measured by means of a Faraday rotation technique (see for example Kane and Troim, 1967) using the frequencies 2.23 MHz and 3.23 MHz. Transmissions from the ground at these frequencies are received in the rocket, and the rotation of the plane of polarisation of the radio waves can be observed to yield information about the electron density distribution in the D-region.

The payloads were launched with Booster II Arcas rockets from White Sands on 29 August 1966 at 0534 and 0632 MST. The peak altitudes were 110 ± 3 and 108 ± 3 km respectively. Because of difficulties with rocket tracking, the altitudes could be determined within only ± 3 km. However, one trajectory relative to the other is uncertain to ± 1 km.

The positive voltage sweep on the ion collector decreased the collected current only by a small amount, of the order 10% at 70 km and 20–30% at 90 km. Near apogee, when $v_i \gtrsim v_R$, the decrease of collected positive current was about 50% for a full positive sweep. It is thus reasonable to assume that the collected current is nearly identical to the ram current given by Equation (1) for $v_i < v_R$; this condition is fulfilled throughout the flight except for a few km below peak.

Figure 2 shows the ion and electron densities for the two flights. The Faraday rotation experiment is expected to give the most reliable absolute measurements and can therefore be used as a reference for the measurement of positive ion densities at higher

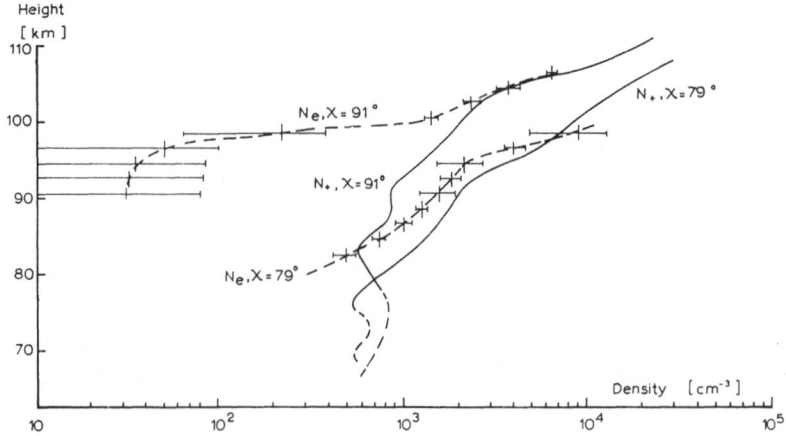

Fig. 2.   Ion and electron densities for solar zenith angles $\chi = 79°$ and $\chi = 91°$.

altitudes where negative ions can be neglected. Figure 2 shows that there is good agreement between the measured values of electron and ion densities at the upper part of the trajectories without introducing any scaling of ion current. Because of possible influences from shock waves at lower altitudes, the ion density profiles are shown as dotted lines below approximately 75 km.

From Figure 2 the corresponding values of $\lambda = N_-/N_e$ are shown in Figure 3. Because of the uncertainties in the electron and ion density measurements the values given in Figure 3 are useful only to establish the orders of magnitude of λ for the two zenith angles involved here.

Measurements of ion and electron densities using similar techniques in the auroral zone have been reported by Folkestad *et al.* (1969) and by Jespersen *et al.* (1968). These authors find for nighttime auroral conditions λ values which lie between the two profiles of Figure 3. This is presumably due to auroral particles increasing the negative ion detachment rate.

From a comparison between the results presented here and the previous results from rocket measurements (Sechrist, 1968) and VLF observations (Thomas and Harrison, 1969), it can be seen that detachment of negative ions starts at and below 80 km for solar zenith angles $\chi = 98°-99°$, whereas detachment starts near ground sunrise for higher altitudes. This indicates that the negative ions are different in these two height ranges; however, it is difficult to conclude what negative ion species are present.

Thomas and Bowman (1969) in a survey of solar radiation during sunrise conclude that the VLF observation of detachment near $\chi = 98°-99°$ can be explained by radiation at wavelengths above 3260 Å, which sets an upper limit on the photon energy necessary for detachment of 3.8 eV.

For $\chi = 91°$ (first flight) and altitudes 80–110 km, where these measurements were made, no significant detachment was observed, even though the same radiation ($\lambda > 3260$ Å) had been present for a long time. This means that the ions in this height range have a larger affinity or the cross section for photodetachment is very small for these wavelengths.

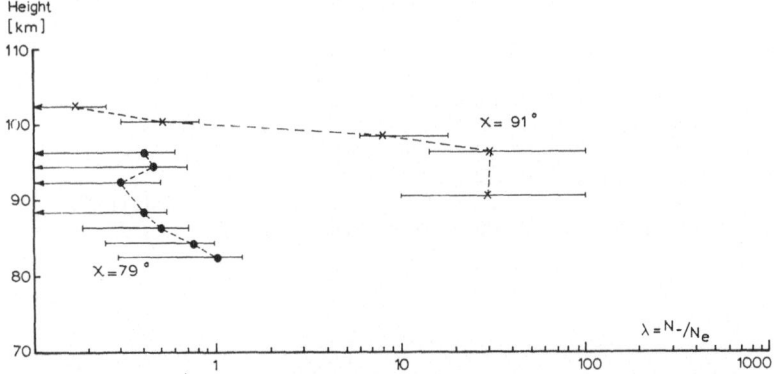

Fig. 3.   Ratio between negative ion and electron concentrations ($\lambda = N_-/N_e$) for solar zenith angles $\chi = 79°$ and $\chi = 91°$.

If we want to find out how the solar radiation changed between the two flights we need to consider:

(a) Solar Lyman-α radiation that 'penetrated' to 95 km for the first flight and to 80 km for the second flight.

(b) Radiation between 2000 Å and 3000 Å, which depends heavily on the O_3 distribution and has a similar solar zenith dependence as solar Lyman-α for a nighttime O_3 distribution.

This means that radiation under both (a) and (b) could give rise to the detachment. The photon energy necessary for detachment must be at least 4–6 eV (corresponding to 2000 Å–3000 Å). It is difficult to argue what negative ions are present on the basis of these numbers because the photodetachment as a function of radiation wavelength is not known for ions of interest for the D-region (for example NO_2^- and NO_3^-).

It is of interest to find out what ion production function is in agreement with the profiles of Figure 2. Solar Lyman-α radiation seems to be the only ionising radiation which changes between the two flights.

In order to calculate production functions for solar Lyman-α for the two flights it is necessary to know the NO distribution. It would then be interesting to compare the production with the densities shown in Figure 2 and derive information about recombination processes. However, because of the existing uncertainty in the determination of NO, this exercise has rather limited value.

Some features are still worth mentioning. The ion density does not change by a large amount between the two flights, and irrespective of the NO distribution chosen, it is found that a presunrise source must give rise to ionisation in addition to cosmic rays. The most likely explanation is diffuse Lyman-α radiation; however, particle precipitation could produce the same effect.

The ratio between negative ions and electrons, λ, is determined with best accuracy for the first flight where it is expected that the values are typical for night conditions. The λ-values for the second flight only give an upper limit for λ.

References

Folkestad, K., Landmark, B., Skovli, G., Kane, J. A.: 1969, *J. Atmospheric Terrest. Phys.* **31**, 835–44.
Jespersen, M., Kane, J. A., Landmark, B.,: 1968 *J. Atmospheric Terrest. Phys.* **30**, 1955.
Kane, J. A. and Troim, J.: 1967, *J. Geophys. Res.* **72**, 1118.
Mechtly, E. A. and Smith, L. G.: 1968, *J. Atmospheric Terrest. Phys.* **30**, 363.
Sechrist, C. F.: 1968, *J. Atmospheric Terrest. Phys.* **30**, 371.
Sonin, A. A.: 1967, *J. Geophys. Res.* **72**, 4547.
Thomas, L. and Bowman, M. R.: 1969, *J. Atmospheric Terrest. Phys.* **31**, 1311–22.
Thomas, L. and Harrison, M. D.: 1970, *J. Atmospheric Terrest. Phys.* **32**, 1–4.

ION DENSITIES IN THE LOWER *D*-REGION MEASURED
WITH A PARACHUTED PROBE

R. JAESCHKE and A. PEDERSEN

Space Science Department, ESTEC, Noordwijk, Holland

A parachute-borne instrument package for performing measurements in the lower ionosphere between 80 and 40 km altitude has been developed by the ESTEC Space Science Department. The parachute probe is a piggy-back on Skylark rockets. It is ejected from the rocket during ascent at about 45 km height and braked by a drag parachute to an apogee of about 80 km. At this height an internal timer releases the drag parachute of 0.75 m diameter and the main parachute of 2.5 m diameter is deployed. The package descends to an altitude of 15 km where two thirds of the rigging lines are cut off by explosive bolts, triggered by a barometric switch. The package then falls rapidly to the ground, thus reducing the potential air traffic risk. A schematic view of the rocket and probe trajectories and of the sequence of events is shown in Figure 1.

A schematic view of the ejection system is given in Figure 2. The instrument package weighing approximately 3 kg is ejected sideways from the rocket by spring force. To avoid any disturbance of the rocket attitude two counterweights are simultaneously ejected in the opposite direction. The package is a cylindrical box 20 cm in height, having an inner diameter of 13.5 cm. The container is divided into two parts, one 9 cm high for the storage of the two parachutes, and the other 11 cm high for the installation of the flight instrumentation and experiments. The package is self-contained, carrying its own telemetry system. A detailed description of the instrument package is given elsewhere (Berkeljon *et al.*, 1970).

Prototypes of the parachute package have been included in two ESRO sounding rocket payloads. These were launched on 24 October 1969 and 22 June 1970. The first flight was successful and the parachute probe operated as expected. However, the rocket motor burnt faster than predicted and ejection of the package occurred at higher altitude than planned. Accordingly the apogee of the package was at about 140 km height instead of 90 km. For the second flight the main rocket motor did not ignite and no results were obtained.

The two flights were regarded largely as technological trials. However, two scientific devices were included in the instrumentation, a Geiger-Müller tube and an ion density probe. Results from the probe measurements in the first round are reported here.

The ion density probe is of the 'blunt probe' type. A disk-shaped collector of 6 cm diameter with 1 cm wide guard ring is mounted at the bottom of the parachute package. For subsonic flow at altitudes below approximately 70 km a thin boundary layer forms in front of the probe. A negative voltage bias on the probe draws the positive ions diffusing into the boundary layer to the probe, whereas positive ions outside the boundary layer flow past the probe and are not detected. The metalised parachute is

Fiocco (ed.), Mesospheric Models and Related Experiments, 279–282. All Rights Reserved.
Copyright © 1971 by R. Reidel Publishing Company, Dordrecht-Holland.

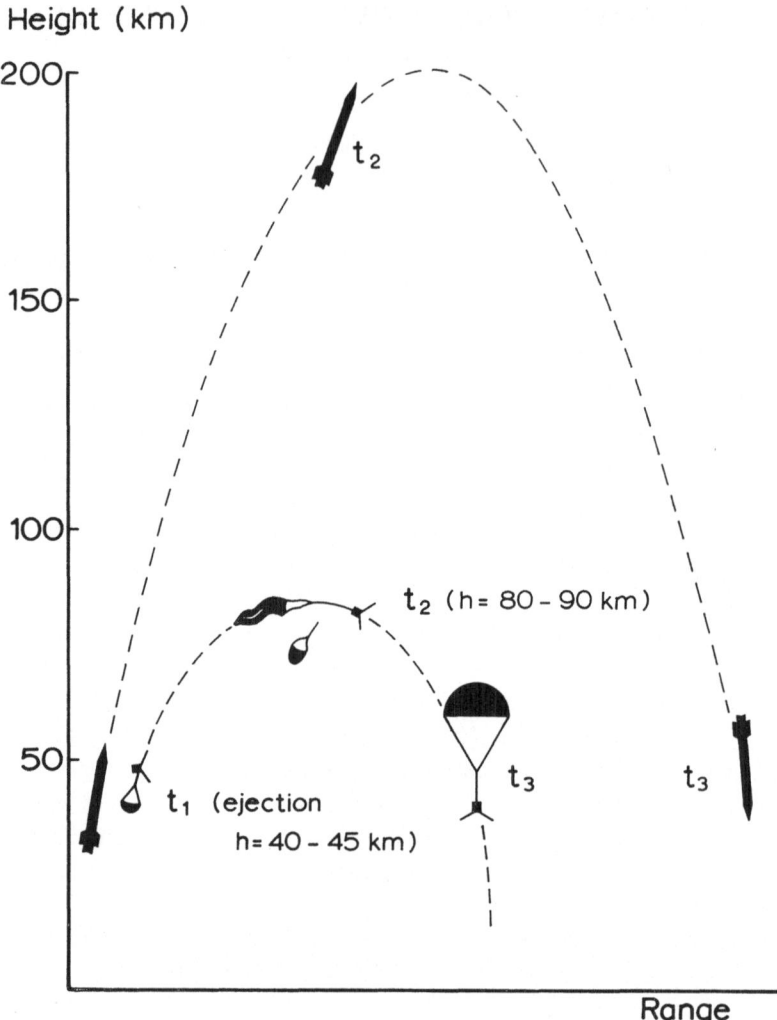

Fig. 1.   Schematic trajectories and sequence of events.

used as return electrode for the probe. The theory of blunt probes for subsonic flow conditions is discussed by Hale *et al.* (1968). They show that the current collected is given by a simple mobility expression

$$di_+ = N_+ \mu_+ eE \, dA,$$

where di_+ is the current collected by an element of the probe surface dA, e is the electric unity charge, E is the electric field at the surface, μ_+ is the mobility, and N_+ is the number density of the positive ions collected.

The ion mobility μ_+, as a function of height, is calculated from

$$\mu_+ = \mu_0 \left(T p_0 / T_0 p \right),$$

using a model atmosphere. μ_0, T_0 and p_0 are values for mobility, temperature and pressure reduced to ground level. The electric field E near the probe surface can be calculated for the simple probe geometry chosen and is found to be proportional to the voltage applied to the probe. In the present case the voltage bias had a triangular wave form, 0 to -4 V with a period of 30 s. From the measured probe voltage to current characteristic the number density N_+ could then be determined.

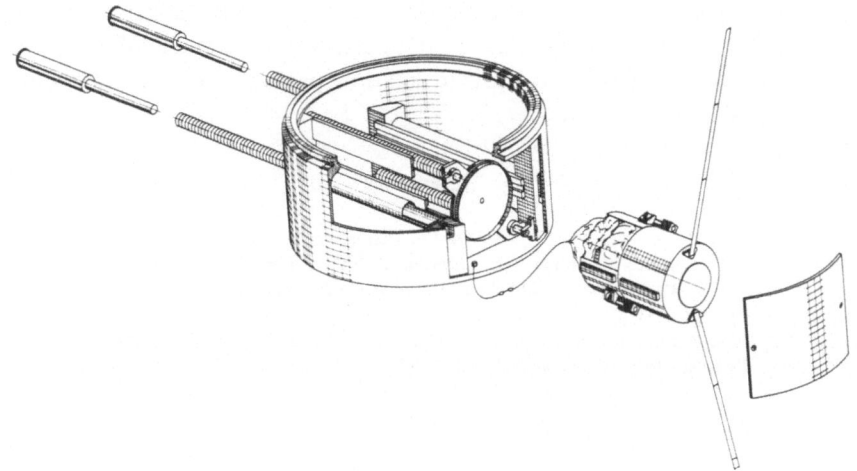

Fig. 2.   Schematic view of instrument package and ejection system.

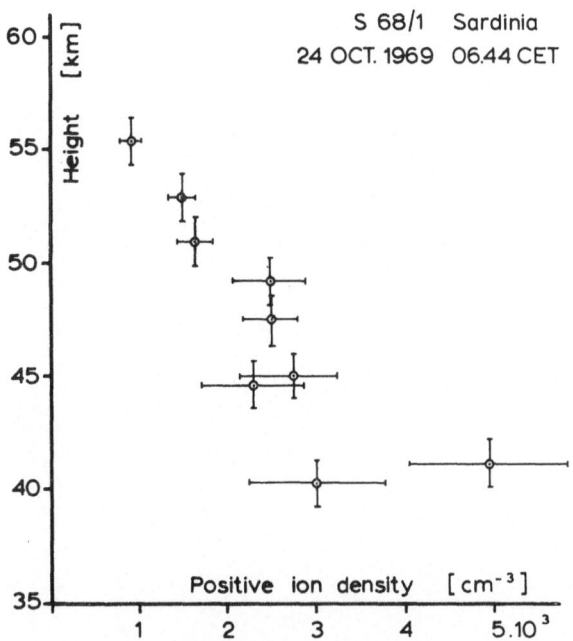

Fig. 3.   Positive ion density versus height.

The result is shown in Figure 3. Data were obtained between 55 and 40 km height. Over this range the ion density increases from 1×10^3 cm^{-3} to about 5×10^3 cm^{-3}. Few measurements in this height region have been reported so far. Data have been published by Hale *et al.* (1968) from White Sands, New Mexico, and by Pedersen (1966) from Andenes, Norway. The ion densities obtained at these two locations were rather similar and the values reported here are in agreement with these earlier results. The similarity of ion densities measured at low and high latitudes is difficult to explain, considering that ion production from cosmic rays should be higher by a factor of 5–10 at the higher latitudes. It is therefore hoped to obtain more information from future flights of the piggy-back instrumentation package to be carried out from Sardinia and Kiruna, Sweden.

References

Berkeljon, R., Englert, G., Jaeschke, R., Köhn, D., Pedersen, A., and Scheper, R.: 1970, 'A rocket-borne, ejectable instrument package with parachute', ESRO TN-102 (ESTEC).
Hale, L. C., Hoult, D. P., and Baker, D. C.: 1968, in *Space Research VIII*, North Holland Publ. Co., Amsterdam, pp. 320–331.
Pedersen, A.: 1966, 'Measurements of ion concentrations in the *D*-region of the ionosphere', Uppsala Ionosphere Observatory, Report No. 15, pp. 9–32.

ION COMPOSITION IN THE *E* AND LOWER *F*-REGION
ABOVE KIRUNA AT SUNSET AND SUNRISE

E. KOPP, P. EBERHARDT, and J. GEISS

Physikalisches Institut, University of Berne, Switzerland

On February 25 and 26, 1970, the ion and neutral composition and the total ion density of the atmosphere between 110 and 220 km were measured simultaneously with a magnetic mass spectrometer (ESRO payloads S 61). The first rocket was launched during sunset and the second at sunrise (solar zenith angles between 97–98°).

NO^+, O_2^+ and O^+ were the major ions between 110 and 220 km. In the evening NO^+ was dominant below 180 km and O^+ above 180 km. In the morning NO^+ was the major ion over the whole altitude range. The evening ion density had a daytime profile. The ion distribution in the morning was comparable to one at night with a remarkable sporadic *E*-structure below 160 km containing 3 thin layers.

(Paper to be published in *Planetary Space Sci.* **18**.)

Fiocco (ed.), Mesospheric Models and Related Experiments, 283. All Rights Reserved.

ROCKET MASS SPECTROMETRY IN JAPAN

N. FUGONO

Radio Research Laboratories, Tokyo, Japan

Summary

In Japan, observations of the upper atmosphere using sounding rockets were started in 1958. The measurement of ion composition by mass spectrometer began in the autumn of 1965. Since that time, seven flight experiments with Bennett type mass spectrometers and one with a new periodic static field type mass spectrometer have been carried out. Five- or three-stage mass spectrometers were used.

The ion composition of the E and F layers was measured. Figure 1 illustrates the profiles of major constituents for the K-10 rocket flight, in the evening on January 14, 1969. The ionogram was taken at the Yamagawa Radio Observatory, about 40 km westward from the launching site. The sporadic E layer appeared at about 100 km but the spectrometer could not detect it.

Scientific observation by satellites now being prepared aim at mass spectrometric composition. One satellite, 'SRATS' (Solar Radiation And Thermospheric Structure), will be orbited by a Mu-series rocket. Its orbit is expected to have 250 km perigee, 2000 km apogee and about 30 deg of inclination.

A three-stage Bennett type mass spectrometer, installed at the side window of the satellite within the orbital plane, will measure only three ionic species, such as H^+, He^+ and O^+.

The 'Ionosphere Observation Satellite' being prepared by the Radio Research Laboratories, Ministry of Posts and Telecommunications, will have a circular orbit of 1000 km height and 70 degrees inclination and will carry two Bennett type mass

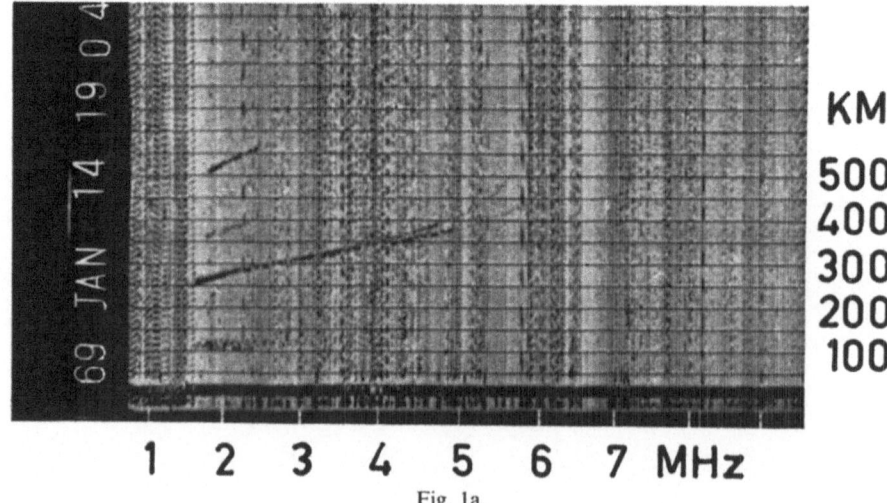

Fig. 1a.

Fiocco (ed.), Mesospheric Models and Related Experiments, 284–285. All Rights Reserved.

spectrometers identical to those in the SRATS satellite, installed along its spin axis but facing opposite directions. A mass range is scanned stepwise in 2 s.

We also plan to measure ion composition of the lower ionosphere with a S-210 rocket carrying a modified quadrupole mass spectrometer. A special type of mass filter will be used (Hayashi and Sakudo, 1968, 1969).

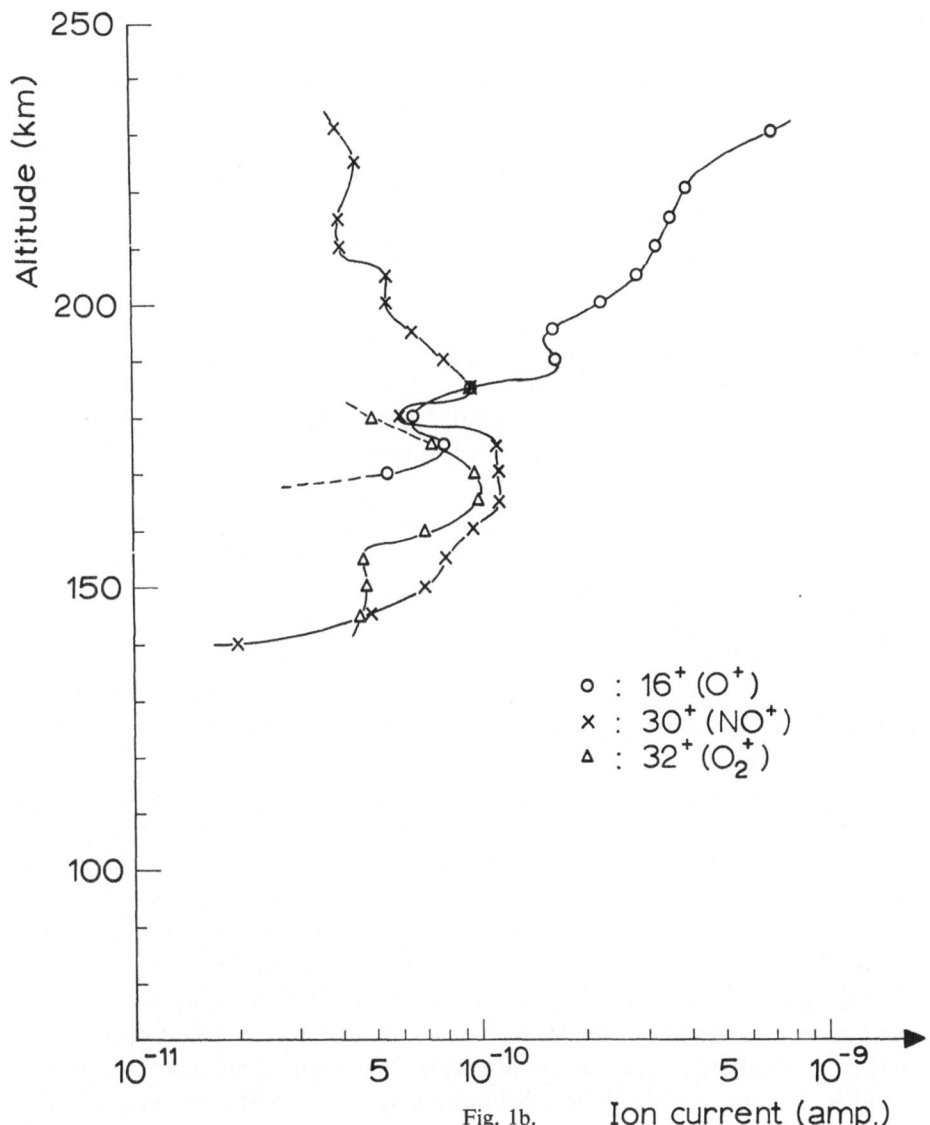

Fig. 1b.   Ion current (amp.)

References

Hayashi, T. and Sakudo, N.: 1968, *Rev. Sci. Instr.* **39**, 7, 958–61.
Hayashi, T. and Sakudo, N.: 1969, *Rev. Sci. Instr.* **40**, 7, 923–4.

MEASUREMENT OF SUBMILLIMETER-WAVE
STRATOSPHERIC EMISSION SPECTRA

JOHN CHAMBERLAIN, W. J. BURROUGHS, and J. E. HARRIES

Division of Electrical Science, National Physical Laboratory, Teddington, Middlesex, England

The submillimeter-wave properties of the atmosphere are determined largely by the spectrum of water vapour. Although oxygen and ozone, for example, have significant features, it is the pure rotation spectrum of monomeric water vapour that dominates all else. As a consequence, solar radiation is detectable at sea level only beyond about 1000 μm and at 3000 m altitude it is detectable above 280 μm with the bulk of the detected power lying in two windows centred at 357 and 454 μm (Gebbie, 1957). The amount of radiation collected depends on

    (a) the length of the atmospheric path;

    (b) the amount of water vapour along it;

    (c) the area of the radiation collector;

    (d) the throughput capability (Jacquinot, 1958) of the spectrometer system;

    (e) the difference between the effective temperature of the source and that of the detecting system.

The spectra of the atmospheric gases can also be observed in emission from the radiation exchange that can be set up between a spectrometric system and the atmosphere. Here, the amount of signal recorded depends on

    (a) the effective optical depth of the atmospheric path;

    (b) the amount of water vapour (and other gases) along it;

    (c) the area of the radiation 'collector' used to view the atmosphere (less important than in (c) above because the 'source', the sky, is extended);

    (d) the throughput of the spectrometer;

    (e) the difference between the effective temperature of the gas and the detecting system.

Hence, spectral information about the atmosphere can be obtained from both absorption and emission submillimeter-wave spectra. The absorption studies can be carried out from mountain sites (Gebbie and Burroughs, 1968; Gebbie *et al.*, 1968; Biraud *et al.*, 1969; Burroughs and Chamberlain, 1971) where a large collector is used but where the proportion of the available observation time that may be used is determined by the whims of the local climate; from aircraft (MacQueen *et al.*, 1968; Eddy *et al.*, 1969) where there is less atmospheric absorption but where, of necessity, a smaller collector is used; or from balloons (Gay *et al.*, 1968) where guidance of the collector can be a problem.

The spectra of the atmospheric gases can also be observed in emission in experiments in which less stringent demands are made on the size of the viewer and practically no guidance is required; but they do need to be made at altitude. Aircraft provide the most efficient platforms but their altitudes are, of course, limited and balloons or

Fiocco (ed.), Mesospheric Models and Related Experiments, 286–291. All Rights Reserved.
Copyright © 1971 by D. Reidel Publishing Company, Dordrecht-Holland.

rockets may be necessary. Experiments have already been carried out using aircraft (Bader *et al.*, 1967; Gebbie *et al.*, 1968a) and balloons. In the case of either emission or absorption studies it is virtually essential to use Fourier transform spectrometry if spectra having medium to high resolution and acceptable signal-to-noise ratio (S/N) are to be obtained (Vanasse and Sakai, 1967).

The spectrum is obtained by numerical Fourier transformation of the output from the detector of a two-beam interferometer (Vanasse and Sakai, 1967). The signal-to-noise ratio in this spectrum is limited by detector noise, by fluctuations in the incident radiation level caused by variable obscuration (for example, clouds), scintillation or poor following, by restrictions of dynamic range brought about by the detection of unwanted radiation and by the effective temperature difference between the source and a representative part of the instrument. Conventional instruments have used amplitude modulation (AM) of the radiation (Gebbie and Burroughs, 1968; Gebbie *et al.*, 1968b; Gebbie *et al.*, 1969a, 1969b; Harries *et al.*, 1969; Burroughs and Chamberlain, 1971) but phase modulation (PM), produced by a sinusoidal jitter of path-difference within the interferometer (Connes *et al.*, 1967; Chamberlain, 1971), relaxes the limitations given above particularly with regard to the signal fluctuations and the inclusion of unwanted radiation.

Experiments have been carried out by NPL in the last 9 months to measure the sky emission using a modular Michelson interferometer (Chantry *et al.*, 1969) fitted with PM (Chamberlain and Gebbie, 1971). The apparatus was mounted in a Comet 2E aircraft of the Royal Aircraft Establishment, Farnborough, and flown near 12000 m altitude, 3000 m above the tropopause. The ambient atmospheric temperature was 222 K, the Golay cell detector temperature was 300 K. Because of the geometry of

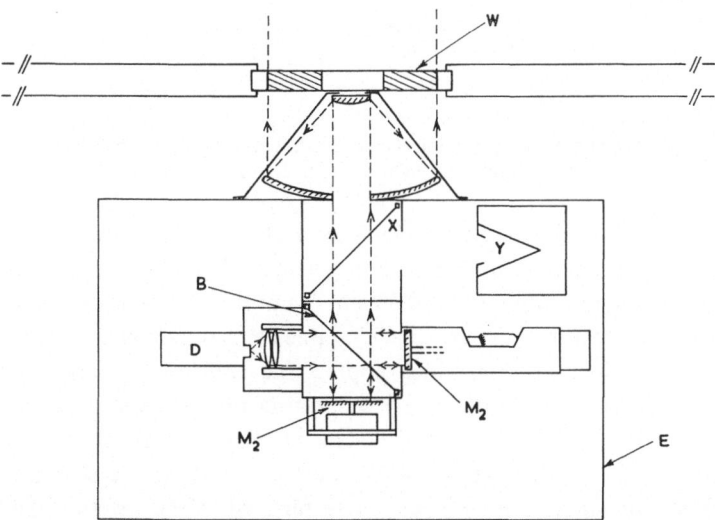

Fig. 1.   Schematic diagram of the installation of the Michelson interferometer plus telescope in the aircraft.

the aircraft, observations were made along an axis about 10° above the horizontal using a 350 mm diameter telescope looking through 4 circular polypropylene windows, 125 mm in diameter and 16 mm thick (Figure 1). The inclination of the observation axis gave an effective emission path about 5.6 times the vertical path. PM interferograms had S/N ratios of about 80 (see Figure 2), which represents more than an order of magnitude improvement over earlier AM measurements (Gebbie *et al.*, 1968a). These ratios are comparable with those of Eddy *et al.* (1969) obtained for solar spectra using AM with a liquid-helium-cooled detector quoted to have a noise equivalent power 1000 times lower than that of a Golay cell (10^{-13} WHz$^{-1/2}$ (Eddy *et al.*, 1969) compared with 10^{-10} WHz$^{-1/2}$ for a Golay cell (Putley and Martin, 1967)). The radiation collector of Eddy *et al.* (1969) was an order of magnitude smaller than ours. Our system could be calibrated using a conical black body (Y) maintained at solid-CO_2 temperature, by inserting mirror X (Figure 2).

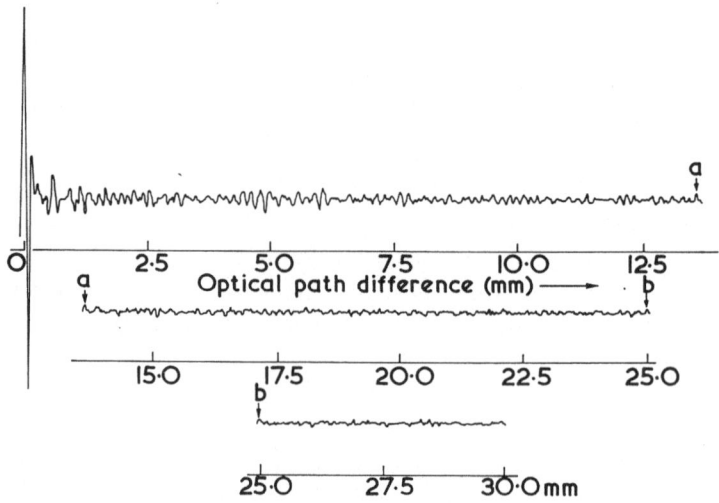

Fig. 2.   Example of a phase-modulated interferogram showing the high signal-to-noise ratio obtained in submillimetre atmospheric emission studies.

In order to indicate the quality of the spectra obtained in this work, the transform of the interferogram shown in Figure 2 is given in Figure 3 for the spectral range 58 to 62 cm^{-1}. The resolution (R) is 0.33 cm^{-1}, but it has been found that for 'strong' lines (i.e. emission features that are 'black' at line centre) the removal of the weighting function $w(x/D)$ does not introduce appreciable spurious features ('feet'), whilst narrowing and, therefore, more clearly separating the principal features (e.g. H_2O line at 59.95 cm^{-1} and O_2 line at 60.46 cm^{-1}). Thus the effective unapodized resolution limit (R_0) is ~0.17 cm^{-1}, while the true width of the strong atmospheric features shown in Figure 3 is ~0.2 cm^{-1}. In fact, when the spectral line-width is of the same order as the instrumental line-width, it has been shown (Dowling, 1967), that the absence of apodization introduces side lobes ('feet') to the observed feature that are completely negligible. It should be noted that the oxygen features in Figure 2 have not previously been observed in atmospheric spectra. From a series of such spectra a number of other oxygen and ozone features have been identified for the first time. Speculative identifications can be made of a number of other features but

ATMOSPHERIC EMISSION SPECTRUM (SINGLE OBSERVATION)
RESOLUTION=0.33 cm⁻¹ ALTITUDE =11.9km ANGLE TO HORIZON=15°

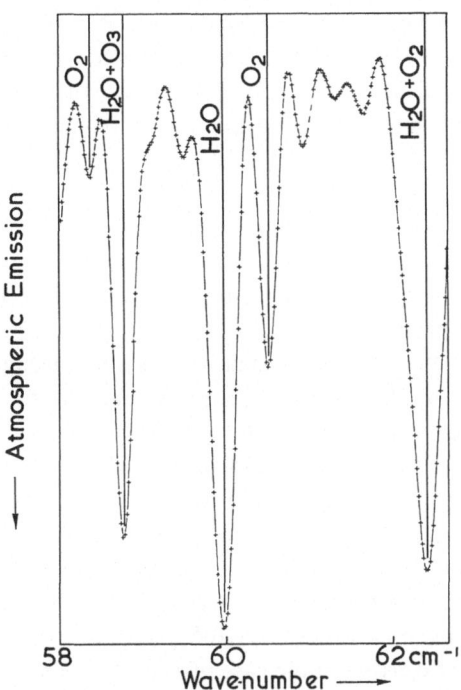

Fig. 3.   Atmospheric emission spectrum obtained from a single observation (see Figure 2), flying at an altitude of 11.9 km. The aircraft was circling, so that the angle of elevation to the horizon of the observed emission path was 15°. Resolution 0.33 cm⁻¹ (unapodized this is effectively 0.17 cm⁻¹). A comparison of the H_2O and O_2 features at 59.95 and 60.46 cm⁻¹ respectively gives a mixing ratio from this spectrum of $(2.8 \pm 0.4) \times 10^{-6}$ g g⁻¹.

more work is required if unambiguous assignments are to be made. It is, perhaps, worth mentioning that there is a feature at 68.7 cm⁻¹ near 68.0 cm⁻¹ where an atomic oxygen feature is predicted (Bates, 1951). However, this feature must be considered in the context of the noise level in the spectra which is not quite low enough, in our opinion, to be very sure about the assignment. Neither do the spectra have a sufficiently high S/N ratio below 20 cm⁻¹ and so there is no additional evidence for the 8 cm⁻¹ $(H_2O)_2$ feature (Gebbie *et al.*, 1969b). However, the Q branch predicted by Viktorova and Zhevakin (1967) to occur at 49.8 cm⁻¹ is well within the observable range but is not apparent in our spectra, although Eddy *et al.* (1969) claim to have observed it and, possibly, two intervening branches at 22 and 35 cm⁻¹. The evidence for the assignment to $(H_2O)_2$ of data we have obtained by the techniques outlined here is given in the following paper (Burroughs *et al.*, 1971).

An important practical application of the data we have recently obtained is to a determination of the mixing ratio of water vapour to air in the stratosphere. This has been evaluated from experimentally measured values of the mixing ratio of water vapour to oxygen determined directly from the relative emission strengths of adjacent water and oxygen lines in the spectra. If certain reasonable assumptions are made (Burroughs, 1969) it can be shown (Burroughs and Harries, 1970) that the mixing ratio is related to the experimentally determined emission strengths, and well-known spectroscopic and physical quantities. Using a number of pairs of adjacent, but resolved, water and oxygen lines a value of the mass mixing ratio

$$\mu = (2.3 \pm 0.3) \times 10^{-6} \text{ g g}^{-1}$$

was found. This value is considerably more accurate than an earlier estimate (Gebbie *et al.*, 1968a) made from submillimeter-wave data because the use of PM has improved the S/N ratio, thus enabling adjacent lines to be used and has removed a number of radiometric calibration problems.

The mixing ratio of water to ozone has also been found using the same technique, and from this a value for the effective integrated path length of ozone above the aircraft of 0.37 ±0.10 cm atm is obtained.

The foregoing outline of some of the results we have obtained using phase-modulated Fourier spectrometry shows that high S/N ratios are obtainable in both interferograms and spectra with the practical convenience of being able to use a detector that does not require cooling. Spectra of high resolution are obtainable in a relatively short time (30 to 40 min). The measurements of stratospheric constituents by this method have fairly small experimental errors ($\sim \pm 10\%$) and are, we believe, relatively free from significant systematic errors in the case of H_2O/O_2 studies, but less accurate for O_3 determinations. Thus we conclude that this technique should be considered in terms of any extensive study of stratospheric constituents using aircraft or balloon-borne equipment.

References

Bader, M., Cameron, R. M., Burroughs, W. J., and Gebbie, H. A.: 1967, *Nature* **214**, 377.
Bates, D. R.: 1951, *Proc. Phys. Soc.* B, **64**, 805.
Biraud, Y., Gay, J., Verdet, J. P., and Zean, Y.: 1969, *Astron. Astrophys.* **2**, 413.
Burroughs, W. J.: 1969, Ph.D. Thesis, London.
Burroughs, W. J. and Chamberlain, J.: 1971, *Infrared Phys.* **11**, 1.
Burroughs, W. J. and Harries, J. E.: 1970, *Nature* **227**, 824.
Burroughs, W. J., Harries, J. E., Chamberlain, J., and Gebbie, H. A.: 1971, this volume, p. 292–98.
Chamberlain, J.: 1971, *Infrared Phys.* **11**, 25.
Chamberlain, J. and Gebbie, H. A.: 1971, *Infrared Phys.* **11**, 57.
Chantry, G. W., Evans, H. M., Chamberlain, J., and Gebbie, H. A.: 1969, *Infrared Phys.* **9**, 85.
Connes, J., Connes, P., and Maillard, J. P.: 1967, *J. Phys.* **28**, C2-120.
Dowling, J. M.: 1967, Air Force Report No. SSD-TR-67-30.
Eddy, J., MacQueen, R., and Lena, P.: 1969, *Atmos. Sci.* **26**, 1318.
Gay, J., Lequeux, J., Verdet, J. P., Turon-Laccarrieu, P., Bardet, M., Roucher, J., and Zeau, Y.: 1968, *Astrophys. Letters* **2**, 169.

Gebbie, H. A.: 1957, *Phys. Rev.* **107**, 1174.

Gebbie, H. A. and Burroughs, W. J.: 1968, *Nature* **217**, 1241.

Gebbie, H. A., Burroughs, W. J., Harries, J. E., and Cameron, R. M.: 1968a, *Astrophys. J.* **154**, 405.

Gebbie, H. A., Chamberlain, J., and Burroughs, W. J.: 1968b, *Nature* **220**, 893.

Gebbie, H. A., Burroughs, W. J., and Bird, G. R.: 1969a, *Proc. Roy. Soc.* A **310**, 579.

Gebbie, H. A., Burroughs, W. J., Chamberlain, J., Harries, J. E., and Jones, R. G.: 1969b, *Nature* **221**, 143.

Harries, J. E., Burroughs, W. J., and Gebbie, H. A.: 1969, *J. Quant. Spectr. Radiative Transfer* **9**, 799.

Jacquinot, P.: 1958, *J. Phys. Rad.* **19**, 223.

MacQueen, R., Eddy, J., and Lena, P.: 1968, *Nature* **220**, 1112.

Putley, E. H. and Martin, D. H.: 1967, in D. H. Martin (ed.), *Spectroscopic Techniques*, North-Holland Publ. Co., Amsterdam, p. 144.

Vanasse, G. and Sakai, H.: 1967, *Prog. Optics*, **6**, 259.

Viktorova, A. A. and Zhevakin, S. A.: 1967, *Soviet Phys. Dokl.* **11**, 1059.

SOME EFFECTS OF DIMERS OF THE WATER
MOLECULE IN THE ATMOSPHERE

W. J. BURROUGHS, J. E. HARRIES, J. CHAMBERLAIN and H. A. GEBBIE*

Division of Electrical Science, National Physical Laboratory, Teddington, Middlesex, England

The possibility of dimers of the water molecule, for which there was considerable evidence on the basis of P-V-T virial coefficient measurements (Rowlinson, 1945), having a significant atmospheric absorption was first proposed theoretically by Viktorova and Zhevakin (1967). They calculated the submillimetre and millimetre spectrum on the basis of a linear or open structure model (Pauling, 1960) of the dimer in order to provide an adequate explanation for the well known excessive absorption observed in atmospheric 'windows' in this region (Dryazin *et al.*, 1966; Ryadov *et al.* 1966; Burroughs *et al.*, 1966). This theoretical work also predicted the existence of a discrete absorption feature in the spectrum at 7.1 cm^{-1}, which being in a 'window' region should be observable in atmospheric studies. At about the same time broad-band observations of the Sun in the millimetre range indicated a feature in the range 7 to 9 cm^{-1} (Gebbie and Burroughs, 1968), which appeared to be correlated with water content. In order to determine whether these effects are indeed due to the dimer of water vapour a number of submillimetre wavelength experiments have been carried out, both in the laboratory, and in the field observing solar radiation attenuated by the atmosphere. The results considered will consist of: (a) spectroscopic studies of the 7.5 cm^{-1} feature showing the necessity to work at low temperatures and low total pressure in order to observe this line, and (b) monochromatic maser absorption measurements showing that the excess absorption in the 'window' regions can be observed at high temperatures and high total pressures. These apparently contradictory properties of the dimer absorption will be shown to be both of interest in making improved measurements of the parameters of this molecule, and also of importance in estimating its effect on the properties of the atmosphere.

Spectroscopic Studies. All the submillimetre spectroscopic studies reported here have used an NPL-Grubb Parsons Michelson interferometer (Chantry *et al.*, 1969) combined either with a helium-cooled InSb detector (Putley, 1963; Rollin, 1961) or a standard Golay cell (Chantry *et al.*, 1969). The first investigation was to observe the transmission of water vapour at various temperatures, and a range of pressures at $0°C$ (Harries *et al.*, 1969). A discrete absorption feature was observed at 7.5 cm^{-1}. Its integrated strength is (to within the $\pm 25\%$ experimental error) directly proportional to the square of the pressure, which indicates that the absorption is caused by the interaction of pairs of water molecules. Furthermore, the temperature study showed, through the use of an Arrhenius plot, that the binding energy between the two molecules

---

* Now at NBS, Boulder, Colorado 80302, U.S.A.

Fiocco (ed.), Mesospheric Models and Related Experiments, 292–298. All Rights Reserved.
Copyright © 1971 by D. Reidel Publishing Company, Dordrecht-Holland.

of each dimeric pair is 6 ± 3 kcal mole^{-1}. This work was carried out using a multipass cell with a maximum path length of 30 m. The detector was a Putley-type helium-cooled InSb system (Putley, 1963).

The fact that the principal feature in the laboratory work did not coincide with the original feature in the solar spectrum led to further work observing solar radiation attenuated by the atmosphere at altitudes of 50 m (Queen Mary College, London) and 3580 m (Jungfraujoch, Switzerland) (Harries and Burroughs, 1970). The solar spectra measured have been ratioed against background results obtained using a Hg lamp, in order to remove instrumental effects. The results obtained are shown in Figure 1, and the conditions are listed in Table I. It is obvious that the atmospheric feature is a doublet (at 7.4 and 8.4 cm^{-1}), and also that at sea level (curves d and e) it is very much reduced in relation to the intensity of monomeric water lines at 6.1 and 10.9 cm^{-1}. These observations can be explained in the following manner.

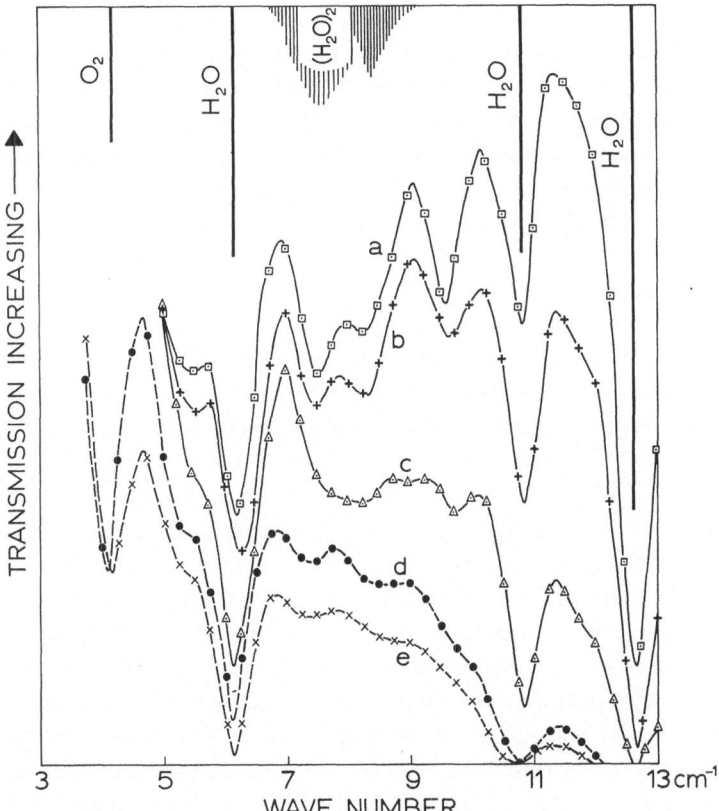

Fig. 1. Ratio of solar spectra against mercury lamp backgrounds for sea level observations (d) and (e) compared with high altitude observations (a), (b) and (c), showing the dependence of the 7.4 and 8.4 cm^{-1} features on increased water vapour absorption combined with changes in temperature, pressure, and atmospheric path-length. Conditions are given in Table I. Resolution is 0.5 cm^{-1}. (The spectrum shown at the top of the diagram is only for identification purposes, and should not be used as a measure of the theoretical strengths of the given features.)

TABLE I

Conditions associated with the spectra shown in Figure 1

Spectrum	Altitude (m)	Temperature (°C)	Mean solar elevation angle (degrees)	Air masses [a]	Precipitable water (mm)
a	3580	− 16.5	48	0.85	1.2
b	3580	− 20.0	48	0.85	1.6
c	3580	− 20.0	11.5	3.08	6.6
d	50	9.0	37	1.66	8.8
e	50	9.0	31	1.94	10.4

[a] With respect to sea-level.

The doublet is not predicted in the original theoretical treatment (Viktorova and Zhevakin, 1967), as this considers only molecules in the ground state. Since the first vibrational state is only 100 cm^{-1} above the ground state this level will be significantly populated ($\sim 40\%$ of the molecules at the atmospheric temperatures considered). The molecular parameters of the dimer will be changed in this vibrational state, and hence it is reasonable to assume that the feature at 8.4 cm^{-1} is the vibrational equivalent of the 7.4 cm^{-1} line. One effect of such an assignment is that the relative strength of the 8.4 cm^{-1} feature should decrease with temperature, which is possibly the case in Figure 1.

The relative decrease in intensity of these features at lower altitudes, despite the increase in total water observed (Table I), can again be explained in terms of the theoretical spectrum (Viktorova and Zhevakin, 1967). This spectrum is calculated on the basis of a symmetric rotor approximation to the slightly asymmetric dimer molecule. This approximation predicts the existence of a series of Q-branches, each formed by the coincidence of a large number of $\Delta J = 0$ transitions for a given $K \rightarrow K + 1$ branch, which occur at 7.1, 22.4, 35.6 and 49.8 cm^{-1} etc. The removal of the symmetric rotor approximation will tend to make the theoretical spectral distribution of the lines more random, and hence tend to blur out any systematic features. This effect will explain why it is much more difficult to observe the higher frequency Q-branches (Harries et al., 1969). The effect of temperature on the feature will be twofold. First, since a Q-branch is composed of a large number of transitions, as the temperature is increased the occupation of higher J-levels will increase, hence more transitions will contribute to the given branch. Second, as the temperature is increased the occupation of the vibrational states will be raised with respect to the ground state, with a corresponding increase in the strength of vibrational satellite lines (e.g. at 8.4 cm^{-1}). Finally, increased atmospheric pressure will further broaden the features, and hence lead to a loss of contrast in the observed spectrum.

The combination of these effects of pressure and temperature, in contrast to the oversimplified theoretical model given by Viktorova and Zhevakin (1967), means that the spectrum of the dimer will be very complex. However, the considerations given briefly above indicate why ground level observations at ambient sea-level conditions have only shown limited evidence of such a relatively weak but extensive absorption feature, rather than the anticipated increase in the strength of the discrete dimer feature observed at altitude (3580 m).

Maser Studies. The principal maser studies aimed at providing evidence of the existence of dimeric absorption have used the two radiations of the HCN maser at 29.71 cm^{-1} and 32.17 cm^{-1} (Burroughs *et al.*, 1969). The absorption at 32.17 cm^{-1} is dominated by two nearby (monomeric) water vapour lines, while at 29.71 cm^{-1} it is primarily the background far-wing contributions that are observed and hence any contribution due to the dimer will be relatively greater in this latter case. A comparison of these two absorptances obviated the need for absolute measurements and enabled a direct comparison to be made of the dependence on pressure of the relative attenuations. The results showed a distinctly different behaviour for the two wavelengths, and analysis of the ratio of the two attenuations indicated that in the far-wing region of absorption a contribution existed, attributable to dimeric water vapour, which largely explained the excess absorption in the 'window' regions. The temperature dependence of the intensity of the new contribution to the background enabled a binding energy of 5.2 ± 1.5 kcal mole^{-1} to be determined. Both this value and the spectroscopic estimate given earlier are in agreement with a prediction of Pauling (1960) for a water dimer with an open linear structure containing a single hydrogen bond.

These maser results indicate that the total atmospheric absorption coefficient (α) at 29.7 cm^{-1} could be expressed in the form (for constant temperature and total atmospheric pressure)

$$\alpha = k_0 + k_1 \varrho + k_2 \varrho^2 \text{ dB km}^{-1} \tag{1}$$

where ϱ = absolute humidity (g m^{-3}); k_0 = absorption due to other atmospheric constituents which for the purposes of the discussion can be assumed to be negligible; $k_1 \varrho$ = absorption due to monomeric water vapour; $k_2 \varrho^2$ = absorption due to dimeric water vapour; also at high values of ϱ (> 10 g m^{-3}) the dimeric contribution should be easily observed. Moreover, recent improved theoretical evaluation of monomeric water vapour absorption (Emery, 1968; Birch *et al.*, 1969) meant that a measurement of $k_1 \varrho$ to compare with theory would be of interest. The atmospheric attenuation at 29.7 cm^{-1} has been measured as a function of ϱ in a 7 m multipass gas cell using path-lengths up to 196 m, in which the humidity, but not the atmospheric pressure, could be controlled. Results obtained at a mean temperature of 298 ± 1.5 K are shown in Figure 2 and a least squares fit gives an expression:

$$\alpha = 3.70 + 8.12 \varrho + 0.12 \varrho^2 \text{ dB km}^{-1}. \tag{2}$$

The value of $k_0 = 3.70$ dB km^{-1} has little meaning as it includes the reflection losses in the cell and cannot be wholly attributed to other atmospheric constituents (collision-induced absorption in nitrogen is the most important, but only gives $k_0 \sim 1$ dB km^{-1} (Gebbie *et al.*, 1963)). The value of $k_1 = 8.12$ dB km^{-1} (g m^{-3})$^{-1}$ is in reasonable agreement with theory (Birch *et al.*, 1969) thus emphasising the importance of the inclusion of the dimer in obtaining agreement between theory and experiment. The value of $k_2 = 0.12$ dB km^{-1} (g m^{-3})$^{-2}$ means that at 29.7 cm^{-1} the contribution of dimeric absorption amounts to $\sim 20\%$ at $\varrho = 15$ g m^{-3}. This measurement is also in agreement, within experimental error, with the dimeric contribution measured in the study at 29.7 and 32.3 cm^{-1} where the total pressure was controlled and ϱ kept constant (Burroughs *et al.*, 1969). In the discussion it will be seen that though this dimer contribution is important in the submillimetre region it becomes very much more significant in the mid-infrared where the monomeric term is much reduced (Bignell, 1970). Finally, it should be emphasised that despite the rapid decrease in dimer concentration with increasing temperature (Zhevakin and Naumov, 1967) estimate an approximately T^{-12} dependence at $T \simeq 300$ K), the compensating effect of the ϱ^2 dependence results in the importance of the atmospheric continuum due to this species being of principal interest at high humidities (and therefore high temperatures). Thus the effect of excess absorption in atmospheric 'window' region must be of greatest concern in the properties of the lower troposphere at low latitudes.

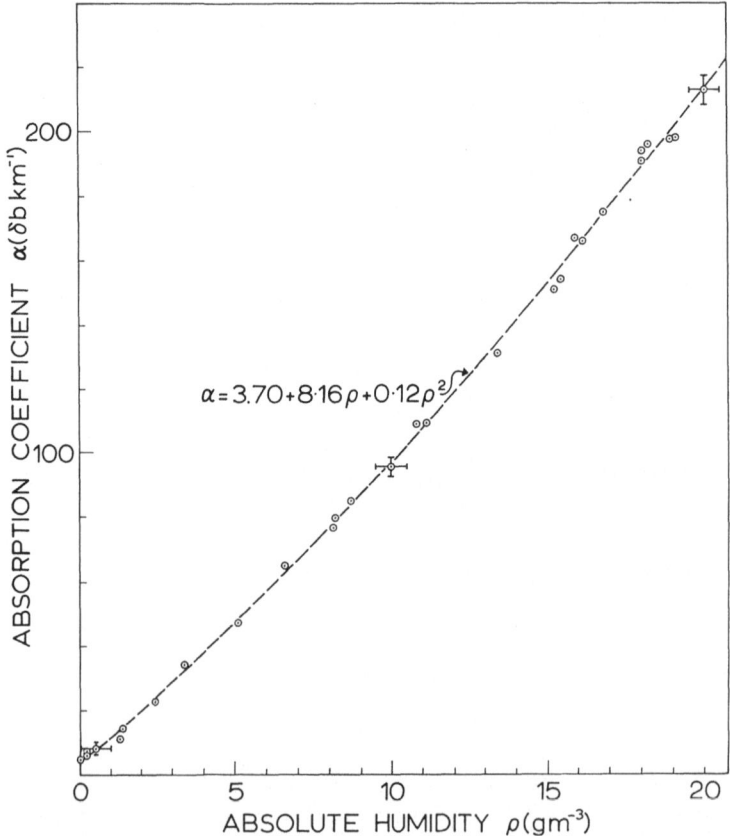

Fig. 2.   The variation of atmospheric absorption coefficient (α) at 29.7 cm^{-1} as a function of absolute humidity (ϱ) showing the effect of dimers in terms of a measurable ϱ^2 dependence.

It can be seen from the results presented here that whilst the discrete feature at $7.5\,\text{cm}^{-1}$ is only easily observed at reduced pressures and temperatures, the more significant continuum absorption is of greatest importance at raised temperatures and high water vapour pressures. It is probable that in such a complex asymmetric top spectrum composed of contributions from molecules in a number of vibrational states the only discrete features will occur at low frequencies and at low temperatures. Possibly the $7.5\,\text{cm}^{-1}$ feature is the only easily observable atmospheric feature, though aircraft studies at an altitude of 12 km have observed a line at $49.5\,\text{cm}^{-1}$ which shows the anticipated behaviour of a dimeric transition (Eddy et al., 1969). For this reason the implications of the dimer will be considered almost exclusively in terms of the continuum absorption. This broad weak absorption will probably extend up to about the binding energy of the molecule ($\sim 1700\,\text{cm}^{-1}$) and spectroscopic studies in the 600 to $100\,\text{cm}^{-1}$ region at high temperatures have attributed the observed excess absorption, compared with theory, to the dimer (Varanasi et al., 1968). This absorption will be of importance in communications, radiative transfer in the atmosphere, and meteorology.

In the field of communications the excess absorption in the millimetre range due to the dimer is empirically well studied (e.g. Dryazin et al., 1966). Apart from the increased attenuation, the effect of humidity fluctuations will produce excess noise due to the ϱ^2 dependence of dimer absorption. Measurements at 337 μm in warm humid conditions have shown very large variations in transmitted signal, that may in part be due to the dimer (D. T. Llewellyn-Jones, private communication).

In radiative transfer and meteorology the importance of the dimer will centre on the 21 to 17 μm and 13 to 8 μm 'window' regions in the atmosphere. Recent work by Bignell (1970) at temperatures $\sim 30\,°\text{C}$ and absolute humidities (ϱ) up to $\sim 25\,\text{g m}^{-3}$ in the 11 to 20 μm region has provided strong confirmation of the dimer hypothesis. In particular, results at $560\,\text{cm}^{-1}$ show a dependence with ϱ which have a much more dominant ϱ^2 term than the $29.7\,\text{cm}^{-1}$ result (Figure 2). This result can be most easily explained in terms of dimers of the water molecule. These observations also show an anomalous decrease of α with increasing temperature for constant ϱ which can be best explained by the exponential decrease of dimer concentration with an increase in temperature. Bignell also makes an extensive survey of measurements in the infrared to find supporting evidence of the effects he has observed, and finds a large amount of agreement in other work. It should be noted that the previous failure to identify the contribution of the dimer rests principally on the fact that most observations were carried over limited ranges of low values of ϱ. From these results he concludes that the effect of the continuum will be noticeable in satellite-inferred 'surface' temperatures as the excess absorption in the 8 to 13 μm 'window' region (especially in humid tropical latitudes) will lead to the soundings being weighted to a greater altitude than anticipated. Consequently a lower temperature will be inferred, and early Nimbus III results may well provide support for this thesis (Wark and Hilleary, 1969; Hanel and Conrath, 1969). Bignell estimates the effect of this species to give an excess radiative cooling of as much as 1 to $2\,°\text{C}$ day^{-1} in the dampest layers of the atmosphere. Radiometric measurements in the tropical Pacific show an excess cooling up to about the 850 mb level of this order (Cox, 1969). This excess cooling of the lower atmospheric layers would tend to stabilize the lower levels in the tropics, and explain why these regions are not 'boiling' with convection (Cox, 1969). If this effect can be assumed to be due to the existence of dimers, the implications of this species in tropical meteorology are of considerable importance.

Finally, it should be noted that in order to obtain improved molecular parameters of the dimer more sensitive spectroscopic techniques will be required. The most obvious possibility is microwave methods, but it should be remembered that the extreme complexity of the spectrum will require water vapour at very low pressures in order to identify single very weak transitions. The large dipole moment of the dimer (~ 3.5 Debye; Viktorova and Zhevakin, 1967) will produce very broad lines ~ 40 MHz Torr^{-1} so that low vapour pressures (~ 1 Torr) are essential. At low pressures the relative concentration of the dimer will be much reduced, so that optimum conditions will probably be obtained at temperatures less than $0\,^{\circ}$C observing saturated vapour in very long effective path-lengths ($\gtrsim 100$ m). However, despite these restrictions it should be possible to observe these transitions with a sensitive microwave spectroscopic system.

References

Bignell, K. J.: 1970, *Quart. J. Meteor. Soc.* **96**, 390.

Birch, J. R., Burroughs, W. J., and Emery, R. J.,: 1969, *Infrared Phys.* **9**, 75.

Burroughs, W. J., Jones, R. G., and Gebbie, H. A.: 1969, *J. Quant. Spectrosc. Radiat. Transfer* **9**, 809.

Burroughs, W. J., Pyatt, E. C., and Gebbie, H. A.: 1966, *Nature* **212**, 387.

Chantry, G. W., Evans, Helen M., Chamberlain, John, and Gebbie, H. A.: 1969, *Infrared Phys.* **9**, 85.

Cox, S. K.: 1969, *J. Atmospheric Sci.* **26**, 1347.

Dryazin, Yu. A., Kyslakov, A. G., Kukin, L. M., Naumov, A. I., and Fedoseev, L. I.: 1966, *Radiofizika* **9**, 1078.

Eddy, J. A., Lee, R. M., Lena, P. J., and MacQueen, R. M.: 1969, *J. Atmospheric Sci.* **26**, 1318.

Emery, R. J.: 1968, *Appl. Opt.* **7**, 1247.

Gebbie, H. A. and Burroughs, W. J.: 1968, *Nature* **217**, 1241.

Gebbie, H. A., Stone, N. W. B., and Williams, D.: 1963, *Mol. Phys.* **6**, 215.

Hanel, R. and Conrath, B.: 1969, *Science* **165**, 1258.

Harries, J. E. and Burroughs, W. J.: 1970, *Infrared Phys.* **10**, 165.

Harries, J. E., Burroughs, W. J., and Gebbie, H. A.: 1969, *J. Quant. Spectrosc. Radiat. Transfer* **2**, 799.

Pauling, L.: 1960, *The Nature of the Chemical Bond*, Cornell Univ. Press, p. 468.

Putley, E. H.: 1963, *Proc. Inst. Elec. Electron. Engrs.* **51**, 1412.

Rollin, B. V.: 1961, *Proc. Phys. Soc.* **77**, 1102.

Rowlinson, J. S.: 1945, *Trans. Faraday Soc.* **45**, 974.

Ryadov, V. Ya. and Furashov, N. I.: 1966, *Radiofizika* **9**, 859.

Varanasi, P., Chou, S., and Penner, S. S.: 1968, *J. Quant. Spectrosc. Radiat. Transfer* **8**, 1537.

Viktorova, A. A. and Zhevakin, S. A.: 1967, *Soviet Phys. Dokl.* **11**, 1059.

Wark, D. Q. and Hilleary, D. T.: 1969, *Science* **165**, 1256.

Zhevakin, S. A. and Naumov, A. P.: 1967, *Radiofizika* **9**, 1213.

ASTROPHYSICS AND SPACE SCIENCE LIBRARY

Edited by

J. E. Blamont, R. L. F. Boyd, L. Goldberg, C. de Jager, Z. Kopal, G. H. Ludwig, R. Lüst,
B. M. McCormac, H. E. Newell, L. I. Sedov, Z. Švestka, and W. de Graaff

1. C. de Jager (ed.), *The Solar Spectrum. Proceedings of the Symposium held at the University of Utrecht, 26–31 August, 1963*. 1965, XIV + 417 pp.

2. J. Ortner and H. Maseland (eds.), *Introduction to Solar Terrestrial Relations. Proceedings of the Summer School in Space Physics held in Alpbach, Austria, July 15–August 10, 1963 and Organized by the European Preparatory Commission for Space Research*. 1965, IX + 506 pp.

3. C. C. Chang and S. S. Huang (eds.), *Proceedings of the Plasma Space Science Symposium, Held at the Catholic University of America, Washington, D.C., June 11–14, 1963*. 1965, IX + 377 pp.

4. Zdeněk Kopal, *An Introduction to the Study of the Moon*. 1966, XII + 464 pp.

5. Billy M. McCormac (ed.), *Radiation Trapped in the Earth's Magnetic Field. Proceedings of the Advanced Study Institute, Held at the Chr. Michelsen Institute, Bergen, Norway, August 16–September 3, 1965*. 1966, XII + 901 pp.

6. A. B. Underhill, *The Early Type Stars*. 1966, XIII + 282 pp.

7. Jean Kovalevsky, *Introduction to Celestial Mechanics*. 1967, VIII + 427 pp.

8. Zdeněk Kopal and Constantine L. Goudas (eds.), *Measure of the Moon. Proceedings of the Second International Conference on Selenodesy and Lunar Topography held in the University of Manchester, England, May 30–June 4, 1966*. 1967, XVIII + 479 pp.

9. J. G. Emming (ed.), *Electromagnetic Radiation in Space. Proceedings of the Third ESRO Summer School in Space Physics, held in Alpbach, Austria, from 19 July to 13 August, 1965*. 1968, VIII + 307 pp.

10. R. L. Carovillano, John F. McClay, and Henry R. Radoski (eds.), *Physics of the Magnetosphere. Based upon the Proceedings of the Conference held at Boston College, June 19–28, 1967*. 1968 X + 686 pp.

11. Syun-Ichi Akasofu, *Polar and Magnetospheric Substorms*. 1968, XVIII + 280 pp.

12. Peter M. Millman (ed.), *Meteorite Research. Proceedings of a Symposium on Meteorite Research held in Vienna, Austria, 7–13 August, 1968*. 1969, XV + 941 pp.

13. Margherita Hack (ed.), *Mass Loss from Stars. Proceedings of the Second Trieste Colloquium on Astrophysics, 12–17 September, 1968*. 1969, XII + 345 pp.

14. N. D'Angelo (ed.), *Low-Frequency Waves and Irregularities in the Ionosphere. Proceedings of the 2nd ESRIN-ESLAB Symposium, held in Frascati, Italy, 23–27 September, 1968*. 1969, VII + 218 pp.

15. G. A. Partel (ed.), *Space Engineering. Proceedings of the Second International Conference on Space Engineering, held at the Fondazione Giorgio Cini, Isola di San Giorgio, Venice, Italy, May 7–10, 1969*. 1970, XI + 728 pp.

p.t.o.

16. S. Fred Singer (ed.), *Manned Laboratories in Space. Second International Orbital Laboratory Symposium.* 1969, XIII + 133 pp.

17. B. M. McCormac (ed.), *Particles and Fields in the Magnetosphere. Symposium Organized by the Summer Advanced Study Institute, held at the University of California, Santa Barbara, Calif., August 4–15, 1969.* 1970, XI + 450 pp.

18. Jean-Claude Pecker, *Experimental Astronomy.* 1970, X + 105 pp.

19. V. Manno and D. E. Page (eds.), *Intercorrelated Satellite Observations related to Solar Events. Proceedings of the Third ESLAB/ESRIN Symposium held in Noordwijk, The Netherlands, September 16–19, 1969.* 1970, XVI + 627 pp.

20. L. Mansinha, D. E. Smylie and A. E. Beck, *Earthquake Displacement Fields and the Rotation of the Earth. A NATO Advanced Study Institute. Conference Organized by the Department of University of Western Ontario, London, Canada, 22 June–28 June, 1969. 1970,* XI + 308 pp.

21. Jean-Claude Pecker, *Space Observatories.* 1970, XI + 120 pp.

22. L. N. Mavridis (ed.), *Structure and Evolution of the Galaxy. Proceedings of the NATO Advanced Study Institute, held in Athens, September 8–19, 1969.* 1971, VII+312 pp.

23. A. Muller (ed.), *The Magellanic Clouds. A European Southern Observatory Presentation: Principal Prospects, Current Observational and Theoretical Approaches, and Prospects for Future Research. Based on the Symposium on the Magellanic Clouds held in Santiago de Chile, March 1969, on the Occasion of the Dedication of the European Southern Observatory.* 1971, XII+189 pp.

24. B. M. McCormac (ed.), *The Radiating Atmosphere. Proceedings of a Symposium Organized by the Summer Advanced Study Institute, held at Queen's University, Kingston, Ontario, August 3–14, 1970.* 1971. XI+455 pp.

26. I. Atanasijević, *Selected Exercises in Galactic Astronomy,* 1971, XII + 143 pp.

27. C. J. Macris (ed.) *Physics of the Solar Corona. Proceedings of NATO Advanced Study Institute on Physics of the Solar Corona, held at Cavouri-Vouliagmeni, Athens, Greece, 6–17 September 1970.* 1971, XII+345 pp.

SOLE DISTRIBUTORS FOR U.S.A. AND CANADA:

SPRINGER-VERLAG NEW YORK, INC., 175 Fifth Ave., New York, N.Y. 10011